ELECTROCHEMISTRY
OF SEMICONDUCTORS

ELECTROCHEMISTRY OF SEMICONDUCTORS

by Viktor A. Myamlin • Yurii V. Pleskov

Institute of Electrochemistry
Academy of Sciences of the USSR, Moscow

Translated from Russian

With a Foreword
C. G. B. Garrett
Bell Telephone Laboratories, Inc.
Murray Hill, New Jersey

℗ Springer Science+Business Media, LLC 1967

Viktor Alekseevich Myamlin, senior scientist at the Institute of Electrochemistry of the Academy of Sciences of the USSR, was born in 1924. After graduating from the Moscow Engineering Physics Institute (MIFI) in 1949, he lectured at that Institute's Department of Theoretical Physics. Myamlin is the author of 26 scientific articles dealing with his research work in nuclear theory, quantum electrodynamics, and electrochemistry of semiconductors as well as co-author (with V. G. Levich and Yu. A. Vdovin) of *A Course in Theoretical Physics*, published in 1962 in the Soviet Union.

Yurii Viktorovich Pleskov, born in 1933, is a graduate of the Department of Electrochemistry of Moscow State University. A senior scientist at the Institute of Electrochemistry of the Academy of Sciences of the USSR, Pleskov has been conducting research on the electrochemistry of semiconductors and oxidation processes in solid electrodes. He is the author of more than 30 scientific papers.

The original Russian text, published by Nauka Press, Moscow, in 1965, for the Institute of Electrochemistry of the Academy of Sciences of the USSR, has been revised and updated by the authors for the American edition.

Виктор Алексеевич Мямлин,
Юрий Викторович Плесков

Электрохимия полупроводников

ELEKTROKHIMIYA POLUPROVODNIKOV

ELECTROCHEMISTRY OF SEMICONDUCTORS

The article by V. A. Tyagai and Yu. V. Pleskov which has been appended to this volume originally appeared in *Elektrokhimiya* Vol. 1, No. 10, pp. 1167-1173 (1965).

Library of Congress Catalog Card Number 66-12887

ISBN 978-1-4899-6248-5   ISBN 978-1-4899-6533-2 (eBook)
DOI 10.1007/978-1-4899-6533-2

© 1967 Springer Science+Business Media New York
Originally published by Plenum Press in 1967.
Softcover reprint of the hardcover 1st edition 1967

Foreword

The serious study of the electrochemistry of semiconductors is hardly more than a decade old, but already several hundred papers have been written on the subject, and in spite of the appearance of several excellent review articles, no general work of reference has hitherto existed. The present translation of Myamlin and Pleskov's book, published in Russian in 1965, should satisfy this need.

The existence of a photovoltaic effect at a semiconductor-electrolyte interface was first reported as long ago as 1839 by E. Becquerel. However, little progress toward a quantitative understanding of the differences between metal and semiconductor electrodes was possible until the advent of semiconductor materials of controlled purity, around the time of the advent of the transistor. Curiously enough, the transistor itself grew out of studies by Brattain and Bardeen of the way in which the surface space-charge region of a semiconductor may be modified by the application of external fields and by changes in chemical environment. The junction transistor, of course, is a bulk device, dependent rather upon the injection and collection of minority carriers at boundaries between regions of different conductivity type within a single semiconductor crystal. The MOS transistor, on the other hand, represents a return to the original concept, except that now the modulating field is applied to the surface across an oxide film instead of through an electrolytic boundary layer. In any case, the development of the transistor has continued to exert a powerful influence on that of semiconductor electrochemistry.

When Brattain and I began our joint investigations of the interface between germanium and an aqueous electrolyte in 1953, we were not in the first place interested in the current flowing across the interface. We hoped, in fact, that as long as the applied voltage was not too large, the interface would be an "ideally polarizable" electrode. The object was to understand the structure of the interface region on both sides of the geometric surface; to be able to say, for example, what fraction of the change in electrode potential under prescribed conditions should be identified with a change in the height of the semiconductor space-charge barrier. We soon found, however, that usually a current does flow, and that the magnitude of the current often depends very substantially on whether the germanium is p-type or n-type. It thus became clear that there were two things to investigate: the structure of the interfacial space-charge region, and the role of the two current carriers in the various electrode reactions. At the time of the publication of our main paper in 1955, we could make no claim to have solved either problem completely.

Since 1955, a great deal of work has been done in both areas, conspicuously in the United States, Germany, and the Soviet Union. Much of the Russian work has been carried out at the Institute of Electrochemistry of the Academy of Sciences of the USSR in Moscow, where Yu. V. Pleskov and his colleagues have conducted a wide range of investigations of germanium and other semiconductor electrodes in aqueous and nonaqueous solvents. V. A. Myamlin, in addition to his contributions to semiconductor electrochemistry, is coauthor, with V. G. Levich (Corresponding Member of the Academy of Sciences and chairman of the theoretical department of the Institute of Electrochemistry) and Yu. A. Vdovin, of a standard text on quantum mechanics, published in 1962.

In the present work, Myamlin and Pleskov give a comprehensive survey of the field of semiconductor electrochemistry, both with regard to theory and experiment. Much of the Russian work is now made available in English for the first time. The introduction contains an elementary exposition of semiconductor principles, aimed primarily at electrochemists unfamiliar with solid state theory. The first chapter covers, broadly, the first of the two topics to which I alluded above: the structure of the space-charge layer. In this area Pleskov and his collaborators have introduced the use of methyl formamide, an electrolyte in which germanium behaves much more nearly as an "ideally polarizable" electrode than it does in water. This work is described in some detail (§§ 13, 18, and 20). The germanium – methyl formamide interface, however, in the state in which it was investigated, still displays the troublesome "fast surface states." Credit goes to Brattain and Boddy for the development of a preparation technique that essentially eliminates fast surface states at the interface between germanium and an aqueous electrolyte. Brattain and Boddy further showed that if copper, silver, or gold ions are introduced into a system prepared in this way, fast surface states reappear, with well-defined energy levels characteristic of each impurity. This highly significant achievement is also well described by Myamlin and Pleskov in Chapter I. Other sections of Chapter I are devoted to silicon (§§ 15 and 18), where the situation is less well understood, to zinc oxide (§ 16a), where Dewald's classic investigation led to an unusually clear and unambiguous understanding, and to cadmium sulphide (§ 16b), which has been investigated in Pleskov's own laboratory, again with clean-cut results.

Chapter II takes up the second of the two questions: the role of the semiconductor current carriers in electrode reactions. Here the situation is still very complex, both in terms of theory and experiment. Myamlin and Pleskov begin by expounding the quantum mechanical theory of Dogonadze, Chizmadzhev, and Kuznetsov (§ 25), which is an elaboration of earlier work due to Gerischer. This lays a formally acceptable groundwork. The more practical question, as always in chemical kinetics, is: what is the rate-limiting step? Sometimes the rate-limiting step is the transport of some of the participating ions from the bulk solution; sometimes it is the transport of holes or electrons from the bulk semiconductor. Either of these possibilities may lead eventually to saturation of the rate of reaction (§§ 27 and 31). In the anodic dissolution of germanium, the rate-limiting step is clearly the transport of holes (§ 32). Generally, these holes have to be generated at recombination centers in the bulk, but the presence in solution of easily-reducible ions such as $K_3Fe(CN)_6$ appears to lead to the injection of hole — electron pairs into the semiconductor, thereby increasing the rate of anodic dissolution (§ 36). The anodic dissolution of silicon is also determined initially by the supply of holes which, however, are generated not in the bulk but in the surface space-charge region (§ § 37 and 38). Later, however, unless the solution is alkaline or contains fluoride ions, the silicon is apt to become covered with a relatively thick oxide film, and the nature of the electrode reaction changes (§ 39). The role of the semiconductor carriers in the cathodic processes — for example the liberation of hydrogen at a germanium cathode — has been a matter of some controversy, but Myamlin and Pleskov support the original view that the rate-limiting step is the transport of conduction-band electrons (§ 34).

The remaining chapters of the book, which are shorter, are devoted to more specialized topics. Chapter III concerns the frequency-dependence of the impedance of the electrolyte interface, a topic to which Pleskov and his colleagues have devoted several research papers. Chapter IV contains a discussion of the special case in which no net current crosses the surface — i.e., corrosion. The last chapter is devoted to an exposition of the utility of various surface techniques for a variety of purposes in semiconductor technology.

Bell Telephone Laboratories, Inc.   C. G. B. Garrett
Murray Hill, New Jersey
January, 1967

Preface to the American Edition

The electrochemistry of semiconductors, the foundation of which was laid in 1955 by the work of W. H. Brattain and C. G. B. Garrett, in ten years has become an independent field of science, lying between electrochemistry and semiconductor physics.

We now have to a considerable extent a physical picture of the surface of a semiconductor in contact with an electrolyte, and some important principles of electrochemical kinetics on semiconductors have been established. Very detailed studies have been made of the electrochemical behavior of germanium which remains, like mercury, in the electrochemistry of metals, a type of model system for investigating the fundamental properties of semiconductor electrodes.

Future progress in this field requires, on the one hand, detailed investigations of the physicochemical aspects of the semiconductor—electrolyte interface and, on the other, extension of the range of semiconductors used as electrode materials. In the present book the authors have given as full an account as possible of the basic theoretical concepts of the electrochemistry of semiconductors and have systematized the voluminous experimental material.

In the preparation of the American edition, in addition to correcting some errors and misprints, we considered it worthwhile to add bibliographies of the most important papers published in 1965 after the appearance of the Russian edition. The article "Most immediate problems in the electrochemistry of semiconductors," which was written by one of us (Yu. P.) to-gether with V. Tyagai, is a logical continuation of our book and is included as an appendix.

We hope that the translation of our work will promote the further extension of scientific exchange in the field of the elec-trochemistry of semiconductors, to whose development Soviet and American physical chemists have made a great contribu-tion.

November 1966

V. A. Myamlin
Yu. V. Pleskov

Preface

A new field of electrochemistry — the electrochemistry of semiconductors — has emerged as a result of the development of the physics of semiconductors and the technology of semiconductor instruments. During the ten years of its existence (from 1954 to 1964) more than 400 papers have been published, most of which are devoted to germanium and silicon. In recent years new semiconductor materials such as the intermetallic compounds, oxides, and sulfides have attracted the attention of investigators.

The electrochemistry of semiconductors is not only of importance because electrochemical methods are used widely for the processing of semiconductor materials. The application of the physical concepts of semiconductors contributes to a fuller understanding of a number of general rules of electrochemical kinetics; in particular, those relating the electrochemical behavior of a solid to its electronic structure. In the electrochemistry of metals, the electronic structure is determined unequivocally by the metal of the electrode. A change from one metal to another produces a simultaneous change in both the chemical nature of the solid and the electronic structure, with the former effect often predominating. In the case of semiconductor electrodes, the introduction of a minute (from the chemical point of view) amount of an impurity produces a drastic change in the electrophysical properties of the electrode material without affecting its chemical nature. We should remember also that the oxide layers appearing on the surface of metal electrodes often have semiconductor properties.

The main characteristics of semiconductor electrodes as compared with those of metal electrodes are connected with the low concentration of free charges in the former. As a result, the electric field extends deep within the volume of the semiconductor and a space charge region is formed close to its surface; the rate of electrode reactions depends on the concentration of free current carriers at the electrode surface; certain essentially new phenomena appear, namely, the dependence of the rate of electrochemical reactions on illumination, recombination processes of free current carriers, etc. In most of the semiconductor materials investigated there are two types of free carriers — free electrons and holes — and both of these participate in electrode processes.

Investigations of single-crystal semiconductor materials formed the basis for the development of the electrochemistry of semiconductors. Polycrystalline semiconductors and, in particular, oxide films of a semiconductor nature on barrier-layer metals (aluminum, tantalum, etc.), are not examined in the present book as it is not certain that the basic concepts of electrochemistry of single-crystal semiconductors applies to them.

The present monograph includes literature up to the end of 1964. The authors have tried to make it suitable for physicists and technologists specializing in the field of surface properties of semiconductors and the production of semiconductor instruments, and for electrochemists. As a rule, the examination of each problem consists of qualitative assessments, followed by a quantitative calculation and an account of experimental material.

Sections 1-12, 17, 19, 24-30, 35, 37, 41-45, and 48 were written by V. A. Myamlin and Sections 13-16, 18, 20-23, 31-34, 36, 38-40, 46, 47, and 49-61 by Yu. V. Pleskov.

The authors would like to thank Yu. Ya. Gurevich, A. M. Kuznetsov, R. M. Lazorenko-Manevich, and V. A. Tyagai for valuable advice and useful comments.

V. A. Myamlin
Yu. V. Pleskov

Contents

Notation

a — light absorption coefficient

$b = U_n/U_p$ — ratio of mobilities of electrons and holes

C — differential capacitance

$C_1, C_0,$ and C_2 — capacitances of space charge region in semiconductor, Helmholtz layer, and diffuse part of double layer in an electrolyte

C_t — capacitance of surface levels

C_n, C_p — probabilities of trapping an electron and a hole by a level

c^0 — concentration of ions in the volume of the electrolyte

c_+, c_- — concentrations of cations and anions in solution

$\vec{D}$ — electrostatic induction

D_n, D_p — diffusion coefficients of electrons and holes

d_0 — thickness of Helmholtz layer

E — strength of electric field

E_s — strength of electric field at surface

E — energy

E_v — energy of upper edge of valence band

E_c — energy of lower edge of conduction band

E_g — width of forbidden gap

E_F — Fermi energy

E_i — energy of center of forbidden band

E_t — energy of surface level

e — absolute value of electronic charge

F — Faraday (unit)

$F(Y, \lambda)$	— function defined by equation (10.9)
$f(E)$	— Fermi distribution function
f_t	— function of the filling of the surface levels by electrons
h	— Planck's constant
i	— free carrier current density
i^0	— exchange current
i_n, i_p	— densities of electron and hole currents
j	$= \sqrt{-1}$
k	— Boltzmann's constant
k	— rate constant of electrochemical reaction
L_p	— diffusion length of holes
L_1	— Debye length in semiconductor
L_D	— Debye length in semiconductor with intrinsic conductivity
L_2	— thickness of diffuse part of double layer in solution
m_n, m_p	— effective masses of electron and hole
m/r	— ratio of electron and hole currents at semiconductor—electrolyte contact with anode solution
N	— Avogadro's number
N_t	— concentration of surface levels
N_D, N_A	— concentrations of donors and acceptors in semiconductor
N_c, N_v	— densities of states in conduction and valence bands
n	— concentration of electrons
p	— concentration of holes
n_i	— concentration of electrons in semiconductor with intrinsic conductivity
n^0, p^0	— equilibrium concentrations of free carriers in neutral volume of semiconductor
n_1, p_1	— concentrations of free carriers at boundary of space charge region and quasi-neutral region
n_s, p_s	— concentrations of free carriers at surface
n', p'	— concentrations of free carriers in bands when the Fermi level coincides with the surface level

n_t	— concentration of electrons in surface levels
Δn, Δp	— deviations of concentrations from equilibrium values
Q_1	— space charge in semiconductor
Q_2	— charge in diffuse part of double layer in electrolyte
Q_t	— charge in surface levels
R	— universal gas constant
R	— resistance
R	— flow of free carriers to recombination centers
s	— rate of surface recombination
T	— absolute temperature
t	— time
U_n, U_p	— mobilities of electrons and holes
V	— Dember photo emf
V_f	— floating potential
v	— thermal velocity of electrons
w	— thickness of semiconductor sample
x	— coordinate
$Y = e\varphi_1/kT$	— potential drop in space charge region in semiconductor in kT/e units
Z	— impedance of electrode
α	— transfer coefficient of cathode reaction
α'	— current multiplication factor
β	— transfer coefficient of anode reaction
γ	— fraction of valence electrons in current of reduction reaction
Γ_n, Γ_p	— excess of electrons and holes at semiconductor surface
ε	— dielectric constant
ε_1, ε_0, ε_2	— dielectric constants of semiconductor, Helmholtz region, and electrolyte
$\eta = \varphi_0 - \varphi_0^0$	— overvoltage in Helmholtz layer
θ	— coverage of surface by adsorbed hydrogen
$\lambda = p^0/n_i = n^i/n^0$ $= (p^0/n^0)^{1/2}$	— degree of alloying of semiconductor

$\bar{\mu}$	— electrochemical potential
$\bar{\mu}_n$	— electrochemical potential of electrons
μ	— chemical potential
ν	— stoichiometric number
ρ	— specific resistance of semiconductor
$\rho(x)$	— electric charge density
σ	— surface conductivity of semiconductor
σ_B	— bulk conductivity of semiconductor in equilibrium
σ_n, σ_p	— trapping cross sections of level for electron and hole
τ	— lifetime of minority carriers
τ_p, τ_n	— lifetimes of holes and electrons
$\varphi(x)$	— electrostatic potential
φ_B	— potential in the bulk of the semiconductor
φ_c	— potential in the bulk of the solution
φ_s	— potential at the surface of the semiconductor
$\varphi_1 = \varphi_s = \varphi_B$	— potential drop in space charge region in semiconductor
φ_0	— potential drop in Helmholtz region
φ	— electrode potential
$\varphi_{1,2}$	— galvanic potential at the semiconductor—electrolyte contact
$\Delta\varphi$	— change in potential
φ_α	— characteristic value of surface potential, defined by equation (19.6)
$\psi' = \varphi(-d_0) - \varphi_c$	— potential drop in diffuse part of double layer in solution
Ω	— characteristic frequency of levels defined by equation (43.6)
ω	— angular frequency of alternating current

Superscripts

0	— equilibrium value
$\sim$	— values corresponding to the transmission of alternating current

Subscripts

1	— space charge region in semiconductor
2	— region of diffuse part of double layer in solution
0	— Helmholtz region
s	— surface
t	— surface states
n	— electrons
p	— holes

Publisher's Note

The following journals cited in this work are available in cover-to-cover translation:

Russian title	English title	Publisher
Doklady Akademii Nauk SSSR	Doklady Chemistry	Consultants Bureau
	Soviet Physics–Doklady	American Institute of Physics
Élektrokhimiya	Soviet Electro-chemistry	Consultants Bureau
Fizika Tverdogo Tela	Soviet Physics–Solid State	American Institute of Physics
Izvestiya Akademii Nauk SSSR: Seriya Khimicheskaya	Bulletin of the Academy of Sciences of the USSR: Division of Chemical Sciences	Consultants Bureau
Pribory i Tekhnika Éksperimenta	Instruments and Experimental Techniques	Instrument Society of America
Uspekhi Khimii	Russian Chemical Reviews	The Chemical Society (London)
Zhurnal Fizicheskoi Khimii	Russian Journal of Physical Chemistry	The Chemical Society (London)
Zhurnal Prikladnoi Khimii	Journal of Applied Chemistry of the USSR	Consultants Bureau
Zhurnal Tekhnicheskoi Fiziki	Soviet Physics–Technical Physics	American Institute of Physics

Electrochemistry of Semiconductors

Introduction

The main purpose of the introduction is to give an elementary description of certain concepts in the physics of semiconductors, such as the energy bands in a crystal, holes, etc. In addition, a number of special problems which to a considerable extent form the basis of further discussion (such as the recombination theory of Shockley and Read) are also examined in the introduction. Thus, the introduction is devoted only to certain problems of semiconductor physics and does not claim to be complete or systematic. *

§ 1. Band Theory of Solids. Concept of Metals, Semiconductors, and Insulators

An electron in an isolated atom may have only certain energies. There are certain, so-called allowed values or levels of energy. Figure 1 gives the energy levels at which an electron in an isolated atom may be found, namely, E_1, E_2, etc.

Let us examined a system consisting of similar atoms. If the atoms are at a distance from each other such that their mutual interaction can be neglected, then each of them has the same set of energy levels. If the atoms are brought closer together, or a crystal lattice is built from them, then the electron levels are split as a result of the interaction between the atoms (see Fig. 1). A particular band of allowed energies from E_1 to E_1', E_2 to E_2', etc., will correspond to each level.

---

* The problems of semiconductor physics are examined in a number of monographs and collections [1-9].

1

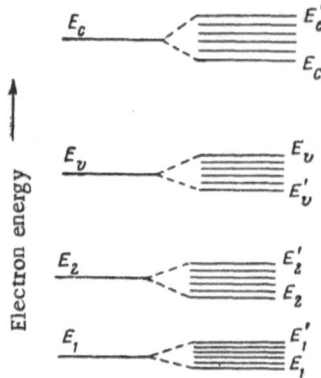

Fig. 1.   Electron energy levels in an
atom and a crystal.

The levels in these bands lie so close to each other that they may be considered as a continuous band. The energy regions between the allowed energy bands (from E_1' to E_2, from E to E_C) are called forbidden energy bands.

Let us examine a crystal in which all the energy bands of the valence electrons are filled completely and do not overlap. A conduction band follows the last-filled valence band. Since, at absolute zero, the former does not contain electrons, the crystal is an insulator at this temperature.

In actual fact, for orderly motion — i.e., an electric current — to be produced by an electric field, the electrons must increase their energy and move from a lower to a higher level.

However, if all the levels in the allowed energy band are occupied, the electrons cannot acquire the additional kinetic energy of orderly motion and there is no flow of electric current. This means that the body is an insulator. The transition of electrons from one band to another is practically excluded (in the case examined of low fields) as the energy the electrons could acquire from the electric field is insufficient for this purpose.

If at 0°K the conduction band is partially filled with electrons, the crystal has the properties of a metal. Orderly motion of the electrons is set up when an electric field is applied. The good electrical conductivity of metals is due to the fact that a sufficiently large number of electrons are in the allowed band and participate in this motion.

If a crystal which is an insulator at absolute zero is heated, then the electrons in the valence band may move into the

Table 1. Widths of Forbidden Bands of a Series of the Most Important Semiconductors and Insulators

Material	Width of forbidden band eV
InSb	0,17
Ge	0,67
Si	1,09
CdSe	1,7
GaAs	1,4
GaP	2,3
CdS	2,4
ZnO	3

free band by using the energy of thermal motion. In this case they would participate in the electrical conductivity. A crystal in which a sufficiently high electrical conductivity is produced by the thermal excitation of the electrons is called a semiconductor with intrinsic conductivity. *

Depending on the magnitude of the electrical conductivity at room temperature, a crystal may show the properties of a semiconductor or an insulator. Materials which have a specific resistance at room temperature within the range of 10^{-2} to 10^9 $\Omega \cdot$ cm are arbitrarily considered to be semiconductors. As will be shown below, the width of the forbidden band has a decisive effect on the electrical conductivity of solids. The width of the forbidden band of insulators is considerably greater than that of semiconductors. Thus, a typical semiconductor such as germanium has a forbidden band width of 0.67 eV, while that of a typical insulator equals several electron volts (for example, 5.6 eV in diamond). The boundary between semiconductors with intrinsic conductivity and insulators lies at a forbidden band width of approximately 1 eV (Table 1).

§ 2. n- and p-Type Semiconductors. Holes and Electrons

The thermal excitation of electrons and their transition into the conduction band produces vacancies (i.e., levels not occupied by electrons) in the energy spectrum of the filled (val-

---

*In contrast with a semiconductor with extrinsic conductivity (see § 2).

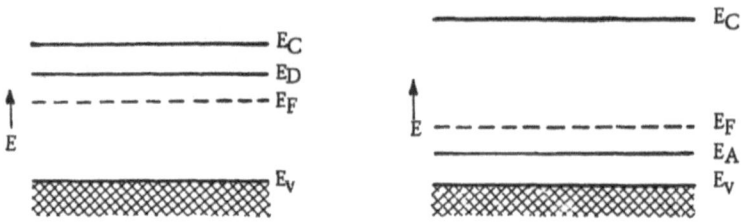

Fig. 2. Structure of the energy levels   Fig. 3. Structure of the energy levels
in an n-type semiconductor   in a p-type semiconductor.

ence) band. If a semiconductor with vacancies in the filled band is placed in an electric field, a drift of electrons occurs. Filling of a vacancy by an electron obviously produces a new vacancy in the filled band. Thus, the drift of electrons along the field is equivalent to the movement of the vacancy in the opposite direction. Thus the transmission of a current by electrons of the valence band may be regarded as the result of the movement of vacancies in this band. Such vacancies are called holes. Thus, a semiconductor has two types of current carrier produced by thermal excitation, namely, electrons in the conduction band and positively charged carriers, i.e., holes, in the valence band. It is evident that the number of free electrons equals the number of holes in an intrinsic semiconductor.

If foreign atoms are introduced into the crystal lattice of a semiconductor, additional energy levels appear within the energy spectrum of the crystal. This is shown schematically in Fig. 2, where the energy level of the impurity E_D is in the forbidden band of the semiconductor. Let us assume that this level lies quite close to the lower edge of the unfilled band. At a temperature other than absolute zero, thermal ionization of the foreign atoms occurs with appreciable probability. When this occurs an electron of the foreign atom moves into the conduction band. The foreign atoms supplying electrons to the free band are called donors, while the semiconductor described is called an extrinsic semiconductor with electron conductivity or an n-type semiconductor. Group V elements such as phosphorus, arsenic, and antimony are usually used as donors in germanium and silicon.

The ionization energy of these atoms is of the order of 0.01 eV, so that at room temperature the donor atoms are completely ionized.

Another example of an extrinsic semiconductor is illustrated in Fig. 3. Should the foreign atoms have a high affinity for electrons and form negative ions readily, then electrons from the filled band are transferred to the foreign atoms. The places vacated in the filled band become holes. The foreign atoms which trap electrons from the valence band are called acceptors, while the semiconductors containing the acceptor impurity are called extrinsic semiconductors with hole conductivity or p-type semiconductors. The most common acceptor impurities for germanium and silicon are Group III elements, namely, boron, aluminum, gallium, and indium.

Thus, high conductivity may be produced in materials with a wide forbidden band by introducing the appropriate impurity.

§3. Statistical Theory of Semiconductors*

Let us now examine the quantitative relations which make it possible to find the concentrations of current carriers in the conduction and valence bands as functions of temperature and the width of the forbidden band.

The distribution of electrons within the levels is described by the Fermi function [10] $f(E)$ which gives the probability of an electron level of energy E being filled by an electron:

---

* Below we present concepts on semiconductors based on the so-called one-electron approximation. It is assumed that a many-electron system, which is represented by a solid, may be reduced to the problem of the motion of one electron in the periodic field of a crystal lattice and in the self-consistent field of the other electrons [10]. Such concepts of the one-electron approximation as a band, effective mass, etc., are being replaced in the many-electron theory being developed at the present time by other concepts such as collective excitation spectrum, quasiparticle, etc.

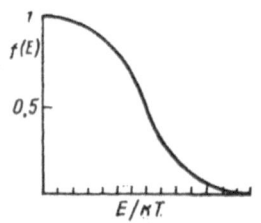

Fig. 4. Relation of the Fermi
function to energy, measured
in kT units.

$$f(E) = \cfrac{1}{1+\exp\left(\cfrac{E - E_F}{kT}\right)} \quad . \tag{3.1}$$

Here, k is Boltzmann's constant and T is the absolute temperature.

The energy level with an energy E_F is called the Fermi level or the chemical potential of the electrons in the system. Figure 4 gives the form of the Fermi function. At T = 0°K, $f(E)$ = 1 when E < E_F and $f(E)$ = 0 when E > E_F. When T > 0°K, the function $f(E)$ changes from 1 to 0 over an energy region with a width of several kT.

If the energy of a level differs considerably from the Fermi energy (E − E_F ≫ kT), then in (3.1) we can neglect unity in comparison with exp (E − E_F/kT), and the Fermi distribution becomes a Boltzmann distribution:

$$f = e^{\frac{E_F - E}{kT}} \quad . \tag{3.2}$$

We then find the electron concentration in the conduction band, i.e., the number of electrons in all the energy levels of this band.

As we know from quantum mechanics, the number of quantum states $\rho(E)dE$ in 1 cm^3 of a crystal in the energy range from E to E + dE in the conduction band is determined by the expression [3]*

* Equation (3.3) may be derived on the basis of various assumptions, for example, for an electron that is moving in a so-called square potential well. It is assumed that the electron is moving in a closed area of space within which there are no forces. Infinitely great forces operate at the boundaries of this area to keep the electron within its limits. This formula may also be derived by a "quasiclassical" approximation [10].

$$P(E)\,dE = a(E - E_c)^{1/2}dE, \quad a = \frac{4\pi\,(2m_n)^{3/2}}{h^3},$$

(3.3)

where m_n is the effective electronic mass, h is Planck's constant, and E_C is the lowest energy level of the conduction band.

The product $f(E)\rho(E)dE$ gives the number of electrons in the energy range from E to $E + dE$. Then the integral

$$\int_{E_c}^{E'_c} \rho(E)f(E)\,dE,$$

(where E'_c is the upper energy level of the conduction band) determines the concentration of electrons n in the conduction band. Using equations (3.1) and (3.3), we obtain

$$n = a \int_{E_c}^{E'_c} \frac{(E - E_c)^{1/2}}{1 + \exp\left(\dfrac{E - E_F}{kT}\right)}\,dE.$$

(3.4)

Two simplifications may be introduced into equation (3.4). Since, usually $E_C - E_F \gg kT$ at room temperature, unity may be neglected over the whole integration range in comparison with $\exp(E - E_F/kT)$:

$$n = a \int_{E_c}^{E'_c} \exp\left(\frac{E_F - E}{kT}\right)(E - E_c)^{1/2}\,dE = a e^{\frac{E_F}{kT}} \int_{E_c}^{E'_c} e^{-\frac{E}{kT}}(E - E_c)^{1/2}\,dE.$$

(3.5)

The integrand in equation (3.5) decreases extremely rapidly with energy and therefore the value of the upper limit E'_c may be replaced by infinity with sufficient accuracy. *

----

*We are assuming that an electron is described by a single effective mass. In any case, this is so for a crystal with a cubic lattice. The effective mass tensor should be introduced in the case of more complex crystals. We should also note that the concept of effective mass describes well the motion of electrons only at the bottom of the energy band. As the value of the integral in equation (3.4) is determined mainly by the value of the integrand expression at the bottom of the band (because of the rapidly falling exponent with the index E/kT), this is a good approximation.

Then the integral in equation (3.5) is readily calculated:

$$n = a \, (kT)^{3/2} \, \frac{\pi^{1/2}}{2} \, e^{-\frac{E_c - E_F}{kT}} = N_c e^{-\frac{E_c - E_F}{kT}} \, ,$$

(3.6)

where the value N_C is determined by the relation

$$N_c = \frac{a}{2}(kT)^{3/2} \pi^{1/2} = 2\left(\frac{2\pi m_n kT}{h^2}\right)^{3/2} \, .$$

(3.7)

The concentration of electrons in the conduction band in relation to temperature and the position of the Fermi level relative to the edge of the conduction band E_C may be determined from equation (3.6). For germanium at room temperature, $N_C = 5 \cdot 10^{19}$ cm^{-3}.

The relation of the density of holes in the valence band to the Fermi level E_F and temperature T may be derived similarly:

$$p = N_v e^{-\frac{E_F - E_v}{kT}} \, ,$$

(3.8)

$$N_v = 2 \left(\frac{2\pi m_p kT}{h^2}\right)^{3/2} \, ,$$

(3.9)

where E_V is the energy of the upper edge of the valence band and m_p is the effective mass of the holes.

§ 4.   Position of the Fermi Level in Intrinsic and Extrinsic Semiconductors

The concentrations of electrons and holes are equal in an intrinsic semiconductor. By using equation (3.6) and (3.8) we obtain the relation

$$n_i = N_c e^{-\frac{E_c - E_F}{kT}} = N_v e^{-\frac{E_F - E_v}{kT}} \, ,$$

(4.1)

where n_i is the concentration of electrons in an intrinsic semiconductor. By solving equation (4.1) with respect to E_F, we find

$$E_F = \frac{E_c + E_v}{2} + \frac{kT}{2} \ln \frac{N_v}{N_c}. \qquad (4.2)$$

By using the relations (3.7) and (3.9) we have

$$E_F = \frac{E_c + E_v}{2} + \frac{3}{4} kT \ln \frac{m_p}{m_n}. \qquad (4.3)$$

For example, in the case of germanium, the effective masses of electrons and holes are close to each other and, as follows from equation (4.3), the Fermi level in germanium with intrinsic conduction is close to the center of the forbidden band.

The concentration of electrons in an intrinsic semiconductor n_i may be found from equations (3.6) and (3.8)

$$n_i^2 = N_c N_v e^{-\frac{E_c - E_v}{kT}} \qquad (4.4)$$

or

$$n_i = \sqrt{N_c N_v}\, e^{-\frac{E_c - E_v}{2kT}}. \qquad (4.5)$$

Expression (4.5) shows that the number of electrons in an intrinsic semiconductor depends on the width of the forbidden band $E_g = E_c - E_v$. For germanium, $E_g \approx 0.7$ eV and, as follows from (4.5), the number of electrons in intrinsic germanium equals $n_i = 2.5 \cdot 10^{13}$ cm^{-3} at room temperature. The concentration of electrons in semiconductors with a wide forbidden band is considerably less.

Let us now examine extrinsic semiconductors. The condition for electroneutrality may be written in the form

$$p + N_D - n - N_A = 0, \qquad (4.6)$$

where N_D and N_A are the concentrations of fully ionized donor and acceptor impurity centers, respectively. *

As a concrete example, let us examine an n-type sample in which the concentration of donor centers considerably exceeds that of acceptor centers: $N_D \gg N_A$; then equation (4.6) becomes

$$p + N_D - n = 0. \tag{4.7}$$

Looking ahead, let us assume that the density of holes in the valence band is considerably less than the density of electrons; then equation (4.7) will be written in the form

$$n = N_D. \tag{4.8}$$

By substituting this value for the concentration of electrons in equation (3.6) we find the Fermi energy:

$$E_F = E_c - kT \ln \frac{N_c}{N_D} = E_c - kT \ln \frac{N_c}{n}. \tag{4.9}$$

The concentration of holes may be determined from equations (3.8) and (4.9):

$$p = \frac{N_c N_v}{N_D} e^{-\frac{E_c - E_v}{kT}}.$$

We should note that the following relation holds:

$$p \cdot n = N_c \cdot N_v\, e^{-\frac{E_c - E_v}{kT}} \equiv n_i^2. \tag{4.10}$$

We may now check the accuracy of the assumption made previously, that $p \ll n$. By substituting $n \approx N_D$ in equation (4.10) we have $p = n_i^2 / N_D \ll n$. The concentrations of electrons and holes in a p-type semiconductor in which $p \approx N$ may be found readily

---

*The case of incomplete ionization of impurities is examined by Shockley [3], for example.

through similar reasoning. We have

$$E_F = E_v + kT \ln \frac{N_v}{N_A}.$$

(4.11)

for the position of the Fermi level.

Electrons and holes carry the electric current in a semi-conductor. However, the two types of carrier do not partici-pate equally in the transmission of a current in semiconductors with extrinsic conduction. The electric current is carried mainly by electrons and hardly at all by the holes in an n-type semiconductor. The opposite is observed in a p-type semicon-ductor. In this connection, the terms "majority" and "mino-rity" current carriers are used. In an n-type semiconductor, the majority carriers are electrons and the minority carriers, the holes, while in a p-type semiconductor the reverse is the case.

§ 5. Surface Electron Levels in a Semiconductor

In addition to the system of energy levels characteristic of the bulk of the semiconductor crystal examined in § 1 there are surface electron levels at the semiconductor–vacuum or semiconductor–electrolyte boundary. They correspond to quan-tum states in which an electron is located at the surface and cannot move into the bulk without exchanging energy with the outer medium.

Tamm [11] was the first to report the existence of sur-face states. The periodicity of the electric fields in a homo-geneous crystal is disrupted at the semiconductor–vacuum boundary. This results in the formation of electronic states lo-calized at the surface, i.e., surface levels. Surface states are also formed by the adsorption of foreign atoms on a semicon-ductor surface (Shockley levels).

The surface levels are characterized by the energy state E^t and concentration N. It is reasonable to examine only the levels lying in the forbidden band. The levels are arbitrarily classified as fast or slow with respect to relaxation time, i.e., the time required to establish an equilibrium concentration of electrons and holes. Fast states have a relaxation time of the order of one millisecond or less; the relaxation time of slow states is from one second to minutes or hours.

The number of electrons n_t in surface states with an energy of E^t is determined by the equation [4]

$$n_t = \frac{N_t}{1 + g\, \exp\left(\dfrac{E_t - E_F}{kT}\right)} = N_t f_t, \text{ where } f_t = \frac{1}{1 + g\, \exp\left(\dfrac{E_t - E_F}{kT}\right)}. \tag{5.1}$$

The properties of the surface level determine the coefficient g. For example, g = 1 if two electrons can be trapped by the level (one electron for each spin orientation). If the surface (donor) level can trap one electron with random spin orientation or remain unfilled, then $g = \tfrac{1}{2}$. Finally, if the surface (acceptor) level can hold two electrons with twin spins or one electron with a random spin, g = 2.

Without limiting its generality we will in future use the filling function of the level f_t in the form

$$f_t = \frac{1}{1 + \exp\left(\dfrac{E_t - E_F}{kT}\right)}. \tag{5.2}$$

Actually, the coefficient g may be written in the form g = exp $(\Delta E_t / kT)$, and equation (5.1) assumes the form

$$n_t = \frac{N_t}{1 + \exp\left(\dfrac{E_t' - E_F}{kT}\right)},$$

where $E_t' = \Delta E_t + E_t$. Such a transformation corresponds to the renormalization of the energy state of the surface level.

If the electrons on the surface levels are in equilibrium with the electrons within the bulk, they have a common Fermi level. The distribution function f_t may be related to the concentration of the electrons within the bulk. For this purpose we

will multiply the numerator and the denominator in equation
(5.2) by $N_c e^{-(-Ec/kT)}$; then

$$f_t = \frac{N_c e^{\frac{E_F - E_c}{kT}}}{N_c e^{\frac{E_F - E_c}{kT}} + N_c e^{\frac{E_t - E_c}{kT}}} \cdot$$

(5.3)

Using equation (3.6), we will write equation (5.3) in the form

$$f_t = \frac{n}{n + n'} ,$$

(5.4)

where, in accordance with equation (3.6), $n' = N_c e^{(Et - Ec/kT)}$
represents the concentration of electrons in the conduction band
when the Fermi level coincides with the surface level. By using equation (4.10), the Fermi distribution function may also be
expressed in terms of the concentration of holes:

$$f_t = \frac{p}{p' + p} ,$$

(5.5)

where p' is the concentration of holes in the valence band when
the Fermi level coincides with the surface electron level.

As we used only the properties of the Fermi distribution
in deriving equations (5.4) and (5.5), these expressions are also
true for bulk impurity levels. We should note that the physicochemical nature of the surface states is still obscure. In particular, we do not know the significance from the chemical
point of view of the process of filling of the surface levels by
electrons described formally by equation (5.1) [12]. In some
cases the specific chemical materials which produce levels on
the surface of a semiconductor have been found; however, the
microscopic picture remains confused (see § 21 for details).

§ 6. Some Problems of Recombination
Phenomena

If the equilibrium of a semiconductor is disrupted by the
introduction of nonequilibrium carriers, the system tends to
return to a state of equilibrium. For example, nonequilibrium

electron-hole pairs are produced by the illumination of a sample. The nonequilibrium electrons pass from the conduction band to free positions, or holes, in the valence band. These processes are called electron and hole recombination processes. With each recombination act, one electron disappears from the conduction band and one hole from the valence band.

Recombination may occur in several different ways. For example, electrons may pass directly from the conduction to the valence band. In this case the excess energy of the electron may be transmitted either to the crystal lattice or be emitted in the form of electromagnetic radiation. Calculations have shown that the second process, that of light emission, is more probable in the case of direct electron transfer.

Another recombination mechanism is possible when there are deep energy levels or traps (close to the center of the forbidden band). Recombination occurs in the following manner. First an electron passes to the level of the trap; the trap then captures a hole and recombination occurs. It has been found that the probability of energy emission in successive portions is considerably higher than the emission of a large amount of energy at once. Therefore, the probability of recombination through a trap exceeds by far that of direct recombination.

Recombination on the surface is a particular case of the second type of process. The surface levels act as traps in this case.

We will now consider the theory of recombination through traps developed by Shockley and Read [13].

Let us examine the following four processes: 1) transfer of electrons from the valence band to the level of the trap; 2) return from the traps to the valence band of some of the electrons that did not have time to recombine; 3) transfer of holes from the valence band to the level of the traps; and, 4) transfer of holes from the level of the trap back to the valence band.

Let us first find the number of electrons from the conduction band captured by the traps during unit time. This value R_c is proportional to the number of electrons in the conduction band and the number of vacancies in the levels. If f_t denotes the fraction of traps occupied by electrons, then the number of vacancies in the levels will be proportional to the value $(1 - f_t)$. Thus

$$R_c = C_n n (1 - f_t), \tag{6.1}$$

where C_n is a certain constant which is proportional to the concentration of traps N_t and independent of the electron concentration:

$$C_n = N_t \cdot \sigma_n \cdot v.$$

Here σ_n is the trapping cross section of the level for an electron and v is the thermal velocity of an electron.

It is obvious that C_n is the probability of trapping an electron when all the traps are vacant.

The number of electrons R_C passing from the trap to the conduction band during unit time is evidently proportional to the number of electrons in the trap levels:

$$R_e = C_n' f_t. \tag{6.2}$$

The constants C_n and C_n' are not independent. In actual fact, $R_e = R_C$ in the equilibrium state. Therefore,

$$C_n' = C_n \frac{n^0 (1 - f_t^0)}{f_t^0}. \tag{6.3}$$

The superscript 0 indicates that the values refer to the equilibrium state.

Let us then assume that all the traps are situated in the same energy level. Then, by using equation (5.4) for C_n', we have

$$C_n' = C_n n', \tag{6.4}$$

where n' is the concentration of electrons in the conduction band when the Fermi level passes through the level of the trap (see p.13). Using equations (6.1), (6.2), and (6.4), we obtain for the overall flow of electrons to the trap level $R_n = R_c - R_e$,

$$R_n = C_n [(1 - f_t) n - n' f_t].$$

(6.5)

Analogous reasoning gives the following expression for the number of holes captured by the traps in unit time:

$$R_p = C_p [f_t p - p' (1 - f_t)],$$

(6.6)

where p' is the number of holes in the valence band when the Fermi level coincides with the trap level.

Let us now examine the case where a steady concentration of nonequilibrium electrons and holes is maintained in the bulk of the semiconductor, for example, with steady illumination of the semiconductor. In this case, $R_n = R_p = R$. Using equations (6.5) and (6.6) we find the degree of filling f_t:

$$f_t = \frac{n C_n + p' C_p}{C_n (n + n') + C_p (p + p')}.$$

(6.7)

Substituting the value found for f_t in equation (6.5), we obtain

$$R = \frac{C_n \cdot C_p (np - n_i^2)}{C_n (n + n') + C_p (p + p')}.$$

(6.8)

Let us write the expression for the concentrations n and p in the form

$$n = n^0 + \Delta n,$$

$$p = p^0 + \Delta p,$$

(6.9)

where n^0 and p^0 are the equilibrium concentrations of electrons and holes, and Δn and Δp the deviations from these concentrations resulting from external effects.

We will then assume that $\Delta n = \Delta p$. * Substituting expression (6.9) for n and p in formula (6.8), we have

---

*The conditions for this assumption to apply are examined in § 28.

$$R = \frac{C_n \cdot C_p \left(n^0 \Delta n + p^0 \Delta p + \Delta n^2\right)}{C_n \left(n^0 + n' + \Delta n\right) + C_p \left(p^0 + p' + \Delta n\right)} . \tag{6.10}$$

Assuming that Δn is small, i.e., that the following inequalities hold:

$$n^0 \gg \Delta n, \quad p^0 \gg \Delta p,$$

from equation (6.10) we find

$$R = \frac{C_n \cdot C_p \left(n^0 + p^0\right) \Delta n}{C_n \left(n^0 + n'\right) + C_p \left(p^0 + p'\right)} . \tag{6.11}$$

We determine the lifetime of the nonequilibrium carriers τ from the relation

$$R = \frac{\Delta n}{\tau} = \frac{n - n^0}{\tau} = \frac{p - p^0}{\tau} . \tag{6.12}$$

Then

$$\frac{1}{\tau} = \frac{C_n \cdot C_p \left(n^0 + p^0\right)}{C_n \left(n^0 + n'\right) + C_p \left(p^0 + p'\right)} . \tag{6.13}$$

Let us examine the relation of the lifetime τ (with the given parameters of the trap) to the concentration of electrons and holes. The relation $n^0 \gg p^0$ holds for an n-type semiconductor. Let us assume that $n^0 \gg n'$, then

$$\frac{1}{\tau} = C_p \equiv \frac{1}{\tau_{p0}} . \tag{6.14}$$

In a p-type semiconductor when $p^0 \gg p'$,

$$\frac{1}{\tau} = C_n \equiv \frac{1}{\tau_{n0}} . \tag{6.15}$$

The times τ_{n0} and τ_{p0} introduced by equations (6.14) and (6.15) are independent of the concentrations of electrons and holes n and p. In the general case, the lifetime τ depends substantially on these concentrations.

§ 7. The Boltzmann Relation and its Applicability when a Current Flows Through a System

The previous paragraphs dealt with semiconductors with a homogeneous distribution of the concentrations of free carriers. Let us now examine the case where there is an electric field in the semiconductor so that the potential depends on the coordinate. Let us find the concentration of electrons n as a function of the potential φ at a given point. When an external electric field is applied, the energy of the electron E will change by a value $e(\varphi - \varphi_B)$, where the potential φ_B is selected as the base line of the electrostatic potential and e is the absolute value of the electronic charge.

Thus, equation (3.2) assumes the form

$$f = e^{\frac{E_F - E + e(\varphi - \varphi_B)}{kT}}.$$

$$(7.1)$$

In accordance with equation (3.5) the concentration of electrons n equals

$$n = e^{\frac{e(\varphi - \varphi_B)}{kT}} a \int_{E_c}^{E_c'} (E - E_c)^{\frac{1}{2}} e^{\frac{E_F - E_c}{kT}} dE.$$

$$(7.2)$$

Integration (as carried out in § 3) gives us

$$n = n^0 \cdot e^{\frac{e(\varphi - \varphi_B)}{kT}},$$

$$(7.3)$$

where n^0 is the concentration of electrons at the point with the potential φ_B.

Analogously, we have the following relation for the concentration of holes:

$$p = p^0 \cdot e^{-\frac{e(\varphi - \varphi_B)}{kT}},$$

$$(7.4)$$

where p^0 is the concentration of holes when $\varphi = \varphi_B$.

Equations (7.3) and (7.4) are called the Boltzmann equations. We shall also give another derivation for these equations as it is more convenient for further discussion. For a definite example we will assume that the electric field is directed along the x axis. Let us examine the distribution of holes.

The electric field applied produces a hole drift current whose density is determined by the relation

$$i_p^{dr} = eU_p pE, \qquad (7.5)$$

where U_p is the mobility of holes and E is the electric field.

Since, generally speaking, the concentration of holes changes with the coordinate, a diffusion current of holes is produced which is proportional to the gradient of their concentration:

$$i_p^{diff} = -eD_p \frac{dp}{dx}, \qquad (7.6)$$

where D_p is the diffusion coefficient of the holes.

The total hole current equals

$$i_p = eU_p pE - eD_p \frac{dp}{dx}. \qquad (7.7)$$

By using Einstein's equation relating the mobility of carriers to the diffusion coefficient,

$$D_p = \frac{kT}{e} U_p, \qquad (7.8)$$

we may rewrite equation (7.7) in the form

$$i_p = U_p \left(epE - kT \frac{dp}{dx} \right). \qquad (7.9)$$

By taking into consideration equation $E = -d\varphi/dx$, we have

$$i_p = U_p \left(-ep \frac{d\varphi}{dx} - kT \frac{dp}{dx} \right). \qquad (7.10)$$

Under equilibrium conditions when the current i_p equals zero, equation (7.10) assumes the form

$$U_p \left(-ep \frac{d\varphi}{dx} - kT \frac{dp}{dx} \right) = 0.$$

(7.11)

The latter equation may be readily integrated:

$$p = Ae^{-\frac{e\varphi}{kT}}.$$

(7.12)

The integration constant A is found from the boundary condition that the concentration equals p^0 at the point with the potential φ_B. Then from equation (7.12) we have

$$p = p^0 e^{-\frac{e(\varphi - \varphi_B)}{kT}}.$$

(7.13)

An expression analogous to (7.9) may be readily derived for the electron current i_n. The total electron current equals

$$i_n = U_n \left(enE + kT \frac{dn}{dx} \right).$$

(7.14)

At equilibrium $i_n = 0$ and integration of equation (7.14) gives the Boltzmann equation (7.3).

The Boltzmann distribution is disrupted when an electric current flows through the semiconductor. However, under certain conditions the following relations hold:

$$|i_p| \ll |eU_p pE|; \ |i_p| \ll \left| eD_p \frac{dp}{dx} \right|.$$

(7.15)

Then in equation (7.9) the current i_p may be neglected as compared with the terms in the right-hand part, and we again arrive at equation (7.11), whose integration gives the Boltzmann distribution. Actual cases in which inequalities (7.15) hold will be examined in Chapter II.

Literature Cited

1.   A. F. Ioffe, Physics of Semiconductors, Izd. Akad. Nauk Akad. Nauk SSSR, Moscow (1957).
2.   Semiconductors in Science and Technology, Vols. 1 and 2, Izd. Akad. Nauk SSSR, Moscow (1957).

3. W. Shockley, Theory of Electronic Semiconductors, Van Nostrand, New York (1950).

4. R. Smith, Semiconductors, Cambridge Univ. Press, New York (1959).

5. R.D. Middlebrook, Introduction to the Theory of Transistors, Wiley, New York (1957).

6. C. Dunlap, Introduction to the Physics of Semiconductors [Russian translation], IL, Moscow (1959).

7. Ya. A. Fedotov, Fundamental Physics of Semiconductor Instruments, Moscow, "Sovetskoe Radio," (1963).

8. C. Kittel, Introduction to Solid State Physics, GITTL, Moscow (1957).

9. Physics of the Surface of Semiconductors, Collection of translations, G. E. Pikus, ed., IL, Moscow (1959).

10. V.G. Levich, Course in Theoretical Physics, Vol. 1, Fizmatgiz, Moscow (1962).

11. I.E. Tamm, Phys. Z. Sowjetunion, 1:733 (1932).

12. F.F. Vol'kenshtein, Electronic Theory of Catalysis Applied to Semiconductors, Fizmatgiz, Moscow (1960).

13. W. Shockley and W.T. Read, Phys. Rev., 87:835 (1952).

Chapter I
The Semiconductor—Electrolyte System in Equilibrium

§8. Equilibrium Electrode Potential. Galvanic Potential

Let us examine an electrochemical cell with a semiconductor electrode.

The cell consists of metal a, a semiconductor electrode, an electrolyte, a reference electrode, and metal a:

Metal a	Semiconductor	Electrolyte	Reference electrode	Metal a

The potential difference between the metal ends a of the cell is called the electrode potential. It is composed of the potential differences at the boundaries of the phases making up the circuit. Thus there is a potential drop between the metal a and the semiconductor, between the semiconductor and the electrolyte, etc. These interphase potential differences are called galvanic potentials. They arise as a result of the transfer of charges across the phase boundaries. Let us examine, for example, the origin of a galvanic potential at the semiconductor—electrolyte boundary. Let us assume that the following reaction can occur on the semiconductor electrode:

$$A^{2-} \rightleftarrows A^- + e^-, \qquad (8.1)$$

i.e., the ion A^{2-} on the electrode donates an electron to the semiconductor and becomes the ion A^-. If an uncharged semiconductor electrode is brought into contact with an electrolyte containing the ions A^{2-} and A^-, there is a transfer of electrons between the ions and the semiconductor. This transfer will continue until equilibrium is established in the system. The transfer of electrons results in the appearance of charges with different signs in the two phases and this produces the potential difference at the phase boundary.

Let us find the relation between the galvanic and electrode potentials and the properties of the semiconductor and electrolyte [1]. In order to calculate the galvanic potential let us write the equilibrium conditions for reaction (8.1)

$$\bar{\mu}_{A^{2-}} = \bar{\mu}_{A^-} + \bar{\mu}_n. \tag{8.2}$$

Here $\bar{\mu}_{A^{2-}}$ and $\bar{\mu}_{A^-}$ are the electrochemical potentials of the ions A^{2-} and A^-, respectively, and $\bar{\mu}_n$ is the electrochemical potential of the electrons in the semiconductor.

Let us separate the chemical potentials μ and the electrostatic potential φ from the electrochemical potentials; then

$$\mu_{A^{2-}} - 2F\varphi_c = \mu_{A^-} - F\varphi_c + \mu_n - F \cdot \varphi_B, \tag{8.3}$$

where $\mu_{A^{2-}}$ and μ_{A^-} are the chemical potentials of the corresponding ions; φ_C and φ_B are the electrostatic potentials in the bulk of the solution and in the bulk of the semiconductor; and F is a faraday.

The galvanic potential $\varphi_{12} = \varphi_C - \varphi_B$ may be found from equation (8.3)

$$F\varphi_{12} = \mu_{A^{2-}} - \mu_{A^-} - \mu_n. \tag{8.4}$$

Equation (8.4) shows directly that the galvanic potential depends on the chemical potential μ_n of the electrons in the semiconductor.

Let us calculate the electrode potential. For this we will divide the potential difference between the ends of the circuit into two: φ_{ac} and φ_{ca}, where φ_{ac} is the potential difference between the ends of the circuit metal a – semiconductor – electrolyte$^-$ and φ_{ca} is the potential difference between the ends of the circuit electrolyte$-$reference electrode$-$metal a. The electrode reaction on the reference electrode and, consequently, the potential difference in the system electrolyte$-$reference electrode, is independent of the properties of reaction (8.1) and those of the semiconductor. To calculate the potential difference φ_{ac} we will use the condition that when there is thermodynamic equilibrium the electrochemical potential of electrons in the metal $\bar{\mu}_a$ equals the electrochemical potential of electrons in the semiconductor $\bar{\mu}_n$. Then relation (8.2) may be written in the form

$$\bar{\mu}_{A^{2-}} = \bar{\mu}_{A^-} + \bar{\mu}_{a}.$$

(8.5)

Relation (8.3) becomes

$$\mu_{A^{2-}} - 2F\varphi_c = \mu_{A^-} - F\varphi_c + \mu_a - F\varphi_a,$$

(8.6)

where φ_a is the potential in the bulk of the metal a. Now the electrode potential φ may be found:

$$\varphi = \varphi_{ac} + \varphi_{ca} = \varphi_c - \varphi_a + \varphi_{ca} =$$

$$= \frac{1}{F}(\mu_{A^{2-}} - \mu_{A^-} - \mu_a) + \varphi_{ca} = \frac{1}{F}(\mu_{A^{2-}} - \mu_{A^-}) + \text{const},$$

(8.7)

where $\text{const} = -(1/F)\mu_a + \varphi_{ca}$ is independent of the properties of the semiconductor.

Thus, the electrode potential depends on the chemical potentials $\mu_{A^{2-}}$ and μ_{A^-} of the ions in the electrolyte solution but, in contrast to the galvanic potential, it is independent of the chemical potential of the electrons in the semiconductor. Consequently, the impurities which are introduced into the semiconductor material and which determine its electrical properties have no effect on the equilibrium electrode potential if they do not change the chemical nature of the semiconductor.

Using equation (8.4) further, we rewrite equation (8.7):

$$F\varphi = F\varphi_{12} + \mu_n + \text{const}.$$

(8.8)

In accordance with formula (4.9), the chemical potential of the electrons in the semiconductor equals*

$$\mu_n = -kTN \ln \left(\frac{N_c}{n_i} \cdot \frac{n_i}{n^0}\right) + NE_c = -kTN \ln \lambda + \text{const},$$

(8.9)

where $\lambda = n_i/n^0$, and n^0 is the electron concentration in the bulk of the semiconductor. Therefore, division of equation (8.8) by F gives

$$\varphi = \varphi_{12} - (kT/e)\ln \lambda + \text{const}',$$

(8.10)

where const' is a constant that is independent of the chemical potential of the electrons in the semiconductor.

The case examined above dealt with an equilibrium reaction of type (8.1) in which the semiconductor was an inert electrode. However, similar results may be obtained when the material of the electrode participates in the equilibrium reaction determining the potential. Such calculations based on thermodynamic values (heats of formation of the appropriate substances) were carried out for certain semiconductor materials which are important from the practical point of view. Expressions for the equilibrium electrode potentials in the system germanium—water are given below [2].†

$$H_2GeO_3 = HGeO_3^- + H^+,$$
$$\log \frac{[H_2GeO_3]}{[HGeO_3^-]} = 8.72 - pH;$$

(8.11)

---

*In chemistry the chemical potential usually refers to a single mole (so that $E_F = \mu_n/N$, where N is Avogadro's number), in solid state physics, to a single particle. We should bear in mind that F = eN.

† Practically the same results were obtained in [3]. The calculations carried out in [4,5] were based partly on incorrect thermodynamic data. In equations (8.11)-(8.30) the concentration is expressed in gram-equivalents per liter and the potential in volts. If not otherwise stated, the values of the electrode potentials are relative to a normal hydrogen electrode.

$$HGeO_3^- = GeO_3^{2-} + H^+,$$

$$\log \frac{[HGeO_3^-]}{[GeO_3^{2-}]} = 12.70 - pH; \qquad (8.12)$$

$$H_2GeO_3 + 4H^+ + 2e^- = Ge^{2+} + 3H_2O,$$
$$\varphi^0 = -0.223 - 0.1182\,pH + 0.0295 \log \frac{[H_2GeO_3]}{[Ge^{2+}]}; \qquad (8.13)$$

$$HGeO_3^- + 5H^+ + 2e^- = Ge^{2+} + 3H_2O;$$
$$\varphi^0 = 0.035 - 0.1477\,pH + 0.0295 \log \frac{[HGeO_3^-]}{[Ge^{2+}]}; \qquad (8.14)$$

$$GeO_3^{2-} + 6H^+ + 2e^- = Ge^{2+} + 3H_2O,$$
$$\varphi^0 = 0.410 - 0.1773\,pH + 0.0295 \log \frac{[GeO_3^{2-}]}{[Ge^{2+}]}; \qquad (8.15)$$

$$GeO_2 \ (\text{hexagonal}) + 4H^+ + 4e^- = Ge + 2H_2O,$$
$$\varphi^0 = -0.009 - 0.0591\,pH; \qquad (8.16)$$

$$GeO \ (\text{black}) + 2H^+ + 2e^- = Ge + H_2O,$$
$$\varphi^0 = 0.089 - 0.0591\,pH; \qquad (8.17)$$

$$GeO \ (\text{brown}) + 2H^+ + 2e^- = Ge + H_2O.$$
$$\varphi^0 = 0.100 - 0.0591\,pH; \qquad (8.18)$$

$$GeO \ (\text{yellow}) + 2H^+ + 2e^- = Ge + H_2O,$$
$$\varphi^0 = 0.256 - 0.0591\,pH; \qquad (8.19)$$

$$GeO_2 \ (\text{hexagonal}) + 2H^+ + 2e^- = GeO \ (\text{brown}) + H_2O,$$
$$\varphi^0 = -0.117 - 0.0591\,pH; \qquad (8.20)$$

$$GeO \ (\text{brown}) + 2H^+ = Ge^{2+} + H_2O,$$
$$\log [Ge^{2+}] = -4.99 - 2\,pH; \qquad (8.21)$$

$$GeO_2 \ (\text{hexagonal}) + H_2O = HGeO_3^- + H^+, \qquad (8.22)$$

$$\log [HGeO_3^-] = -10.1 + pH;$$
$$Ge^{2+} + 2e^- = Ge, \qquad (8.23)$$

$$\varphi^0 = 0.247 + 0.0295 \log [\text{Ge}^{2+}];$$

$$H_2GeO_3 + 4H^+ + 4e^- = Ge + 3H_2O,$$
$$\varphi^0 = 0.012 - 0.0591\,pH + 0.0148 \log [H_2GeO_3]; \tag{8.24}$$

$$HGeO_3^- + 5H^+ + 4e^- = Ge + 3H_2O,$$
$$\varphi^0 = 0.141 - 0.0738\,pH + 0.0148 \log [HGeO_3^-]; \tag{8.25}$$

$$GeO_3^{2-} + 6H^+ + 4e^- = Ge + 3H_2O,$$
$$\varphi^0 = 0.328 - 0.0886\,pH + 0.0148 \log [GeO_3^{2-}]; \tag{8.26}$$

$$H_2GeO_3 + 2H^+ + 2e^- = GeO \text{ (brown)} + 2H_2O,$$
$$\varphi^0 = -0.076 - 0.0591\,pH; \tag{8.27}$$

$$HGeO_3^- + 3H^+ + 2e^- = GeO \text{ (brown)} + 2H_2O,$$
$$\varphi^0 = 0.182 - 0.0886\,pH + 0.0295 \log [HGeO_3^-]; \tag{8.28}$$

$$GeO_3^{2-} + 4H^+ + 2e^- = GeO \text{ (brown)} + 2H_2O,$$
$$\varphi^0 = 0.557 - 0.1182\,pH + 0.0295 \log [GeO_3^{2-}]; \tag{8.29}$$

$$Ge + 4H^+ + 4e^- = GeH_4,$$
$$\varphi^0 = -0.422 - 0.0591\,pH - 0.0148 \log P_{GeH_4}. \tag{8.30}$$

Figure 5 gives a graph of potential against pH plotted from the equations given above. Similar calculations and graphs are given for silicon in [3, 5, 6].

Up to the present time only a few cases have been known where a semiconductor electrode in a solution containing an oxidation–reduction system acquires an equilibrium potential. Normally the steady potential of a semiconductor electrode differs from the equilibrium value. There are two reasons for this. One of them is of a fundamental nature. As will be shown in § 25, the exchange currents at semiconductor electrodes are many orders less than those at metal electrodes. The low value of the exchange current makes it difficult to de-

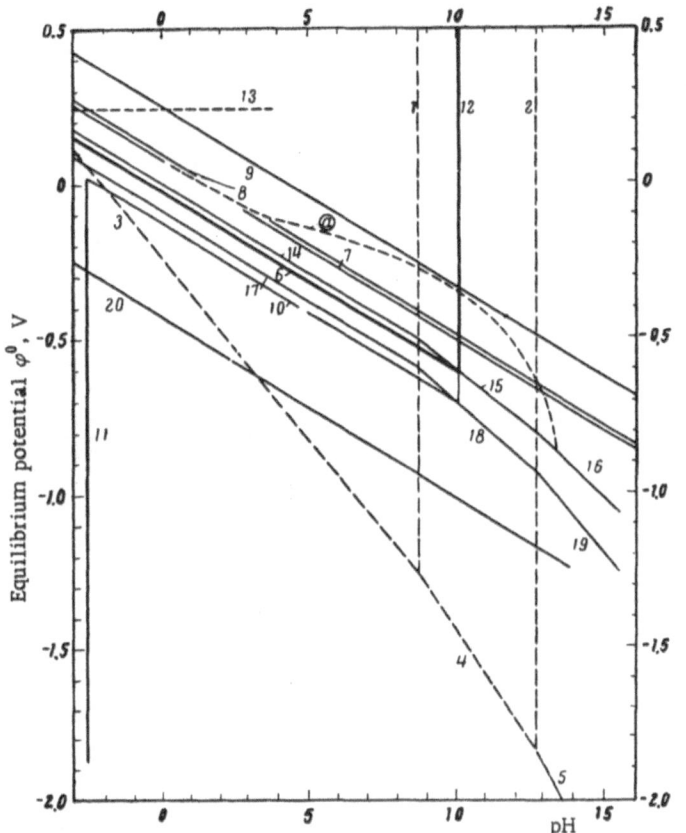

Fig. 5. Graph of potential against pH for the system germanium—
water. The numbers of the curves refer to the reaction equilibrium:
1) (8.11); 2) (8.12); 3) (8.13); 4) (8.14); 5) (8.15); 6) (8.16); 7)
(8.17); 8) (8.18); 9) (8.19); 10) (8.20); 11) (8.21); 12) (8.22); 13)
(8.23); 14) (8.24); 15) (8.25); 16) (8.26); 17) (8.27); 18) (8.28); 19)
(8.29); 20) (8.30). a) Relation of steady potential of germanium
electrode to pH measured experimentally [2].

termine and measure the reversible potential. Over a wide
range of potentials, the electrode behaves as an ideally polarizable one. For various incidental reasons, adsorption processes and electrostatic charging affect the electrode potential.

The other reason, which is of a particular nature, is the fact that the semiconductor materials studied most up to now, namely, germanium, silicon, cadmium, sulfide, and gallium arsenide, undergo corrosion in aqueous solutions. Their potentials are corrosion and not equilibrium potentials (see Chapter IV).

The equilibrium oxidation—reduction potential in the $Fe(CN)_6^{3-}/Fe(CN)_6^{4-}$ system was apparently observed on an electrode of ZnO [7], CdS [8], and polytetracyanoethylene (as a film deposited on glass) [9].

§ 9. Structure of the Electrical Double Layer and Distribution of the Galvanic Potential at the Electrolyte — Semiconductor Interface

The relation of the galvanic potential to the properties of the solution and the semiconductor was examined in the previous section. Now, taking the galvanic potential as given, let us examine its distribution between the phases. There are potential drops in three regions (Fig. 6), namely, in the dense part of the ionic double layer, which is also called the Helmholtz layer, in the space charge region in the semiconductor, and in the diffuse part of the double layer in the electrolyte, or the Gouy layer.* The dense part of the double layer is formed of ions immediately adjacent to the semiconductor. The thickness of the Helmholtz layer is determined by the radius of the ions and is of the order of 10^{-8} cm. The ions in the diffuse part of the double layer may move. Two processes determine the distribution of anions and cations in it: the electric field in the Gouy layer, which produces a nonuniform distribution of the concentrations of the ions, and diffusion processes which

---

*The simplest case is examined here. In particular, we will ignore potential differences in the oxide film or in the layer of oriented dipolar molecules (for example, of the solvent) adsorbed on the electrode surface. Following the terminology used in semiconductor physics and electrochemistry, the charge in the semiconductor will be called the volume or space charge, while the charge in the electrolyte will be the charge of the diffuse part of the double layer.

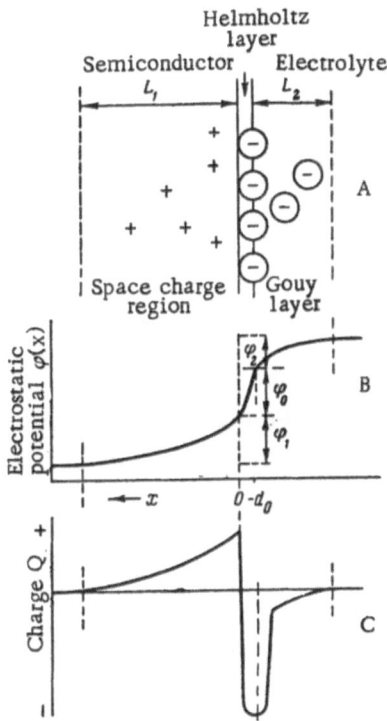

Fig. 6. Structure of the double layer (A) and distribution of the potential (B) and charge (C) at the semiconductor—electrolyte interface.

tend to equalize the concentrations. An equilibrium distribution of the ions is established as a result of these two mutually opposing processes.

A similar picture is found in semiconductors. Depending on the sign of their charge and the direction of the electric field, the free carriers in a semiconductor are either attracted to or repelled from the phase boundary. An equilibrium concentration is established and a space charge layer is formed under the effect of electric forces and diffusion.

Let us now turn to the quantitative estimation of the distribution of the galvanic potential between the three regions mentioned above. For this, let us write Poisson's equation

$$\frac{d^2\varphi(x)}{dx^2} = -\frac{4\pi}{\varepsilon}\rho(x),$$

$$(9.1)$$

where $\varphi(x)$ is the potential at point x; $\rho(x)$ is the charge density at the same point, and ε is the dielectric constant of the medium.

The x axis is perpendicular to the phase boundary with the semiconductor situated in the region x > 0 and the electrolyte, in the region x < $-d_0$ (d_0 is the thickness of the Helmholtz layer). The density of the charge in the semiconductor is expressed by

$$\rho(x) = e[-n(x) + p(x) + N_D - N_A] \; , \tag{9.2}$$

where n, p, N_D, and N_A are concentrations of electrons, holes, donor levels, and acceptor levels, respectively. It is assumed that the donors and acceptors are completely ionized. At equilibrium the distribution of the free carrier concentration in the electric field of the semiconductor is determined by Boltzmann's law*

$$n(x) = n^0 e^{\frac{e[\varphi(x)-\varphi_B]}{kT}} \; ; \; p(x) = p^0 e^{-\frac{e[\varphi(x)-\varphi_B]}{kT}} . \tag{9.3}$$

Here n^0, p^0, and φ_B are, respectively, the concentrations of electrons and holes and the potential in the bulk of the semiconductor in the region beyond the space charge (when $x \to \infty$).†

Now Poisson's equation appears as

$$\frac{d^2\varphi}{dx^2} = -\frac{4\pi e}{\varepsilon_1}[p^0 e^{-\frac{e(\varphi-\varphi_B)}{kT}} - n^0 e^{\frac{e(\varphi-\varphi_B)}{kT}} + N_D - N_A], \tag{9.4}$$

where ε_1 is the dielectric constant of the semiconductor.

---

*We should note that at the present time there is insufficient basis for using the Boltzmann distribution in writing equation (9.3). In actual fact, as § 7 indicates, in the derivation of Boltzmann's formula it is assumed that the electrons move in an external field and do not interact with each other. In the case examined here, the electrons and holes, enriching the region close to the contact, themselves produce a field in which they move. In addition, their interaction energy is quite high. The basis commonly accepted for using the Boltzmann distribution in this case is the assertion that the electrons form a certain self-consistent field which may be substituted in the Boltzmann formula. The limits of applicability of this approximation are not clear.

†The sign of the electrostatic potential used in this book is in accordance with usage in semiconductor physics, namely, a positive shift in the potential corresponds to a downward bend in the energy bands, i.e., a decrease in the electron energy. The rule commonly used in electrochemistry, namely, that a positive shift in the potential corresponds to the transfer of positive charges from the electrode into the solution, is used in selecting the sign of the electrode potential. Thus, the more positive the electrode potential and the space charge, the more negative the electrostatic potential in the space charge region.

We will subsequently examine an n-type semiconductor for which condition $N_D \gg N_A$ and $n^0 \gg p^0$ holds. Then the expression for the charge density ρ may be written as

$$\rho = e\left[-n(x) + N_D\right]. \tag{9.5}$$

In order to obtain a qualitative picture, it is sufficient to examine the case of small changes in potential; we will assume that the relation $e(\varphi - \varphi_B)/kT \ll 1$ holds for all regions of the semiconductor.

Then the exponents in equation (9.4) may be expanded into a series and only the first term of the expansion in φ used. In this case the expression for the charge density appears as

$$\rho(x) = e\left\{-n^0\left[1 + \frac{e\left[\varphi(x) - \varphi_B\right]}{kT}\right] + N_D\right\}. \tag{9.6}$$

There is no space charge in the bulk of the semiconductor so $-n^0 + N_D \approx 0$. With these considerations taken into account, expression (9.6) becomes

$$\rho(x) = -\frac{e^2 n^0}{kT}\left[\varphi(x) - \varphi_B\right].$$

Poisson's equation (9.4) may now be written as

$$\frac{d^2\varphi}{dx^2} = \frac{4\pi e^2 n^0}{\varepsilon_1 kT}\left[\varphi(x) - \varphi_B\right]. \tag{9.7}$$

By integrating equation (9.7), we find

$$\left(\frac{d\varphi}{dx}\right)^2 = \frac{4\pi e^2 n^0}{\varepsilon_1 kT}\left[\varphi(x) - \varphi_B\right]^2. \tag{9.8}$$

For the field strength at the semiconductor surface $E_S = (-d\varphi/dx)\big|_{x=0}$ we have the expression

$$E_s = \frac{\varphi_1}{L_1}, \tag{9.9}$$

where $\varphi_1 = \varphi(0) - \varphi_B$ is the potential drop in the space charge region of the semiconductor, while

$$L_1 = \sqrt{\frac{\varepsilon_1 kT}{4\pi e^2 n^0}}.$$

(9.9a)

The value L_1 is called the Debye length. *

On the basis of equation (9 9), the space charge region may be considered arbitrarily as a plane–parallel capacitor of thickness L_1 in which the field E_1 is constant. The relation of φ_1 to the dimensions of the charged (Debye) region determines the strength of the electric field at the contact $E_S = E_1$.

If the potential drop in the space charge region φ_1 equals zero, then in this case the boundaries of the energy bands are not deformed by the electric field. The electrode potential φ_{fb} when $\varphi_1 = 0$ is called the "flat band potential."

Let us now examine the problem of the potential distri-bution on an electrolyte. The charge density in a uni-unival-ent electrolyte equals

$$\rho = e(c_+ - c_-),$$

(9.10)

where c_+ and c_- are the cation and anion concentrations which at equilibrium are related to the potential $\varphi(x)$ by Boltzmann's equation:

$$c_+(x) = c^0 e^{-\frac{e[\varphi(x) - \varphi_c]}{kT}}; \quad c_-(x) = c^0 e^{\frac{e[\varphi(x) - \varphi_c]}{kT}}.$$

Here, φ_C and c^0 are the potential and the concentration of anions and cations in the bulk of the electrolyte, respectively.

Thus, Poisson's equation becomes

$$\frac{d^2\varphi}{dx^2} = -\frac{4\pi e}{\varepsilon_2} c^0 [e^{-\frac{e(\varphi - \varphi_c)}{kT}} - e^{\frac{e(\varphi - \varphi_c)}{kT}}],$$

(9.11)

where ε_2 is the dielectric constant of the electrolyte.

---

*It would have been more correct to call the value L_1 after Gouy, who was the first to obtain an equation of the type (9.9) for an electrolyte (see below) [10].

When $e(\varphi - \varphi_C)/kT \ll 1$ and the exponents are expanded into a series, equation (9.11) may be written as

$$\frac{d^2\varphi}{dx^2} = \frac{8\pi e^2}{\varepsilon_2 kT} c^0 (\varphi - \varphi_c).$$

A comparison of this equation with equation (9.7) shows that they are analogous. By repeating the calculation we arrive at the conclusion that the value

$$L_2 = \sqrt{\frac{\varepsilon_2 kT}{8\pi e^2 c^0}} \qquad (9.12)$$

determines the thickness of the diffuse part of the double layer in the electrolyte, while the potential drop in this region equals $\psi' = \varphi_C - \varphi (x = -d_0)$.

Thus, the distributions of the diffuse charge in the semiconductor and in the electrolyte are quite similar. The corresponding Debye length determines the magnitude of the charged region.

The electric field E_0 is assumed to be constant in the Helmholtz layer so the potential drop equals $\varphi_0 = E_0 d_0$.

Taking into consideration equation (9.9), we have for the galvanic potential

$$\varphi_{12} = \varphi_1 + \varphi_0 + \psi' = E_1 L_1 + E_0 d_0 + E_2 L_2, \qquad (9.13)$$

where E_2 is the strength of the electric field in the solution when $x = -d_0$. Let us assume then that there are no charges at the phase boundaries between the three regions examined (in particular, there are no surface states in the semiconductor). Then the equation of electrostatic inductions will hold,

$$\varepsilon_0 E_0 = \varepsilon_1 E_1 = \varepsilon_2 E_2, \qquad (9.14)$$

where ε_0 is the dielectric constant of the Helmholtz layer. Using equations (9.9) and (9.14), we may readily find the relations of the potential drops:

$$\frac{\varphi_1}{\psi'} = \frac{L_1 \varepsilon_2}{L_2 \varepsilon_1} ; \quad \frac{\varphi_0}{\psi'} = \frac{d_0 \varepsilon_2}{L_2 \varepsilon_0} .$$

$$(9.15)$$

Thus, the ratio of the potential drops in the semiconductor and the electrolyte is determined by the ratio of the corresponding Debye lengths with an accuracy up to the ratio of the dielectric constants. Normally the Debye lengths in a semiconductor with a low free carrier concentration and in an electrolyte differ by several orders. Actually, using equation (9.9a), we find that for germanium with a donor level concentration of $N_D = n^0 = 10^{14}$ cm^{-3} and a dielectric constant of $\varepsilon_1 = 16$, the Debye length $L_1 \approx 10^{-4}$ cm. The thickness of the diffusion layer in the electrolyte L_2 decreases with an increase in the solution concentration and is insignificant in highly concentrated solutions. For example, in a centinormal aqueous electrolyte solution ($c^0 = 10^{19}$ ions per 1 cm^3, $\varepsilon_2 = 81$), the order of magnitude of the Debye length $L_2 \approx 10^{-6}$ cm.

As has been said previously, the thickness of the Helmholtz layer is of the order of 10^{-8} cm. It follows from equation (9.15) that in a sufficiently concentrated solution the galvanic potential drops mainly in the semiconductor while the potential drop in the electrolyte and in the Helmholtz layer may be ignored. The Debye length is of considerable magnitude in weak electrolyte solutions. For example, in a 10^{-7} M solution (with a concentration of 10^{14} ions per cm^3), $L_2 \approx 10^{-4}$ cm, i.e., it becomes comparable with the Debye length in a semiconductor.

In metal electrodes with an electron concentration per unit volume of $n^0 \approx 10^{22}$ cm^{-3}, all the charge is concentrated in the surface layer which has a thickness of the order of 10^{-8} cm, and therefore the potential drop in a metal is low in comparison with the potential drop in the electrolyte and the Helmholtz layer.

Thus, the semiconductor—solution interface may be regarded as a capacitor whose two plates have a diffuse struc-

Table 2. Relation of Potential Drop in Helmholtz
Layer to Potential Drop in Space Charge Region
in Semiconductor

φ_1, V	φ_{0c}, V	φ_{0l}, V	φ_1, V	φ_{0c}, V	φ_{0l}, V
—0.375	$3 \cdot 10^{-2}$	$2 \cdot 10^{-1}$	0	$<10^{-4}$	$1 \cdot 10^{-1}$
—0.30	$7 \cdot 10^{-3}$	$2 \cdot 10^{-1}$	0.025	$<10^{-4}$	$5 \cdot 10^{-2}$
—0.225	$2 \cdot 10^{-3}$	$2 \cdot 10^{-1}$	0.050	$<10^{-4}$	$2 \cdot 10^{-2}$
—0.150	$<10^{-4}$	$2 \cdot 10^{-1}$	0.075	$7 \cdot 10^{-4}$	10^{-2}
—0.1	$<10^{-4}$	$2 \cdot 10^{-1}$	0.100	10^{-3}	$5 \cdot 10^{-3}$
—0.075	$<10^{-4}$	$1.9 \cdot 10^{-1}$	0.150	$3 \cdot 10^{-3}$	$<10^{-3}$
—0.050	$<10^{-4}$	$1.8 \cdot 10^{-1}$	0.225	$2 \cdot 10^{-3}$	$<10^{-3}$
—0.025	$<10^{-4}$	$1.4 \cdot 10^{-1}$			

ture.* The electric properties of this contact are determined mainly by the diffuse charge in the semiconductor.

We have examined in detail the rules for the distribution of the galvanic potential at low values of it. As will be shown in § 10, this picture also remains true to a considerable extent for high potential drops. However, with an increase in the galvanic potential, the ratio of potential drops in the Helmholtz layer and in the space charge layer begins to increase and with a high space charge the potential drop in the Helmholtz layer may become comparable with the potential drop in the space charge region. In a highly concentrated electrolyte the potential drop may always be neglected in comparison with the corresponding value in the semiconductor or in the Helmholtz layer.

Let us now consider the effect of surface levels on the potential distribution. As will be shown in the next paragraph, the potential difference in the Helmholz layer φ_0 may be written as

---

*Rise [11] put forward the hypothesis that the charge on a metal electrode also has a diffuse structure. However, this idea has not been confirmed experimentally.

$$\varphi_0 = \varphi_{01} + \varphi_{0c}, \tag{9.16}$$

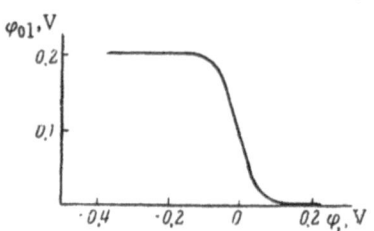

Fig. 7. Potential difference in Helmholtz layer φ_{01} in the presence of surface levels in relation to the potential difference in the space charge region φ_1 (see Table 2).

where φ_{01} is the potential difference in the Helmholtz layer produced by surface levels; and φ_{0c} is the potential difference in the Helmholtz layer caused by the space charge in the semiconductor in the absence of surface levels.

Table 2 gives the values for φ_{0c} and also for φ_{01} at a concentration of surface donor levels $N_t = 10^{13}$ cm^{-2} and with the condition that the Fermi level coincides with the surface levels in the semiconductor. The calculation was carried out for electronic germanium ($n^0/n_i = 10$). Figure 7 illustrates the table.

Table 2 shows that the potential drop in the Helmholtz layer φ_{0c}, produced by the space charge, is considerably smaller than both the potential drop φ_1 in the space charge region in the semiconductor and φ_{01}.* In the range of potentials $\varphi_1 < -0.1$ V, the potential drop in the Helmholtz layer in the presence of levels φ_{01} has a constant value which is comparable with that of the potential drop in the space charge region φ_1. The levels are fully charged at these potentials and their charge does not change with a change in the potential drop φ_1 in the semiconductor. When $\varphi_1 > -0.1$ V, the donor level is filled with electrons and its charge falls. Accordingly, the potential drop in the Helmholtz layer produced by the surface levels, also starts to decrease and soon becomes negligibly small.

As Table 2 and Fig. 7 show, when the overall potential difference at the phase boundary changes, the main change in

*It will be shown in § 10 that the potential drop φ_{0c} becomes comparable with φ_1 when there is a very considerable curvature of the energy bands in the semiconductor.

the Helmholtz potential φ_{0l} occurs at 0.075 V $> \varphi_1 > -0.1$ V. In this region the changes in the potential differences φ_1 and φ_{0l} are comparable. When $\varphi_1 > 0.075$ V and $\varphi_1 < -0.1$ V, a change in the galvanic potential produces a change mainly in the potential difference in the space charge region φ_1. Thus, the effect of the surface level on the potential change in the Helmholtz layer is appreciable only over a narrow range of potentials (a few kT/e), in which there is a change in the degree of filling of the level with electrons.

As will be shown in § 10, the change in the potential in the Helmholtz layer predominates over the change in the potential in the space charge region in the semiconductor if the surface level density satisfies the relation

$$N_t > \frac{\varepsilon_0 kT}{\pi e^2 d_0}$$

(9.17)

(for example, for germanium $N_t > 10^{13}$ cm^{-2}).

In concluding this section, let us examine the applicability of the concept of "equilibrium in a semiconductor" to electrochemical systems. In practice only a few electrodes are ideally polarizable. Normally a current passes through the electrode−electrolyte phase boundary (including cases when the current equals zero in the external circuit, for example, during corrosion). Nonetheless, the equilibrium formulas for the space charge as well as for surface conduction and the differential capacitance (see below) remain valid even when there is a current as long as that current does not exceed a certain critical value. The criterion of "smallness" of current is formulated in different ways for different cases. For example, the equilibrium is hardly disturbed in the case of minority carriers whose concentration is no lower in the Debye region than in the neutral bulk of the semiconductor if the current transmitted through the system remains much weaker than the maximum current of the minority carriers (see § 27). These formulas may also be used under nonsteady conditions, for example, with varying potential, if the rate of its change is low in comparison

with the relaxation rate of the space charge, * and the varying
current remains weak (see above). If these conditions are fulfilled, we may consider that the free carriers in the Debye region are distributed according to Boltzmann's law.

§ 10.   Calculation of Potential Distribution
at the Electrolyte – Semiconductor Interface

Poisson's equation (9.4) must be solved to determine the
electric field and course of the potential drop in the electrolyte
or semiconductor at any potential. We will calculate the potential drop in the semiconductor† then, analogously, we will
solve Poisson's equation for the electrolyte and find the relation between the potential drops in the semiconductor, electrolyte, and Helmholtz layer.

To solve Poisson's equation (9.4) we will multiply the
right- and left-hand parts of the equation by $d\varphi/dx$ and rewrite
(9.4) in the form

$$d\varphi \frac{d^2\varphi}{dx^2} = \frac{1}{2} d \left(\frac{d\varphi}{dx} \right)^2 =$$

$$= -\frac{4\pi e}{\varepsilon_1} \left[N_D - N_A + p^0 e^{-\frac{e(\varphi-\varphi_B)}{kT}} - n^0 e^{\frac{e(\varphi-\varphi_B)}{kT}} \right] d\varphi. \tag{10.1}$$

Integration with the limiting condition $d\varphi/dx = 0$ when $\varphi = \varphi_B$
gives us

$$\left(\frac{d\varphi}{dx} \right)^2 = -\frac{8\pi e}{\varepsilon_1} \left[(N_D - N_A)(\varphi - \varphi_B) - \frac{kT}{e} p^0 \left(e^{-\frac{e(\varphi-\varphi_B)}{kT}} - 1 \right) - \right.$$

$$\left. -\frac{kT}{e} n^0 \left(e^{\frac{e(\varphi-\varphi_B)}{kT}} - 1 \right) \right]. \tag{10.2}$$

---

*With a change in potential the relaxation time of a space charge composed of ma-
jority carriers (the so-called Maxwell time) is determined by the specific resistance of the semiconductor. If the space charge is composed of minority carriers
then its relaxation time $t_d \approx L_1^2/D$, as will be shown in § 42. For example, for
germanium, which has a high resistance, $t_d \approx 10^{-12}$ sec.

†We will be following [12] in this part of § 10.

The derivative $d\varphi/dx$ equals

$$\frac{d\varphi}{dx} = \pm \sqrt{\frac{8\pi kT}{\varepsilon_1}} \left[-(N_D - N_A)y + p^0(e^{-y} - 1) + n^0(e^y - 1)\right], \tag{10.3}$$

where y represents

$$y = \frac{e(\varphi - \varphi_B)}{kT}. \tag{10.4}$$

For the electric field at the surface we obtain the expression

$$E_s = \pm \sqrt{\frac{8\pi kT}{\varepsilon_1}} \left[-(N_D - N_A)Y + p^0(e^{-Y} - 1) + n^0(e^Y - 1)\right]. \tag{10.5}$$

The following symbol is introduced here:

$$Y = \frac{e(\varphi_s - \varphi_B)}{kT}. \tag{10.6}$$

The plus sign appears in front of the root when $Y > 0$ and the minus when $Y < 0$. In actual fact, when $Y > 0$, the potential $\varphi_s > \varphi_B$. In this case the potential φ falls with an increase in x, i.e., the derivative $d\varphi/dx$ is negative while the electric field $E = -d\varphi/dx$ is positive. Using analogous considerations, we find that when $Y < 0$ the electric field E is negative.

From the electrical neutrality of the bulk of the semiconductor it follows that $-N_D + N_A = p^0 - n^0$. Using the symbol*

$$\lambda = \left(\frac{p^0}{n^0}\right)^{1/2}, \tag{10.7}$$

we obtain

$$\frac{-N_D + N_A}{n_i} = (\lambda - \lambda^{-1}).$$

Using the latter relation, we may write the expression for the electric field (10.5) as

---

*From equation (4.10) it also follows that $\lambda = p_0/n_i = n_i/n^0$.

$$E_s = \pm \sqrt{\frac{8\pi kTn_i}{\varepsilon_1}} \cdot F(\lambda, Y),$$

(10.8)

where the function $F(Y, \lambda)$ has the form

$$F(Y, \lambda) = [\lambda(e^{-Y} - 1) + \lambda^{-1}(e^{Y} - 1) + (\lambda - \lambda^{-1})Y]^{1/2}.$$

(10.9)

The function $F(Y, \lambda)$ is tabulated in the book [13] and is shown graphically in [14-16].

The Debye length for a semiconductor with intrinsic conductivity L_D is given by

$$L_D = \sqrt{\frac{\varepsilon_1 kT}{4\pi e^2 n_i}} \cdot$$

(10.10)

Thus the expression for the electric field E_S will be written in the form

$$E_s = \frac{\sqrt{2kT}}{e} \frac{1}{L_D} F(Y, \lambda).$$

(10.11)

Under these conditions we obtain the following expression for the space charge in the semiconductor*

$$Q_1 = -\frac{\varepsilon_1 E_s}{4\pi} = \mp \frac{\varepsilon_1 kT}{2\sqrt{2}\pi e L_D} F(Y, \lambda).$$

(10.12)

The calculation given above is based on the assumption that the charge at the semiconductor—electrolyte boundary is distributed continuously and does not consist of separate point charges. This approximation is normally quite good. In the next approximation we should consider the discreteness of the charges in the ionic part of the double layer and of the charges in the surface states. (This problem also exists in the electrochemistry of metals; see, for example, [17].)

---

*The basis for equation (10.12) will be given in § 12.

Let us now find the surface charge produced by the surface states in a semiconductor. For concreteness we will assume that there are donor levels on the surface which are positively charged in the free state and neutral when occupied by electrons. The Fermi distribution (5.2) determines the number of electrons in the levels. We should also consider that the difference $E_t - E_F$ depends on the magnitude of the potential drop φ_1 in the space charge region of the semiconductor. Actually, the energy of an electron in the surface level equals $E_t^0 - e\varphi_1$, where E_t^0 is the energy of the surface level when there is no space charge in the semiconductor. The charge on the surface levels is determined by the value

$$Q_t = e(N_t - n_t) = \frac{eN_t}{1 + \exp \dfrac{(E_F - E_t^0 + e\varphi_1)}{kT}}.$$

$$(10.13)$$

Let us now determine the magnitude of the potential drop in the Helmholtz layer, which equals $\varphi_0 = d_0 E_0$. The electric field is determined by the equation

$$\varepsilon_0 E_0 = \varepsilon_1 E_s + 4\pi Q_t, \qquad (10.14)$$

where E_S is given by expression (10.11).

Using equations (10.11) and (10.13), we obtain the potential drop in the Helmholtz layer:

$$\varphi_0 = d_0 E_0 = \frac{4\pi e N_t d_0}{\varepsilon_0\left(1 + \exp \dfrac{E_F - E_t^0 + e\varphi_1}{kT}\right)} \pm \frac{\sqrt{2}\,\varepsilon_1 kT d_0}{eL_D \varepsilon_0} F(Y, \lambda).$$

$$(10.15)$$

There is no need to solve Poisson's equation for the electrolyte (9.11) again in order to calculate the potential drop in the electrolyte. Instead let us introduce $N_D = N_A = 0$ and $n^0 = p^0$ and substitute c^0, φ_C, and ε_2 for n^0, φ_B, and ε_1, respectively, in equation (9.4). Then equations (9.11) and (9.4) coincide. Therefore, a solution for equation (9.11) may be written immediately by carrying out the appropriate transformations in equation (10.9). For the electric field E_2 at the point $x = -d_0$ in the electrolyte we have

$$E_2 = \pm \sqrt{\frac{8\pi kT c^0}{\varepsilon_2}} F_2(Y_2) = \pm \frac{\sqrt{2}\,kT}{eL_2} F_2(Y_2),$$

(10.16)

where F_2 is determined by expression

$$F_2(Y_2) = [e^{-Y_2} + e^{Y_2} - 2]^{1/2},$$

(10.17)

and

$$Y_2 = \frac{e}{kT}[\varphi(-d_0) - \varphi_c] = \frac{e}{kT}\psi'.$$

We should then note that with low potential drops in the electrolyte ($Y_2 \ll 1$), the function F_2 may be written approximately in the form

$$F_2 \approx Y_2,$$

(10.18)

while with high potential drops $|Y_2| \gg 1$

$$F_2(Y_2) = e^{\frac{Y_2}{2}}.$$

(10.19)

To obtain the relation between the values of the potential drops in the Helmholtz layer and electrolyte, we use the relation

$$\varepsilon_2 E_2 = \varepsilon_0 E_0.$$

(10.20)

Using equations (10.20) and (10.16), we obtain

$$\varphi_0 = d_0 E_0 = \frac{d_0 \varepsilon_2 E_2}{\varepsilon_0} = \pm \frac{\sqrt{2}\,kT d_0 \varepsilon_2}{eL_2 \varepsilon_0} F_2(Y_2).$$

(10.21)

Thus, equation (10.15) makes it possible to determine the potential drop in the Helmholtz layer if the potential drop in the semiconductor is known and equation (10.21) makes it possible to relate the potential drop in the Helmholtz layer to the corresponding value in the electrolyte.

Let us now consider the relations obtained. Let us first examine the change in the potential drop in the Helmholtz layer dY_0 when the potential drop in the semiconductor changes by dY as a result of electrode polarization. Differentiation of equation (10.15) gives

$$dY_0 = \left\{ \frac{4\pi e^2 N_t\, d_0 \exp\left(\dfrac{E_F - E_t^0}{kT} + Y\right)}{kT\varepsilon_0 \left[1 + \exp\left(\dfrac{E_F - E_t^0}{kT} + Y\right)\right]^2} \pm \right.$$

$$\left. \pm \frac{\varepsilon_1 d_0}{\sqrt{2}\, L_D \varepsilon_0} \frac{-\lambda e^{-Y} + \lambda^{-1}e^Y + \lambda - \lambda^{-1}}{F} \right\} dY. \quad (10.22)$$

When examinining an n-type semiconductor without surface levels ($N_t = 0$) we will assume that the relation (10.22) holds at such high negative values of Y that the function F [equation (10.9)] may be expressed in the form

$$F = \lambda^{1/2}\, e^{\frac{|Y|}{2}} . \tag{10.23}$$

Then equation (10.22) may be written as

$$\frac{dY_0}{dY} = \frac{\varepsilon_1 d_0}{\sqrt{2}\, \varepsilon_0 L_D} \lambda^{1/2} e^{\frac{|Y|}{2}} . \tag{10.24}$$

Since $p_s = p^0 e^{|Y|}$, then by eliminating potential Y, we obtain

$$\frac{dY_0}{dY} = \frac{\varepsilon_1 d_0}{\sqrt{2}\, \varepsilon_0 L_D} \sqrt{\frac{p_s}{n_i}} . \tag{10.25}$$

For an estimate let us assume a hole concentration on the germanium surface equal to $p_s = 10^{18}$ cm^{-3}. Then $dY_0/dY \approx 0.2$. We then come to the conclusion that with very high potential drops in the semiconductor the change in the potential drop in the Helmholtz layer becomes comparable with that in the semiconductor. With even higher potential drops ($p_s > 2 \cdot 10^{19}$ cm^{-3})

we should take into account the degeneracy of the electron gas on the surface and use other formulas to estimate the potential distribution (see below).

In another limiting case the number of surface levels N_t is so great that the second term in equation (10.22) may be neglected. (This indicates that the contribution of the space charge to the potential drop in the Helmholtz layer is small in comparison with that of the surface levels.) Then

$$\frac{dY_0}{dY} = \frac{4\pi e^2 N_t \, d_0 \exp\left(\dfrac{E_F - E_t^0}{kT} + Y\right)}{kT\varepsilon_0 \left[1 + \exp\left(\dfrac{E_F - E_t^0}{kT} + Y\right)\right]^2}.$$

(10.26)

With weak ionization of the donor levels, (10.26) may be rewritten in the form

$$\frac{dY_0}{dY} = \frac{4\pi e^2 N_t \, d_0}{kT\varepsilon_0 \exp\left(\dfrac{E_F - E_t^0}{kT} + Y\right)} = \frac{4\pi e d_0}{kT\varepsilon_0} Q_t,$$

(10.27)

where $Q_t = eN_t \exp\{[(E_F)/kT] - Y\}$ is the surface level charge.

The surface levels will contribute considerably to the potential drop in the Helmholtz layer in accordance with equation (10.22) with the condition that:

$$\frac{4\pi e^2 N_t \, d_0 \exp\left(\dfrac{E_F - E_t^0}{kT} + Y\right)}{kT\varepsilon_0 \left[1 + \exp\left(\dfrac{E_F - E_t^0}{kT} + Y\right)\right]^2} > 1.$$

(10.28)

For convenience in the investigation of the inequality (10.28) let us determine the potentials at which the following relation holds:

$$\frac{4\pi e^2 N_t \, d_0 \exp\left(\dfrac{E_F - E_t^0}{kT} + Y\right)}{kT\varepsilon_0 \left[1 + \exp\left(\dfrac{E_F - E_t^0}{kT} + Y\right)\right]^2} = 1.$$

(10.29)

Let us introduce the symbol

$$\exp\left(\frac{E_F - E_i^0}{kT} + Y\right) = x.$$

Then equation (10.29) may be written as

$$x^2 + x\left(2 - \frac{4\pi e^2 N_t d_0}{\varepsilon_0 kT}\right) + 1 = 0. \tag{10.30}$$

Equation (10.30) has the following roots:

$$x_{1,2} = \exp\left(\frac{E_F - E_i^0}{kT} + Y_{a,b}\right) = -1 + \frac{2\pi e^2 N_t d_0}{kT\varepsilon_0} \pm$$

$$\pm \sqrt{\frac{4\pi^2 e^4 N_t^2 d_0^2}{\varepsilon_0^2 k^2 T^2} - \frac{4\pi e^2 N_t d_0}{\varepsilon_0 kT}} . \tag{10.31}$$

This result indicates the following. If the relation

$$N_t < \frac{\varepsilon_0 kT}{\pi e^2 d_0} \tag{10.32}$$

holds, then for any potential Y the change in the potential drop in the Helmholtz layer is always less than that in the semiconductor. However, if the reverse of the inequality (10.32) holds, then there is a certain range of potentials for Y_a to Y_b within which the change in the potential drop in the Helmholtz layer remains greater than the change in the potential drop in the semiconductor. The width of this region may be found by using equation (10.31).

Let us estimate the change in potential drop in the diffuse part of the double layer in the electrolyte. Let us first examine the case where there are no surface levels. From the equation for electrostatic inductions $\varepsilon_1 E_s = \varepsilon_2 E_2$ and equations (10.11) and (10.16), we obtain

$$\frac{\varepsilon_1}{L_D} F(Y, \lambda) = \frac{\varepsilon_2}{L_2} F_2(Y_2). \tag{10.33}$$

Let us assume that at first the potential drop is small in the electrolyte so that $Y_2 \ll 1$. In this case, $F_2(Y_2)$ is determined by equation (10.18). At the same time, let the potential drop in the semiconductor be so high that equation (10.23) holds. Then the relation (10.33) will be written in the form

$$-\frac{\varepsilon_1}{L_D} \lambda^{1/2} e^{-\frac{|Y|}{2}} = \frac{\varepsilon_2}{L_2}|Y_2|.$$

(10.34)

For the change in potential we have

$$\frac{\varepsilon_1 \lambda^{1/2}}{2L_D} e^{-\frac{|Y|}{2}} = \frac{\varepsilon_2}{L_2} \frac{d|Y_2|}{d|Y|},$$

(10.35)

hence

$$\frac{d|Y_2|}{d|Y|} = \frac{L_2 \varepsilon_1}{2L_D \varepsilon_2} \lambda^{1/2} e^{-\frac{|Y|}{2}} = \frac{|Y_2|}{2} \ll 1.$$

(10.36)

Thus, we come to the conclusion that if the potential drop in the semiconductor satisfies the condition

$$|Y| < 2 \ln \frac{2L_D \, \varepsilon_2}{L_2 \varepsilon_1 \, \lambda^{1/2}},$$

(10.37)

then the change in the potential drop in the semiconductor remains greater than that in the electrolyte.

When the reverse of inequality (10.37) holds, equation (10.33) is written in the form

$$\frac{\varepsilon_1}{L_D} \lambda^{1/2} e^{-\frac{|Y|}{2}} = \frac{\varepsilon_2}{L_2} e^{-\frac{|Y_2|}{2}},$$

(10.38)

while we obtain the following expression for the change in the potential drops:

$$d|Y| = d|Y_2|.$$

(10.39)

In this case the potential applied from outside is divided approximately equally between the space charge region in the conductor and the diffuse part of the double layer in the electrolyte.

Let us then determine the relation of the change in the potential drop in the electrolyte dY_2 to the change in the potential drop in the Helmholtz layer dY_0 in the presence of levels. Let their number be so great that $|Y_2| \gg 1$ and $Y_0 \gg 1$. Using equation (10.23) we rewrite equation (10.21) in the form

$$|Y_0| = \frac{\sqrt{2}\, e_2 d_0}{L_2 e_0}\, e^{\frac{|Y_2|}{2}}.$$

(10.40)

Differentiation of equation (10.40) gives

$$|dY_0| = \frac{e_2 d_0}{\sqrt{2}\, L_2 e_0}\, e^{\frac{|Y_2|}{2}}\, |dY_2|$$

or

$$d|Y_2| = \frac{2|dY_0|}{|Y_0|}.$$

(10.41)

Consequently, if the potential drop in the Helmholtz layer is great $(Y_0 \gg 1)$ we may neglect the change in the potential drop in the electrolyte on polarization of the electrode.

The calculations given above were carried out on the assumption that the position of the Fermi level at the semiconductor surface satisfies the condition

$$(E_v + 3kT) \leqslant E_F \leqslant (E_c - 3kT).$$

(10.42)

In this case the free carriers in the whole space charge region obey Boltzmann statistics. If the inequality (10.42) does not hold, i.e., the Fermi level approaches the edges of the forbidden band, the concentration of free carriers must be calculated using the Fermi−Dirac distribution. This case (which is called degeneracy of the free carriers on the surface) was examined by Green [18, 19]. The expression for the electric field on the surface becomes

$$E_s = \frac{2kT}{e\pi^{1/4} L_D} \exp\frac{E_c - E_t}{2kT} \left[\frac{2}{3} F_{1/2}(G)\right]^{1/2}.$$

(10.43)

Here E_i is the energy at the center of the forbidden band, while $F_{3/2}(G)$ represents the function

$$F_{3/2}(G) = \int_0^\infty \frac{\delta^{3/2}}{1 + \exp(\delta - G)} d\delta,$$

(10.44)

where

$$G = \frac{E_F - E_c}{kT} + Y.$$

The function $F_{3/2}(G)$ was tabulated in [18, 20].*

The space charge may be found from the formula

$$Q_1 = -\frac{\varepsilon_1 E_s}{4\pi}$$

(10.45)

(see p. 57).

Green investigated the distribution of the potential at the semiconductor–electrolyte boundary in the case of degeneracy of the free carriers on the surface [19].

It should be noted that with strong fields at the boundary, when the thickness of the space charge region is comparable with the mean distance between charges and the length of an electron wave in a semiconductor (10^{-6} cm), the applicability of Maxwell–Boltzmann or Fermi–Dirac statistics for describing the electron gas in the space charge region becomes doubtful and, consequently, the applicability of the space charge

---

*In calculations with equation (10.43) it should be remembered that, according to (3.6),

$$\frac{E_F - E_c}{kT} = \ln n^0 - \ln N_c.$$

theory described above is also doubtful. These difficulties are discussed in [18, 21].

§ 11. Surface Conductivity

Under the influence of an electric field, the concentration of electrons and holes in the Debye region of a semiconductor differs from that in its bulk. This phenomenon can be observed when measuring the conductivity of the semiconductor. As the conductivity is proportional to the concentration of free carriers, the conductivity in the space charge region changes in relation to the applied field. In order to characterize the difference between the conductivity of the space charge layer and that in the bulk of the semiconductor, a value called surface conductivity is introduced:

$$\sigma = eU_n\Gamma_n + eU_p\Gamma_p; \tag{11.1}$$

here U_n and U_p are the mobilities of the electrons and holes, respectively, and Γ_n and Γ_p, the "excess" electrons and holes, respectively, on the surface.

By an excess of free carriers on the surface we mean the number of them in the space charge region at a given potential in excess of the number there would have been if there were no field (when $\varphi_S = \varphi_B$). The values Γ_n and Γ_p are determined by the relations

$$\Gamma_n = \int_0^\infty (n-n^0)\,dx; \quad \Gamma_p = \int_0^\infty (p-p^0)\,dx. \tag{11.2}$$

Before undertaking exact calculations, let us examine qualitatively the relation of the surface conductivity to the applied potential. Let us write σ in the form

$$\sigma = A \cdot g(Y), \quad \text{where } Y = \frac{e(\varphi_s - \varphi_B)}{kT}. \tag{11.3}$$

The function g was calculated in the work by Garrett and Brattain [12] and is given in Fig. 8 for an n-type semiconductor.

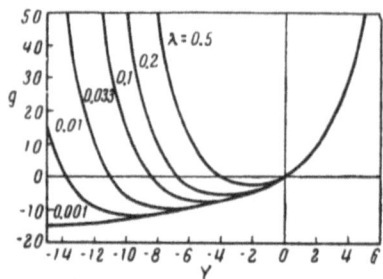

Fig. 8. Integral of the surface conductivity [equation (11.5)] in relation to the potential drop in the space charge region Y at various values of the parameter λ (for an n-type semiconductor) [12].

When $Y = 0$ (flat band potential) the concentrations n and p are everywhere equal to n^0 and p^0, respectively; by using formula (11.2) we find that Γ_n and Γ_p equal zero, i.e., $\sigma = 0$.

In accordance with equations (11.2) and (11.3), when $Y > 0$, the electron concentration at the contact n_s exceeds the electron concentration in the bulk n^0 and, therefore, Γ_n increases with potential while the hole concentration at the contact decreases. This enrichment of the surface in electrons produces an increase in the surface conductivity with potential. Analogously, an increase in σ at high negative potentials is produced by an increase in Γ_p, i.e., an enrichment of the contact in holes.

As Fig. 8 shows, there is a certain intermediate range of potentials ($Y < 0$) in which the surface conductivity is negative. This is explained in the following way. At low negative potentials the electron concentration in the space charge layer decreases and the value $n - n^0$, and, consequently Γ_n, also becomes negative. Under these conditions the value $p - p^0$ is positive and, consequently, $\Gamma_p > 0$. However, a change in the hole concentration of an n-type semiconductor proceeds more slowly than a change in its electron concentration and this leads to negative surface conductivity. The dependence of Γ_n, Γ_p, and σ on potential is shown schematically in Fig. 9. Obviously, a negative surface conductivity does not imply the appearance of negative electrical conductivity in the sample. Within the meaning of the concept "surface conductivity" used here, the relation $\sigma < 0$ means that the electrical conductivity in the space charge region is less than the bulk electrical conductivity of the semiconductor.

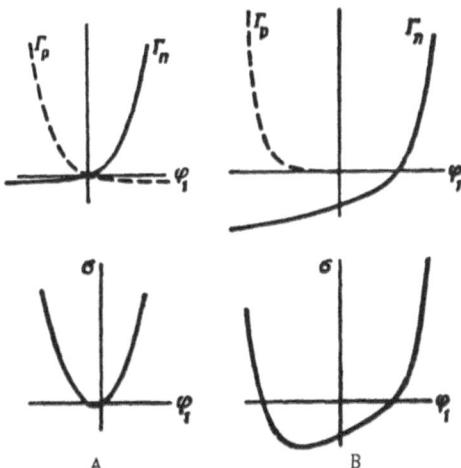

Fig. 9. Qualitative relation of the excess electron Γ_n and hole Γ_p charges and surface conductivity σ to the potential drop in the space charge region. A) Semiconductor with intrinsic conductivity; B) semiconductor with extrinsic conductivity.

Let us now derive an expression for surface conductivity. For this we need to arrange the integrals in equation (11.2). It is convenient to use a new independent variable for the integration. Let us write Γ_n in the form

$$\Gamma_n = \int_0^\infty (n - n^0)\,dx = \int_{\varphi_s}^{\varphi_B} (n - n^0)\frac{d\varphi}{\frac{d\varphi}{dx}} = n^0 \int_{\varphi_s}^{\varphi_B} \frac{(e^{\frac{e(\varphi - \varphi_B)}{kT}} - 1)}{\frac{d\varphi}{dx}}\,d\varphi.$$

Similar rearrangements need to be carried out for the value Γ_p. Then we may write the following expression for the surface conductivity:

$$\sigma = eU_n n^0 \int_{\varphi_s}^{\varphi_B} \frac{(e^{\frac{e(\varphi - \varphi_B)}{kT}} - 1)\,d\varphi}{d\varphi/dx} + eU_p p^0 \int_{\varphi_s}^{\varphi_B} \frac{(e^{-\frac{e(\varphi - \varphi_B)}{kT}} - 1)\,d\varphi}{d\varphi/dx}.$$

The derivative $d\varphi/dx$ is determined by equations (10.8) and (10.9). Let us express the ratio of electron and hole mobilities by $b = U_n/U_p$, and then the expression for surface conductivity becomes

$$\sigma = U_p n_i kT \int_Y^0 \frac{\lambda(e^{-y} - 1) + b\lambda^{-1}(e^y - 1)}{E} \, dy.$$

By using an equation of the type (10.9) for the transformation of the latter formula, we obtain an expression for surface conductivity:

$$\sigma = \frac{1}{\sqrt{2}} eU_p n_i L_D \lambda^{-1/2} g, \qquad (11.4)$$

where the function g equals*

$$g = \lambda^{1/2} \int_Y^0 \frac{\lambda(e^{-y} - 1) + b\lambda^{-1}(e^y - 1)}{F(y, \lambda)} \, dy. \qquad (11.5)$$

By comparing equations (11.3) and (11.4) we see that the constant A in equation (11.3) is determined by the expression

$$A = \frac{1}{\sqrt{2}} eU_p n_i L_D \lambda^{-1/2}.$$

It was assumed in this theory that the mobility of the current carriers is the same on the surface as in the bulk. This approximation is not always correct. Actually, if the thickness of the space charge layer is comparable with the free path of the carriers, then their collision with the boundary surface produces further scattering of free carriers and their mobility decreases in comparison with their mobility in the bulk. This effect, which has been studied by Schrieffer [22], results in a surface conductivity which is lower than that calculated by equation (11.4), particularly at high values of $|Y|$.†

———————
*The values of the function g are tabulated in [13].

†The surface mobility in a semiconductor was calculated in [23].

It is interesting to compare the surface conductivity in the space charge region in a semiconductor with that in the diffuse region of the double layer in the electrolyte. Estimates show that the surface conductivity of an electrolyte may be neglected [24]. In actual fact, the order of magnitude of the surface conductivity in an electrolyte, produced by the movement of ions in the diffuse section of the ionic layer, may be written as

$$\sigma_2 \sim eU_i\Gamma_i,$$

where U_i and Γ_i are the mobility and excess concentration of ions in the diffuse part of the double layer. As the contact is electrically neutral, $-\Gamma_i = \Gamma_p - \Gamma_n$. However, the mobility of the ions is considerably less than that of electrons and holes $U_i/U_n \approx 10^{-5}$, so that the surface conductivity of the electrolyte is considerably less than that of the semiconductor.

§ 12.   Differential Capacitance

Let us now calculate the differential capacitance of an electrolyte−semiconductor contact, * which is determined by the equation† :

$$C = \frac{dQ_2}{d\varphi_{12}} = -\frac{dQ_1}{d\varphi_{12}},$$

(12.1)

where $\varphi_{12} = \varphi_c - \varphi_B$ is the total potential drop at the phase boundary and Q_2 is the ionic charge in the electrolyte.

It is obvious that the condition of electrical neutrality of the system results in the equality in absolute value of the charges of the semiconductor and ionic layers of the double layer.

---

*The differential capacitance of a metal−semiconductor contact was examined in [25].

†See also § 41 for the determination of the differential capacitance. For the sake of brevity, from now on we will use the term capacitance for differential capacitance.

It is convenient to calculate the value

$$\frac{1}{C} = -\frac{d\varphi_{12}}{dQ_1} = -\frac{d\varphi_1}{dQ_1} - \frac{d\varphi_0}{dQ_1} - \frac{d\psi'}{dQ_1} = \frac{1}{C_1} + \frac{1}{C_0} + \frac{1}{C_2}. \tag{12.2}$$

Thus formally, the contact capacitance is reduced to three capacitances in series: C_0, C_1, and C_2, i.e., the capacitances of the Helmholtz layer, the space charge region in the semiconductor, and the diffuse part of the double layer in the electrolyte, respectively.

To estimate the order of magnitude of these capacitances, let us examine the simplest case where there are no surface levels in the semiconductor and the potential drop between the phases is low, so that we may use the results given in § 9. The use of formula (9.13) gives us an expression for the galvanic potential at the electrolyte−semiconductor contact

$$\varphi_{12} = \frac{\varepsilon_1 E_1 L_1}{\varepsilon_1} + \frac{\varepsilon_0 E_0 d_0}{\varepsilon_0} + \frac{\varepsilon_2 E_2 L_2}{\varepsilon_2}.$$

The expression for the semiconductor charge may be written in the form *

---

* Actually, let us write Poisson's equation in the semiconductor in the following way:

$$dE_1 = \frac{4\pi}{\varepsilon_1} \rho(x)\, dx.$$

Let us integrate this equation from x = 0 to x = ∞. Taking into account that there is no electric field in the bulk of the semiconductor, we have

$$-E_s = \frac{4\pi}{\varepsilon_1} \int_0^\infty \rho(x)\, dx.$$

The integral $\int_0^\infty \rho(x)\, dx$ represents the total charge in the semiconductor per unit surface, i.e., Q_1. Hence,

$$-E_s = \frac{4\pi}{\varepsilon_1} Q_1.$$

The last equation relates the charge in the semiconductor to the electric field at the contact.

$$Q_1 = -\frac{\varepsilon_1 E_s}{4\pi} .$$

(12.3)

By using formulas (12.1) and (9.14) we obtain the following expression for the differential capacitance:

$$\frac{1}{C} = 4\pi \left(\frac{L_1}{\varepsilon_1} + \frac{d_0}{\varepsilon_0} + \frac{L_2}{\varepsilon_2} \right) .$$

Comparison with relation (12.2) gives us the differential capacitances

$$C_1 = \frac{\varepsilon_1}{4\pi L_1}, \quad C_0 = \frac{\varepsilon_0}{4\pi d_0}, \quad C_2 = \frac{\varepsilon_2}{4\pi L_2} .$$

(12.4)

Thus, the capacitances are inversely proportional to the thicknesses of the corresponding regions. Taking into account the order of magnitude of the Debye lengths, we find that the capacitance of the whole system is determined by the capacitance of the space charge region in the semiconductor C_1.

Let us now calculate the capacitances of the semiconductor and electrolyte at any potential. As was shown in § 10, the charge in the semiconductor per unit area of the contact is determined by expression (10.12) in the absence of surface levels. According to equation (12.2) the capacitance of the semiconductor layer may be obtained by differentiation of equation (10.12):

$$C_1 = \frac{\varepsilon_1}{4\sqrt{2}\,\pi L_D} \frac{|-\lambda e^{-Y} + \lambda^{-1} e^{Y} + \lambda - \lambda^{-1}|}{[\lambda\,(e^{-Y} - 1) + \lambda^{-1}\,(e^{Y} - 1) + (\lambda - \lambda^{-1})\,Y]^{1/2}} .$$

(12.5)

At very positive potentials for which $\lambda^{-1} e^{Y} \gg (\lambda - \lambda^{-1})Y$, the relation of the capacitance to the potential may be represented by the following expression:

$$C_1 = \frac{\varepsilon_1 \lambda^{-1/2}}{4\sqrt{2}\,\pi L_D} e^{\frac{|Y|}{2}} .$$

(12.6)

At high negative potentials,

$$C_1 = \frac{\varepsilon_1 \lambda^{1/2}}{4\sqrt{2}\,\pi L_D} e^{\frac{|Y|}{2}} .$$

(12.7)

A comparison of equations (12.6) and (12.7) shows that, in general, the curves are not a symmetrical function of the potential as $\lambda^{-1} \neq \lambda$. The capacitance curve is symmetrical only in the case of an intrinsic semiconductor ($\lambda = 1$).

We will need another approximate expression for the relation of the capacitance to the potential. Let us examine a heavily doped n-type semiconductor ($\lambda^{-1} \gg \lambda$). In the region of negative potentials when $|Y| \gg 1$, the expression for the capacitance may be written in the form

$$C_1 = \frac{\varepsilon_1}{4\sqrt{2}\pi L_D} \frac{\lambda e^{-Y} - \lambda^{-1}}{[\lambda e^{-Y} - \lambda^{-1}(Y+1)]^{1/2}}.$$

If the potential also satisfies the condition

$$\lambda e^{-Y} \ll \lambda^{-1}, \tag{12.8}$$

then the expression for the capacitance becomes

$$C_1 = \frac{\varepsilon_1 \lambda^{-1/2}}{4\pi \sqrt{2} L_D} \frac{1}{\sqrt{-(Y+1)}}. \tag{12.9}$$

Substituting the values of L_D and λ from equations (10.10) and (10.7) in equation (12.9), we have

$$C_1 = \frac{1}{2} \frac{\sqrt{\varepsilon_1 e n^0}}{\sqrt{2\pi}\sqrt{-\varphi_1 - \dfrac{kT}{e}}}. \tag{12.10}$$

Thus we see that in this potential range the capacitance is related linearly to the inverse square root of the potential drop in the space charge region. Under these conditions, the concentration of both the holes and free electrons in the space charge region is lower than the concentration of donor impurities in the sample. The space charge consists of ionized donors and is called the "depletion layer" (or "Schottky layer")

Equation (12.10) may be conveniently written in the form

$$\frac{1}{C_1^2} = \frac{8\pi}{\varepsilon_1 e n^0}\left(-\varphi_1 - \frac{kT}{e}\right). \tag{12.11}$$

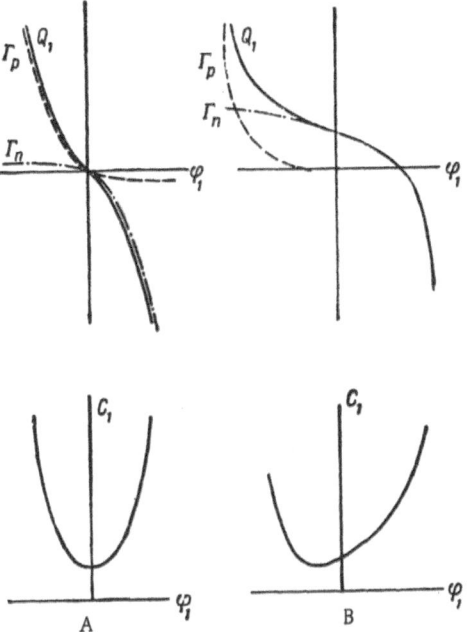

Fig. 10. Qualitative relation of the excess charge
of electrons Γ_n, excess charge of holes Γ_p, total
space charge Q_1, and capacitance C_1 of the semiconductor layer of the double layer to the potential drop in the space charge region. A) Semiconductor with intrinsic conductivity; B) semiconductor with extrinsic conductivity.

As equation (12.11) shows, $-\varphi_1 = kT/e$ when $1/C^2 = 0$.
Therefore, extrapolation of the relation of the inverse root of
the capacitance to the electrode potential to $1/C^2 \rightarrow 0$ gives a
potential which differs by kT/e from the flat band potential (φ_1
= 0). Such an extrapolation is a convenient method for determining the flat band potential from measurements of the differential capacitance of the depletion layer.

The relation of Q_1 of the space charge and the capacitance
C_1 to the applied potential is illustrated in Fig. 10. The curve
is symmetrical close to the minimum in the case of "intrinsic"
germanium. The minimum is much broader in the case of

samples with extrinsic conductivity. When the surface is enriched in majority carriers (enriched or accumulative layer) or
in minority carriers (inverse layer), the capacitance is an exponential function of the potential. In the intermediate range of
potentials (depletion layer) the capacitance changes with potential relatively slowly.

Let us now find the capacitance of the semiconductor
electrode at the flat band potential, i.e., when $Y = 0$. Equation (12.5) gives us [26]:

$$C_1(Y = 0) = \sqrt{\frac{4\pi e^2}{\varepsilon_1 kT}} \sqrt[4]{4n_i^2 + (N_D - N_A)^2}.$$

(12.12)

It may be shown that for a semiconductor with intrinsic
conductivity, the capacitance minimum corresponds to the flat
band potential. In the case of an n-type semiconductor electrode the potential minimum is more positive than the flat band
potential, while that of a p-type semiconductor is more negative.

We will need another expression for the semiconductor
capacitance

$$C_1 = -\frac{dQ_1}{d\varphi_1} = -\frac{d}{d\varphi_1} \int_0^\infty \rho(x)\,dx = -\frac{d}{d\varphi_1} \int_{\varphi_s}^{\varphi_B} \frac{\rho(\varphi)\,d\varphi}{d\varphi/dx} =$$

$$= -\frac{\rho(\varphi_s)}{E_s} = \frac{e(n_s - p_s - N_D)}{E_s}.$$

(12.13)

An expression for the capacitance of the diffuse ionic
layer in the electrolyte C_2 may be derived directly from equation (12.5) (see [27]).

This conversion is accomplished by replacing n_i by c^0,
Y by $e\psi'/kT$, and ε_1 by ε_2. In this case we select $\lambda = 1$. After
these conversions we obtain an expression for C_2 by using
equation (12.5):

$$C_2 = \frac{\varepsilon_2}{4\pi \sqrt{\dfrac{\varepsilon_2 kT}{2\pi e^2 c^0}}} \cdot \frac{\left| e^{-\frac{e\psi'}{kT}} - e^{\frac{e\psi'}{kT}} \right|}{\left[e^{-\frac{e\psi'}{kT}} - 2 + e^{\frac{e\psi'}{kT}} \right]^{1/2}}$$

$$= \frac{\varepsilon_2}{8\pi L_2} \frac{\left| \operatorname{sh} \dfrac{e\psi'}{kT} \right|}{\operatorname{sh} \dfrac{e\psi'}{2kT}} = \frac{\varepsilon_2}{4\pi L_2} \coth \frac{e\psi'}{2kT}. \tag{12.14}$$

Let us examine the capacitance of the surface levels C_t. Differentiation of equation (10.13) for the charge on the levels gives us an expression for C_t:

$$C_t = \frac{e^2}{kT} \frac{N_t \exp\left(\dfrac{E_F - E_t^0}{kT} + Y\right)}{\left\{1 + \exp\left(\dfrac{E_F - E_t^0}{kT} + Y\right)\right\}^2}. \tag{12.15}$$

As the total charge of the semiconductor layer of the double layer equals the sum of Q_1 and Q_t, the capacitance is also determined by the sum of C_1 and C_t.

Relation (12.15) shows that at high negative values of Y, when $|(E_F - E_t^0)/kT| \ll |Y|$, the capacitance of the surface levels changes with potential according to the equation

$$C_t \approx \frac{e^2}{kT} N_t e^Y, \tag{12.16}$$

while at high positive values of Y the capacitance C_t depends on the potential in the following way:

$$C_t \approx \frac{e^2}{kT} N_t e^{-Y}. \tag{12.17}$$

Thus, the capacitance $C_t \to 0$ when $|Y| \to \infty$. The relation of the capacitance of the levels to the potential is represented by a curve with a maximum.

With a high surface charge at the electrode, for example with the number of surface levels $N = 10^{14}$ cm^{-2}, the value $e^2 N_t / kT \approx 10^8$ cm^{-1}. Then, in accordance with equation (12.17) even at potentials $Y = e\varphi_1/kT \approx 1$ the capacitance of the surface levels becomes comparable with that of the Helmholtz layer.

Equations (12.5) and (12.15) for the capacitance of the space charge region and surface states were derived with the assumption that the charge on the levels and the space charge are determined by equilibrium formulas. This imposes certain limitations on the frequency of the alternating current. There is dispersion of capacitance at high frequencies. This problem will be dealt with in Chapter III.

§ 13.   Potential Distribution: Germanium. Potential Drop in the Space Charge Layer and Its Relation to the Bulk Properties of Germanium

The potential drop in the space charge region may be determined by investigating such properties of the space charge as surface conductivity, differential capacitance, and photopotential. *

a.   Measurement of Surface Conductivity. Let us examine some of the results obtained by measuring the surface conductivity. Figure 11 gives the experimental relation of surface conductivity σ to the potential φ of a germanium electrode in a solution of KBr in methylformamide [28]. The latter was chosen as the solvent as no electrochemical reactions occur in the system germanium–methylformamide over a wide range of potentials. Therefore, under these conditions the germanium electrode is similar in its properties to an

---

*See § 17 regarding the photopotential. In comparing these methods, we should note that the results of measuring the surface conductivity are strongly affected by the escape of current through the semiconductor–electrolyte boundary, which usually occurs in real systems (see § 23). Another possible source of error is the decrease in mobility of free carriers in the space charge region compared with their mobility in the bulk of the semiconductor because of the additional scattering of free carriers by the surface [22]. The methods using capacitance and the photopotential do not have these limitations. However, both of them, and particularly the former, are sensitive to the presence of fast surface electron states at the interface being studied.

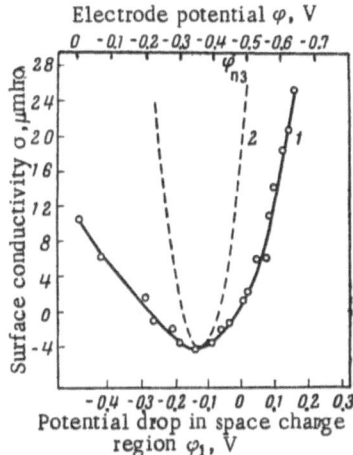

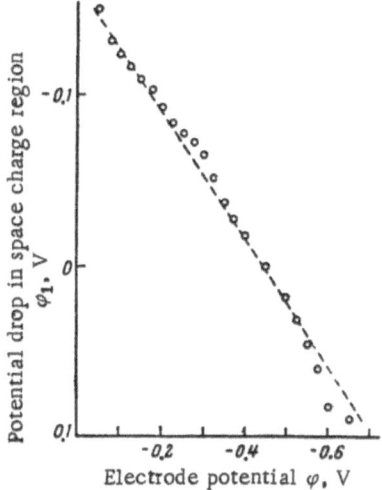

Fig. 11. Experimental relation of surface conductivity σ to electrode potential φ (1) and theoretical relation of surface conductivity σ to potential drop in space charge layer φ_1 (2) for n-type germanium (30 Ω·cm), KBr solution in methylformamide [28].

Fig. 12. Relation of potential drop in space charge region to electrode potential (calculated from Fig. 11).

ideally polarizable electrode and this facilitates the measurement of the electrophysical properties of its surface. The $\sigma - \varphi$ curve was plotted in the following way: the electrode was kept at each potential value for several minutes until the surface conductivity reached a steady value. Then σ was measured.

The same figure gives a theoretical curve for the relation of the surface conductivity to the potential drop in the space charge region φ_1, calculated by equation (11.4). A comparison of the curve shows that the theoretical $\sigma - \varphi_1$ curve is narrower than the experimental $\sigma - \varphi$ relation (with the same scales for φ and φ_1). Consequently, with a given change in the surface conductivity the change in the potential drop in the space charge layer $\Delta\varphi_1$ is always less than the change in the electrode potential $\Delta\varphi$.

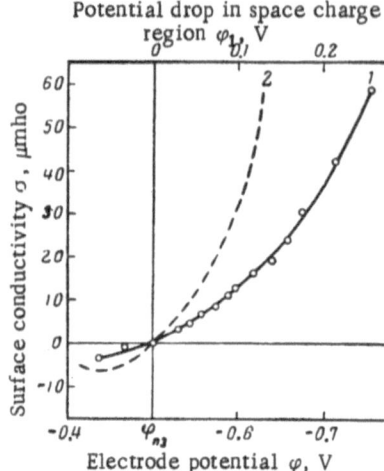

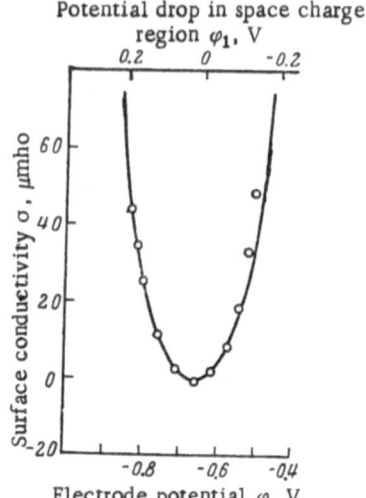

Fig. 13. Relation of surface conductivity σ to electrode potential φ(1) and theoretical relation of surface conductivity σ to potential drop in space charge region φ_1 (2). 0.01 N aqueous NaOH solution, n-type germanium (17 $\Omega \cdot$ cm) [30].

Fig. 14. Relation of the surface conductivity σ of germanium to electrode potential φ with rapid changes in polarization (curve plotted within 20 sec) [31]. Points, experimental values; solid line, theoretical relation of surface conductivity to potential drop in space charge region φ_1.

This indicates that not only φ_1 but also the potential drop φ_0 in the ionic part of the double layer changes with a change in the electrode potential.

Figure 12 gives the relation of φ_1 to φ calculated from Fig. 11 by comparing the values of φ and φ_1 at the same value of the surface conductivity. Figure 12 shows that $\Delta\varphi_1$ and $\Delta\varphi_0$ are comparable in magnitude and are several tenths of a volt in the potential range investigated. *

---

* We should emphasize once more that the potential drop in the space charge region may be measured as an absolute value, while the potential drop in the Helmholtz layer, as in the case of metal electrodes, may be measured with accuracy only to some unknown constant [29].

The values found for $\Delta\varphi_0$ in these experiments are tens
of times greater than those calculated with the assumption that
the electrical double layer is formed only by the space charge
in the semiconductor and electrostatically adsorbed ions from
the solution. The high value of the potential drop in the ionic
layer may be the result of either a high charge concentrated
in the surface states or oriented dipoles (for example, solvent
molecules) at the surface whose degree of orientation changes
with the electrode potential. The number of surface states or
dipoles per unit surface must be close in order of magnitude
to the number of surface germanium atoms. We will return to
this problem in § 14.

Similar results were also obtained in aqueous solutions
(Fig. 13) [30]. However, in this case there was anodic solution of germanium during anode polarization, and this limits
the possibility of using this method to the region of weak anode
and cathode polarization. A complete curve of σ against φ
with a minimum was obtained in [31]. It was found that with a
sufficiently rapid change in electrode potential (the whole curve
was plotted within 20 sec) the experimental curve coincided
with the theoretical curve of σ against φ_1 over a wide range of
potentials (Fig. 14). This is because the processes which produce a potential change in the ionic part of the double layer
and are connected with adsorption, surface oxidation, change
in degree of orientation of dipole solvent molecules at the electrode, etc., proceed relatively slowly. The relaxation time of
the potential drop in the ionic double layer on a germanium
electrode was of the order of 10 sec. Therefore, with rapid
changes in polarization there is a change only in the potential
drop in the space charge region. The method of measuring surface conductivity was also used in [32].

In a number of papers the change in surface conductivity
was studied when an electrode that had previously been in a
gas atmosphere was immersed in a solution. Thus, when germanium was immersed in aqueous solutions its surface conductivity (with a steady potential) changed over a period of
several hours and this may have been caused by a structural

change in the Helmholtz region (for example, by hydration and partial solution of surface oxides) [33,34]. On the other hand, immersion in n-heptane did not change σ and, evidently, produces no change in the structure of the double layer [35]. These experiments are reminiscent of the so-called "Bardeen—Brattain cycle" in which the surface charge of a semiconductor is changed by successively changing the gas atmosphere [36]. However, we should note that reliable data may be obtained only by measuring the relation of the surface conductivity to potential over a sufficiently wide range of potentials. This is necessary for a comparison of experimental and calculated curves to determine the sign and value of φ_1.

   b.   Measurement of Differential Capacitance.   Measurement of the differential capacitance may also be used to determine the distribution of potential at the interface. We must first ensure that the value being measured is actually the capacitance of the space charge in the semiconductor (and is not produced by surface levels, diffusion processes, etc.). For this the experimental curve is compared with that calculated from equation (12.5).

   In the case of germanium, both curves have a minimum and this simplifies their comparison. Here, the main difficulty is connected with the determination of the true surface of the electrode. It is usually accepted that the mirror-like surface of germanium obtained by etching in an SR-4 mixture (see § 23) has a roughness coefficient of 1.3-1.4 [37].*   The absence of a frequency dependence of the capacitance indicates that the capacitance is produced by the space charge (certain complications arising during the measurement of the capacitance of the inversion layer at high frequency will be examined in Chapter III).

---

*Microirregularities whose dimensions are less than the thickness of the space charge layer have practically no effect on the capacitance value. This problem and also the edge effect, have been investigated by Goodman [38].

Bohnenkamp and Engell were the first to measure the capacitance of a germanium electrode [39]. However, in this, as in most of the subsequent work, the capacitance of the space charge alone could not be measured because of the presence of fast electron states on the surface which contributed to the capacitance being measured right up to the maximum frequency obtained. Brattain and Boddy were able to overcome this difficulty as they evidently were able to produce a germanium — electrolyte surface boundary which did not contain an appreciable number of fast surface states [32].

For this the electrode surface and the solution had to be freed from impurities (probably heavy metals). The first of these conditions could be achieved readily by preliminary anode etching of the electrode directly in the solution in which the measurements were to be carried out. It was shown previously that such treatment decreases the number of recombination centers at the germanium — solution boundary [40, 41].* The solution (0.1 N K_2SO_4, pH 7.4) was purified by storage above germanium powder (see § 23). Under these conditions, the relation of capacitance to potential shown in Fig. 15,† was obtained in [32] and the capacitance at its minimum found to be 1.3 times greater than that calculated by equation (12.5). As the ratio of the true surface of germanium to the apparent surface after etching in SR-4 was the same value, Brattain and Boddy concluded that the capacitance measured was determined wholly by the space charge in the semiconductor and the number of fast surface states which could have contributed to the capacitance did not exceed $3 \cdot 10^9 / cm^2$. Regrettably, the frequency dependence of capacitance was not investigated in this work‡ and this conclusion must be treated with care.

---

*It was also noted in [39, 42] that anode etching of germanium directly before measurement increase the reproducibility of the capacitance measurements.

†In Figs. 15 and 17-21 the electrode potentials are relative to a saturated calomel electrode.

‡A pulse method was used in [32] for the measurement of capacitance (see § 23). The pulse time was usually 5 μsec.

Brattain and Boddy interpreted their measurements with the help of the equivalent circuit of the electrode illustrated in Fig. 16. Here, R_B is the ohmic resistance of the electrode and solution, C_1 is the capacitance of the space charge region, and R is the resistance of the electrode reaction. The nature of the resistance R_s', which is small and becomes appreciable only at high anode polarization of the electrode, remains unclear. Some hypotheses as to its origin will be put forward in §47.*

A comparison of the experimental $C - \varphi$ curve (Fig. 15) with the theoretical $C - \varphi_1$ curve calculated by equation (12.5), showed (as in the case of surface conductivity, see Fig. 11), that the steady experimental curve was broader than the theoretical curve. The results of measuring capacitance using a series of samples with different specific resistances (in particular, those shown in Fig. 15), are given in Fig. 17 as the relation of $\varphi_1 - (kT/e) \ln \lambda$ † to the electrode potential. As follows from equation (8.10) up to a certain constant value, the electrode potential equals $\varphi_{12} - (kT/e) \ln \lambda$, where the galvanic potential φ_{12} equals the sum of the potential drops in the space charge region and in the ionic part of the double layer, i.e., $\varphi_{12} = \varphi_1 + \varphi_0$. If, with a change in the electrode potential, only the potential drop in the space charge region changes, then the relation of $\varphi_1 - (kT/e) \ln \lambda$ to φ should be expressed by a straight line with a slope equal to unity (see [43]).

Figure 17 shows that the slope of the line differs considerably from unity as in the case when an analogous value is determined from measurements of the surface conductivity (see Fig. 12). Consequently, the capacitance method indicates a potential drop φ_0 in the ionic region of the double layer which is

---

*This equivalent circuit lacks the capacitance of the Helmholtz layer (which, generally speaking, is in series with the circuit in Fig. 16). This is not surprising if one considers that the capacitance measured is 2-3 orders less than the capacitance of the Helmholtz layer (20 $\mu F/cm^2$). (The complete equivalent circuit of the semiconductor electrode is given in §46.)

†The value $\varphi_s^0 = \varphi_1 - (kT/e) \ln \lambda$ is called the surface potential of the semiconductor.

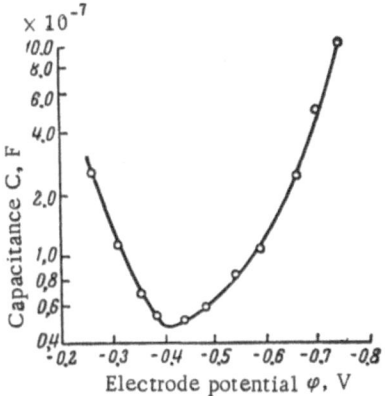

Fig. 15. Capacitance of germanium electrode (n-type, 42 Ω · cm) in 0.1 N K_2SO_4 solution (pH 7.4) [32].

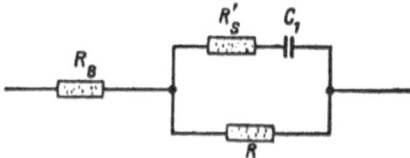

Fig. 16. Equivalent circuit of germanium electrode [32].

comparable with the potential drop in the space charge region and changes with polarization of the electrode. Thus, although there are no fast states, the electrode surface evidently contains slow states which do not contribute to the differential capacitance but determine the potential drop in the ionic double layer.

As in the case of surface conductivity, the capacitance was measured by a method of rapid changes of the electrode potential such that the potential drop in the ionic layer had insufficient time to change. In [32], the electrode potential changed suddenly from the starting value (corresponding to no current or to a weak anode current) to a certain new value. The new values of the potential and capacitance were determined within 1 sec of the switch over. With this method of measurement, the relation shown in Fig. 17 becomes the relation illustrated in Fig. 18. At potentials of ⁻0.6 to ⁻0.8 V, the slope of the curve of $\varphi_1 - (kT/e)\ln\lambda$ against φ equals unity, i.e., with rapid polarization the change in the electrode potential equals the change in φ_1. (This relation is not maintained with very strong anode polarization.)

Depending on the initial potential and current, solution pH, and other conditions, the line in Fig. 18 shifts along the potential axis, i.e., the initial potential drop in the ionic double layer changes (§ 14 is devoted to this problem).

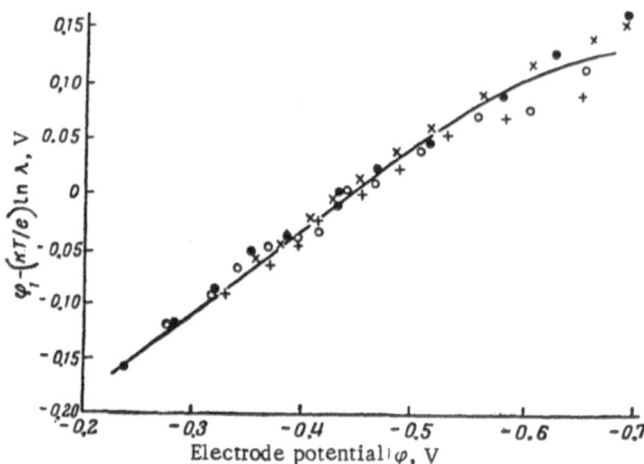

Fig. 17. Relation of $\varphi_1 - (kT/e)\ln \lambda$ to the germanium electrode
potential φ with a slow change in potential [32]. +) 4.75 Ω · cm,
n-type; ●) 42.2 Ω · cm, n-type; ○) 22 Ω · cm, n-type; x) 7.5 Ω
· cm, p-type.

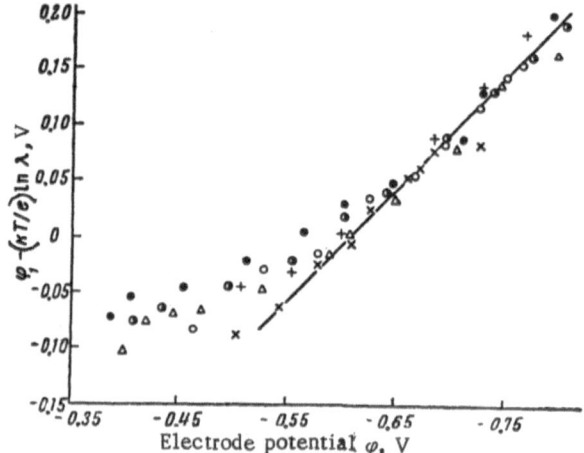

Fig. 18. Relation of $\varphi_1 - (kT/e)\ln \lambda$ to the germanium electrode
potential φ with a rapid change in potential (initial value of φ was
that of the steady potential) [32]. ○ and x) 7.5 Ω · cm, p-type (2
series of experiments); ◑ and +) 42.2 Ω · cm, n-type (2 series of
experiments);△) 22 Ω · cm, n-type; ●) 25.5 Ω · cm, p-type.

Later Boddy [44] developed a method for measuring the capacitance with a rapid and continuous change in the electrode potential. The experimental curve which was plotted within 10 sec coincides with the theoretical over a wide range of potentials as well as at the minimum point.

A series of papers [39, 42, 45, 46] has been devoted to the measurement of the differential capacitance of a germanium electrode and the results agree qualitatively with those given above. The $C - \varphi$ curves for high-resistance germanium have a minimum but the value of the capacitance at the minimum is 2-3 times as great as the calculated value. Evidently, in most cases, this is not caused by the difference in the geometric and the true surface of the electrode. Part of the capacitance was probably due to fast electron surface states. In all of the above papers it was noted that the potential drop in the Helmholtz layer contributed appreciably to the general change in potential (the measurements were carried out with slowly changing polarization).

The concentration of free carriers increased when the semiconductor was heated or illuminated. This produced an increase in capacitance. As in this case the concentrations of excess electrons and holes are the same ($\Delta p = \Delta n$), with an increase in Δp and Δn the relative difference between the total concentrations of holes and electrons produced by the presence of donors or acceptors in the sample decreases. Therefore, the conductivity of the sample when heated or illuminated approaches the intrinsic conductivity while the $C - \varphi$ curves for n- and p-types become symmetrical as for "intrinsic" germanium [39].

§ 14.   Potential Distribution: Germanium. Nature of Components of Interphase Potential Difference.

The experimental material given in the previous section confirms that the basic concepts presented in § 9 on the structure of the double layer and potential distribution at the semiconductor–electrolyte boundary are valid for a germanium

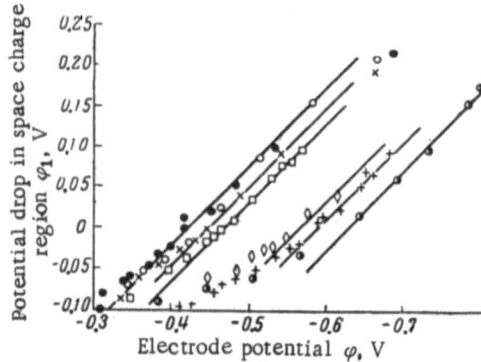

Fig. 19. Relation of φ_1 to the potential of a germanium electrode φ with a rapid change in potential. The
parameter is the initial potential value, which is determined by the anode current: ●) 50 $\mu A/cm^2$; ○) 10
$\mu A/cm^2$; ×) 5 $\mu A/cm^2$; □) 1.5 $\mu A/cm^2$; ◇) 0.5 μA
per cm^2; +) 0.15 $\mu A/cm^2$; ◐) 0.

electrode. We will now examine certain characteristics of a
germanium electrode that were not included within the frame-
work of the general theory developed in § 9.

It was shown in § 13 that after the usual treatment of the
surface of the germanium electrode (etching in aqueous solutions) the value of the interphase potential difference* is much
higher than it would be if the electrical double layer were formed only of free carriers in the semiconductor and the electrostatically adsorbed ions from solution. Let us first examine
the experimental material on the relation of the interphase potential difference to the electrode potential, pH, crystallographic orientation of the surface, etc.

a. Relation of the Interphase Potential
Difference to Various Factors. To obtain the relation shown in Fig. 18, a current was passed to change the

_____

*By interphase potential difference we mean the galvanic potential excluding the
potential difference in the Debye region in the semiconductor and the ψ'-poten-
tial difference in the electrolyte.

electrode potential from a steady value to some new value and during this change the capacitance was measured rapidly (in 1 sec). The potential drop in the space charge region φ_1 was determined from the capacitance. It was found [32] that the position of the line (see Fig. 18) of φ_1 against φ depended on the initial value chosen for the potential during these measurements. Different initial potentials (corresponding to different anode currents) gave the set of lines shown in Fig. 19. Here the initial potential is characterized by the value of the anode current. A change in current from 0 to 10 $\mu A/cm^2$ displaced the curves along the potential axis by 0.18 V but a further change in the initial potential had no effect on the relation of φ_1 to the electrode potential. The relation found may be expressed by the equation

$$\varphi_1 - (kT/e)\ln \lambda = \varphi + const, \tag{14.1}$$

where the constant depends on the initial value of the potential (or the current).

Special experiments showed that the change in the constant is not related to concentration polarization in the solution. Thus, the only explanation for the shift in the lines of φ_1 against φ is a change in the interphase potential difference with a change in the electrode potential selected initially. This potential difference remains the same at all points on each of the $\varphi_1 - \varphi$ lines in Fig. 19, as it does not have time to change within the short time required to measure the capacitance. Only with high anode polarization does the interphase potential difference change in less than 1 sec and the relation of φ_1 to φ deviates from a straight line.

Thus, there are two limiting states for the surface of a germanium electrode in an aqueous solution and two potential drops at the phase boundary correspond to them. One of these is a steady potential (the current equals zero) and the second is an anode current equal to or greater than 10 $\mu A/cm^2$. There is a continuous transition between them.

Lines of φ_1 against φ for these two surface states obtained for electrodes with different crystallographic orienta-

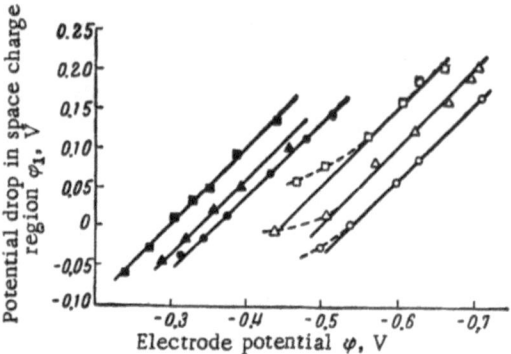

Fig. 20. Relation of φ_1 to the potential of a germanium electrode φ with a rapid change in potential [47]. The parameter is the crystallographic orientation of the surface: ■, □) (111); ▲,△) (110); ●, ○) (100). Initial value of anode current, ■, ▲, ●) 45 $\mu A/cm^2$; □, △, ○) 0.

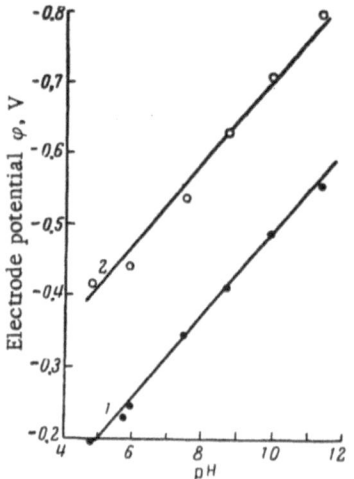

Fig. 21. Relation of germanium electrode potential φ corresponding to the condition $\varphi_1 - (kT/e)\ln \lambda = 0$ to the solution pH [48]. Initial value of anode current: 1) 45 $\mu A/cm^2$; 2) 0.

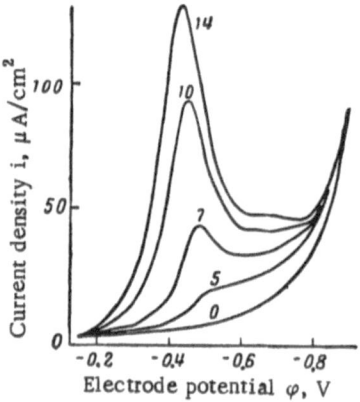

Fig. 22. Cathode polarization curves for a germanium electrode (p-type, 3 $\Omega \cdot cm$) in 48% HF solution, plotted after preliminary anode oxidation at 0.15 V (relative to a hydrogen electrode in same solution) [54]. The oxidation time (in secs) is indicated on the curves.

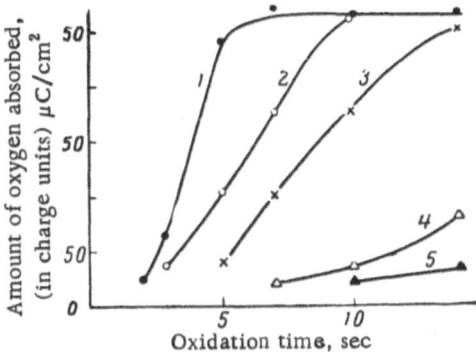

Fig. 23. Relation of amount of oxygen absorbed (in units of electric charge) on the surface of a germanium electrode (p-type, 3 Ω · cm) in 48 % HF solution to the potential and time of preliminary oxidation [54]. Oxidation potential: 1) 0.2 V; 2) 0.17 V; 3) 0.15 V (see Fig. 22); 4) 0.12 V; 5) 0.1 V.

tions of the surfaces, namely, (100), (110), and (111) are given in Fig. 20 [47]. The interphase potential difference changes markedly from one crystal face to another.

Finally, the potential difference φ_0 depends on the solution pH. By comparing the electrode potentials when $\varphi_1 - (kT/e)\ln \lambda$ = const (for example, when const = 0), and when there is the same initial potential and crystallographic face, but different pH values, Boddy and Brattain [48] obtained the relation shown in Fig. 21. When the pH was changed by one unit, the potential difference φ_0 changed by 59 mV (other authors of earlier work found a value close to this [31, 42]).

It was found that the magnitude of the interphase potential difference depends on the degree of oxidation of the surface. Galvanostatic cathode charging curves were used in the first measurements of the amount of oxygen absorbed. It was shown that in sulfuric acid solutions with a steady potential, as well as after preliminary weak anode oxidation, a germanium electrode is covered by about one monolayer of oxygen [49-51].

According to [52] and [53], this amount increases to tens of monolayers with strong anode oxidation in KOH and H_2SO_4 solutions. *

We should note that the cathode charging curves for a germanium electrode are, as a rule, very uninformative. The potential changes monotonically with an increase in the amount of electricity. The curves show no lags in potential growth produced by the formation of phase oxides of constant composition. †

A more efficient method was that of plotting cathode oscillopolarograms with a linearly increasing voltage [54]. Figure 22 shows such polarograms plotted for 48% hydrofluoric acid solution (in this electrolyte germanium oxides are readily soluble and a surface with the minimum degree of oxidation may be obtained).‡ The lower curve was plotted immediately after preliminary cathode polarization of the electrode and the rest, after preliminary anode oxidation. The bulk of the surface oxides was reduced at a potential of −0.4 V. The amount of oxygen adsorbed during anode polarization and removed by cathode polarization may be determined from the area bounded by the lowest polarogram and one of the upper ones. The results of such an investigation relative to potential and anode oxidation time are shown in Fig. 23. During strong anode polarization a steady state is established when the rates of formation and solution of the surface oxide are equal. In cathode polarization of an oxidized electrode about one monolayer of oxygen is reduced. When an electrode has been kept for a long

---

*Here and elsewhere we mean the amount of surface oxides which are reduced by cathode polarization.

†In [52, 53] the authors found steps on the charging curves which they explained by the formation of a surface oxide of definite composition (GeO). The height of these steps corresponded to one monolayer. There are objections to this explanation. In actual fact, the hypothesis that a monolayer of an oxide of constant composition exists on the surface on top of tens of monolayers of a bulk oxide of variable composition lacks sense.

‡In Figs. 22-24, the potential is relative to a hydrogen electrode in the solution investigated.

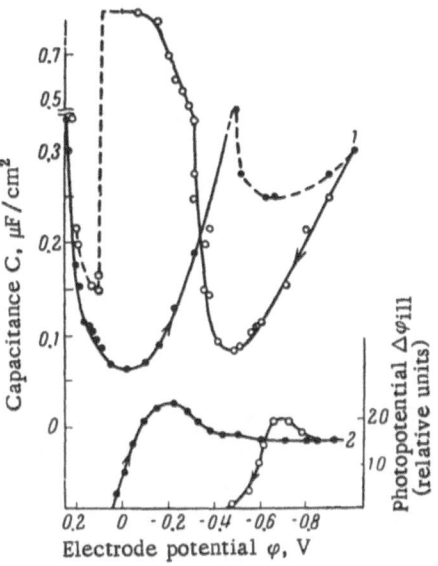

Fig. 24. Relation of capacitance C (curves 1) and the photopotential $\Delta\varphi_{ill}$ (curves 2) to the potential of a germanium electrode φ (p-type, 3 $\Omega \cdot$ cm) in 48% HF solution [54]. Arrows show direction of potential change during plotting. Dotted line shows unstable sections of the curves.

time in an HF solution at a steady potential, regardless of preliminary treatment, only tenths of a monolayer of oxygen cover the surface.

Thus, by using cathode and anode polarization of a germanium electrode in an HF solution, it is possible to achieve two limiting cases in which the amount of oxygen absorbed differs by one monolayer (10^{15} atoms per cm^2).* The transition from one state to the other with the oxidation and reduction of the electrode surface is accompanied by a shift in the curves of

---

*A certain amount of adsorbed oxygen apparently remains on the surface of a germanium electrode even after strong cathode polarization [54].

the differential capacitance (as well as of the photopotential) by about 0.6 V (Fig. 24) and this can only be associated with a change in the interphase potential difference φ_0 (with a constant potential difference in the space charge region φ_1).* If the integral capacitance of the Helmholtz region is taken as equal to $10 \ \mu F/cm^2$, then a 0.6 V change in potential corresponds to a $10^{-5} \ C/cm^2$ change in charge and this is a factor of approximately ten less than the amount of electricity required for reduction of the oxygen adsorbed on the surface. Consequently, if the surface charge is produced by oxygen atoms, then there is one electronic charge for every 10 oxygen atoms.

   b.   The Nature of the Interphase Potential Difference.   Let us now examine the interpretation of the experimental material given above. A high value of $d\varphi_1/d\varphi_0$ may be produced by a sufficiently high concentration of slow surface states.† Let us estimate the charge on these levels, using the value $d\varphi_0/d\varphi_1 \approx 1$ measured in aqueous solutions and in methylformamide [28, 30]. From equation (10.27) it follows that

$$Q_t = \frac{d\varphi_0}{d\varphi_1} \frac{\varepsilon_0 kT}{4\pi e d_0} .$$

(14.2)

Estimation of Q_t on the basis of experimental data gives a value of $3 \cdot 10^{-6} \ C/cm^2 \cdot V$, i.e., $2 \cdot 10^{13}$ electronic charges per cm^2 per volt.

   We should note that the estimation is very approximate as it was assumed that there is only one type of level with a definite energy on the surface. In this case the levels would have

----

*On the $C - \varphi$ curve plotted after anode polarization (oxidized surface), the capacitance at the minimum was practically equal to the theoretical capacitance in the space charge region (considering the roughness of the surface). A certain increase in capacitance at the minimum after cathode polarization, which was less noticeable on low-resistance samples, was explained by the formation of fast surface states during cathode polarization which could possibly have resulted from the adsorption of anions or molecules of hydrofluoric acid.

†These levels must be slow as they have no appreciable effect on the differential capacitance of an electrode measured with an ac current.

been effective only over the narrow range of potentials over which their filling changes. In actual fact the ratio $d\varphi_0/d\varphi_1$ ≈ 1 over the whole range of potentials investigated. This may be explained by a set of levels with energies lying in the forbidden band of the semiconductor. The calculation given above explains formally the effect of screening of the bulk of the semiconductor by surface electron states. However, very little is known of the nature of these states.

To explain the formation of a surface charge and the dependence of $\Delta\varphi_0$ on the solution pH, Gerischer [42] put forward the hypothesis that the surface oxide was hydrated and the OH-groups formed were partly dissociated according to the following equation [for the (100) face]:

$$H_2O + {>}Ge{<}^{OH}_{OH} \rightleftarrows {>}Ge{<}^{O^-}_{OH} + H_3O^+. \qquad (14.3)$$

For the equilibrium of equation (14.3) together with the preceding oxidation of germanium, the equilibrium potential must equal

$$\varphi^0 = \text{const} + 0.059\,\text{pH}. \qquad (14.4)$$

Thus, a change in pH by unity produces a change in the potential difference in the Helmholtz layer of 59 mV. This agrees with experimental data when $5 < \text{pH} < 11$ (Fig. 21).

Let us examine the relation of the interphase potential difference to the amount of oxygen adsorbed on the surface at a constant pH. Oxygen adsorption makes the potential of the flat bands more positive (Fig. 24). This means that during adsorption the electrode is charged positively relative to the solution. At the same time, a negative charge is produced on the electrode surface on dissociation of the surface oxide, according to equation (14.3). The potential difference in the true ionic double layer therefore has a sign opposite to that of the change in the interphase potential difference at the boundary which is connected with oxygen adsorption. We should therefore conclude [54] that the potential difference $\Delta\varphi_0$ produced

by oxygen is the sum of two parts: the dipole potential difference $\Delta\varphi_{01}$, produced by the polar character of the Ge—O bond,*
and the potential difference $\Delta\varphi_{02}$ in the ionic double layer
formed by the dissociation of the surface oxide:

$$
\begin{array}{c}
\diagdownO^{-}H_3O^{+}\\
\diagup\\
\underset{Ge}{}^{(+)}\\
\diagup\diagdown\\
{}^{(-)}OH
\end{array}
$$

$$\overleftrightarrow{\Delta\varphi_{01}}\quad\overleftrightarrow{\Delta\varphi_{02}}\tag{14.5}$$

These components of the interphase potential difference
have opposite signs and the former has the higher absolute
value $|\Delta\varphi_{01}| > |\Delta\varphi_{02}|$. Therefore, in the adsorption and de-
sorption of oxygen at a constant pH the predominating effect is
that of the surface dipole whose positive end is oriented toward
the electrode. A change in pH (with a constant degree of oxidation) has an effect only on $\Delta\varphi_{02}$. We may assume that it is
mainly the structure and the charge of the ionic region of the
double layer which change with a change in the electrode potential. Thus, the charges of the GeOOH⁻ groups screen the bulk
of the semiconductor from the external electric field and act
as (slow) surface states.†

Another explanation was proposed by Boddy and Brattain [48], who related the interphase potential difference to the
oriented adsorption of polar water molecules:

---

*Strictly speaking, the potential difference $\Delta\varphi_{01}$ may be related not only to the
polarity of the Ge—O bond, but also to the effect of oxygen adsorption on the potential difference produced by the extension of the electron cloud beyond the
limits of the crystal lattice of germanium.

†The charge of the surface states is formally described by equation (10.13). We
should keep in mind that with a change in the electrode potential, not only the
filling of the levels by electrons may change, but also their concentration N_t (as
a result of adsorption and desorption).

$$\text{>Ge<}^{OH\cdots\cdot O<^{H}_{H}}$$

$$(14.6)$$

The degree of orientation (and, consequently, the value of φ_0) depends on the current passing through the electrode. It was proposed that at a current density above $10 \ \mu A/cm^2$, the orientation of molecules in this layer was completely disrupted. The transition from $i = 0$ to $i \geq 10 \ \mu A/cm^2$ was accompanied by a change in $\Delta\varphi_0$ of 0.15-0.18 V. This hypothesis was not checked experimentally. To confirm it the kinetics of relaxation of $\Delta\varphi_0$ and of the current with a change in potential must be investigated simultaneously. We should note that the experimental data in [48] on the effect of the electrode potential on φ_0 (Fig. 19) may be explained by assuming that the degree of oxidation of the germanium surface changes during anode polarization.

The difference in the value of φ_0 for different crystallographic faces may be related to a difference in the electronic structure of the crystal surface (displacement of the electron cloud relative to the surface germanium atoms) as well as to a difference in the number of germanium atoms per cm^2 of the surface and, consequently, a difference in the number of oxygen atoms or water molecules adsorbed. Areas of 8, 11, and 13.8 $Å^2$ correspond to each germanium atom on the (100), (110), and (111) faces, respectively. The number of surface germanium atoms and adsorbed groups per cm^2 is inversely proportional to these values. The difference in φ_0 in going from (100) to (110) is 21 mV and from (100) to (111), 61 mV [47].

As the potential difference at the phase boundary depends on a whole series of factors, the flat band potential of a semiconductor electrode determined, for example, from the $\sigma - \varphi$ or $C - \varphi$ relations should be referred to definite measurement conditions (solution pH, degree of oxidation of the electrode, etc.).

   c.   Structure of the Electrical Double Layer on a Germanium Electrode.   The data given

in paragraphs a and b make it possible to draw the following
picture of the structure of an electrical double layer at the
germanium−solution boundary. In alkaline or weakly acid solutions the electrode surface is charged negatively as a result
of the partial dissociation of the surface oxide according to
equation (14.3). The charge of the surface states Q_t, the
space charge in the semiconductor layer Q_1, and the charge of
the electrostatically adsorbed ions in the outer layer of the
double layer Q_2 are in equilibrium with each other, so that

$$Q_t + Q_1 + Q_2 = 0. \tag{14.7}$$

The relation between the values Q_1 and Q_2 is determined by the
capacitances of the space charge region in the semiconductor
C_1 and in the electrolyte C_2 and the capacitance of the surface
states C_t. We may assume that the screening of the surface
charge occurs in such a manner that the energy of the system
is minimal. On the basis of equation (14.7) we may write

$$\frac{dQ_t}{d\varphi_1} + \frac{dQ_1}{d\varphi_1} + \frac{dQ_2}{d\varphi_2}\frac{d\varphi_2}{d\varphi_1} = 0, \tag{14.8}$$

hence

$$C_t + C_1 + C_2\frac{d\varphi_2}{d\varphi_1} = 0. \tag{14.9}$$

Consequently, the derivative $d\varphi_2/d\varphi_1$ equals

$$\frac{d\varphi_2}{d\varphi_1} = -\frac{C_1 + C_t}{C_2}. \tag{14.10}$$

Considering that

$$dQ_2 = -C_2 d\varphi_2,$$
$$dQ_t = -C_t d\varphi_1, \tag{14.11}$$

we obtain an expression for the derivative dQ_t/dQ_2 which
characterizes the relation of the charges in the surface states
and in the diffuse part of the ionic double layer:

$$\frac{dQ_t}{dQ_2} = \frac{C_t}{C_2}\frac{d\varphi_1}{d\varphi_2} = -\frac{C_t}{C_1+C_t}.$$

$$(14.12)$$

If $C_t \gg C_1$, then $dQ_2 \approx -dQ_t$, i.e., the surface charge is screened mainly by the electrostatically adsorbed ions from solution. As a result a much greater density of ionized surface states may be obtained on the nondegenerate surface of the semiconductor at the germanium−electrolyte boundary than at the germanium−gas boundary. For example, on a germanium electrode in an alkali solution, $Q_t \approx 10^{14}$ electronic charges per cm^2, while the space charge Q_1, calculated, for example, from the differential capacitance, does not exceed 10^{11} electronic charges per cm^2. On a "dry" germanium surface where the condition $Q_1 = -Q_t$ holds, the value of the space charge (10^{14}) required to neutralize the charge in the surface levels would be the same, and this would lead to the degeneracy of the free carriers on the surface.

At the flat band potential the space charge Q_1 equals zero, but the electrode has a space charge which is neutralized by the charge of the ions in the outer layer of the double layer, so that $Q_t = -Q_2$. In this case, the structure of the double layer on germanium is reminiscent of the structure of the double layer on a mercury electrode at the potential of the maximum on the electrocapillary curve with specific adsorption (see, for example, [27]). The flat band potential should not be mistaken for the "true" zero point, * when there are no charges on the electrode surface.

Measurement of capacitance obviously cannot give us direct data on the value of the surface charge. In actual fact, on a nondegenerate surface, $C_1 \ll C_0$ and, therefore, the capacitance measured under these conditions is determined by the

---

*By the "true" zero point we mean the potential of the maximum on the electrocapillary curve on a nonoxidized electrode in the absence of specific adsorption.

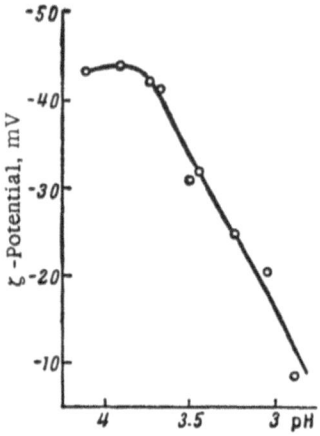

Fig. 25. Relation of the ζ-potential at a germanium—electrolyte boundary to pH. n-Type germanium (10^{-2} $\Omega \cdot$ cm) [58].

space charge in the semiconductor and not its surface charge.* Neither can the surface conductivity or other properties of the space charge be used to determine Q_t. Electrokinetic measurements are most suitable for this purpose.

The potential drop in the diffuse part of the double layer in the electrolyte has a value close to that of the electrokinetic or ζ-potential. The latter may be determined by various methods, using the nature of electrokinetic phenomena (see, for example, [56]). For example, the ζ-potential in water and some organic solvents was calculated from measurements of the streaming potentials of a liquid flowing through finely ground germanium powder [57]. In water and in methanol the ζ-potential was 20-25 mV. Similar values were obtained for dilute aqueous solutions by Sparnaay [58], whose calculations were based on observations of the electroosmotic flow of a liquid through germanium powder. The ζ-potential equals 40-45 mV in millinormal KNO_3 solution; it falls with an increase in concentration.

---

*In [55], a minimum (at −0.6 V) was found on the $C - \varphi$ curve of a germanium electrode in a 0.1 N HCl solution. On the basis of the usual concepts of metal electrodes, the authors [55] related it to the zero point of germanium. Before the measurements the electrodes were subjected to cathode polarization for an hour at a current density of 10^{-2} A/cm^2 and, as will be demonstrated below, this produces a sharp change in the surface properties of germanium which cannot be controlled properly. The nature of the capacitance minimum discovered in [55] remains unclear. It is possible that the measurement results were distorted by the effect of a pseudocapacitance which resulted from the liberation of hydrogen.

Thus, the value of the ζ-potential at the semiconductor–solution boundary is approximately the same as that at the metal–solution boundary, i.e., the structure of the diffuse charge in the electrolyte for such systems is essentially the same. As a rule, all the measurements on semiconductor electrodes examined above were carried out in concentrated solutions in which the potential drop in the diffuse layer of the solution was negligibly small.

The relation of the ζ-potential on a germanium electrode to the solution pH is given in Fig. 25. We may assume approximately that $\zeta \approx \varphi_2$ (see, for example, [27]). Consequently, the value of the ζ-potential reflects the value of the charge in the diffuse section of the ionic layer, which approximately equals Q_t in a dilute solution (10^{-4}–10^{-3} N). Extrapolation of the curve in Fig. 25 to its intersection with the abscissa axis gives us $\zeta = 0$ at pH 2.5 and, consequently, $Q_t = 0$. As was mentioned above, the surface charge on a germanium electrode is usually related to either acidic or basic dissociation of a surface oxide of the type GeO_2. The absence of dissociation at pH $\approx$ 2.5 may be explained by the fact that in this solution the amphoteric surface oxides are at the isoelectric point (according to data from polarographic measurements of solubility, the isoelectric point of GeO_2 is close to pH 2.7).

Q_2 may be determined from the value of the ζ-potential (assuming that $\zeta = \varphi_2$). On the other hand,

$$Q_2 = Q_t = - eN_t\alpha,$$

where N_t is the number of surface acid groups and α is their degree of dissociation.

The latter value is related to the dissociation constant K_1 and concentration of hydrogen ions by the following expression:

$$K_1 = [H^+]\frac{\alpha}{1-\alpha}. \tag{14.13}$$

Hence we have

$$Q_t = -eN_t\alpha = -\frac{eN_t}{1+[\text{H}^+]\cdot K_1^{-1}} = -Q_2. \tag{14.14}$$

By determining Q_2 from equation (14.14) for several values of the ζ-potential (Fig. 25), Sparnaay [58] calculated the dissociation constant of the surface groups $K_1 \approx 10^{-6}$ g-equiv per liter, which is close to the first dissociation constant of germanic acid H_2GeO_3.

Finely ground germanium powder was used for the electrokinetic measurements. However, the surface of germanium after cleaving has specific properties and evidently shows little resemblance to the actual surface of a germanium electrode. The method of measuring contact angles is probably more suitable for the investigation of the latter. It was shown qualitatively in [59] that the contact angle at the germanium— aqueous solution—air boundary depends on the electrode potential. It may be found that the contact angle method will prove useful for the measurement of the surface charge on semiconductor electrodes.

§15.   Distribution of Potential: Silicon

In contrast to germanium, there have been relatively few investigations of silicon. This is partly explained by the fact that silicon is a more difficult subject from the experimental and theoretical points of view as in aqueous solutions it is normally covered by an oxide film whose thickness and composition depends on the composition of the electrolyte and conditions of preliminary treatment. Silicon oxides are soluble in hot concentrated alkalis and in solutions containing fluorine ions. Under these conditions the thickness of the oxide layer is evidently small. The thickness of the oxide layer is small in hydrofluoric acid solutions and the surface oxides are partly reduced by cathode polarization [60].

The most definite data on the potential distribution at the silicon—electrolyte boundary were obtained by Harten [61] who measured the surface conductivity of a silicon electrode in re-

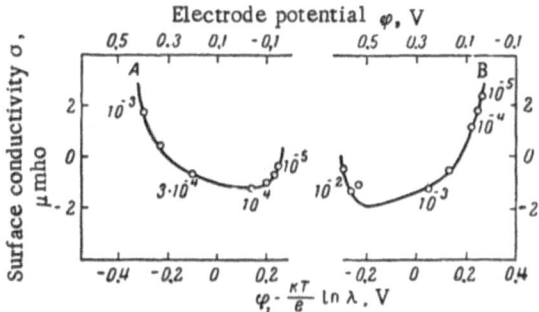

Fig. 26. Relation of the surface conductivity σ of silicon to the electrode potential φ in 1 N H_2SO_4 solution with $CeNH_4(SO_4)_2$ added (concentrations are given in the figure) [61]. A) p-Type, 350 Ω · cm; B) n-type, 250 Ω · cm.

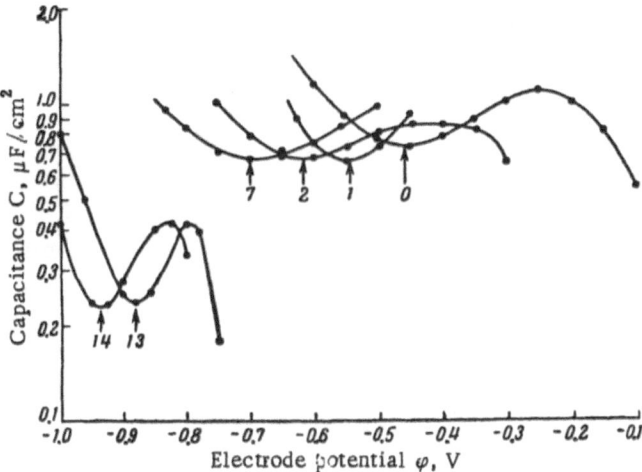

Fig. 27. Relation of capacitance C to electrode potential φ of a silicon electrode (n-type, 8 Ω · cm). The parameter is the solution pH (the pH values are given on the curves) [63].

lation to the potential (Fig. 26).* The electrode potential was changed by changing the concentration of the oxidizing agent $CeNH_4(SO_4)_2$ in an H_2SO_4 solution, and not by passing a current. The curves in Fig. 26 were calculated by equation (11.4). The scale is the same for the electrode potential φ and for the potential drop in the space charge region φ_1. The coincidence of the theoretical relation of σ to φ_1 and the experimental relation of σ to φ justifies the conclusion that $\Delta\varphi = \Delta\varphi_1$, i.e., the change in the electrode potential is concentrated wholly in the space charge region, while the potential difference in the ionic region of the double layer, if it has an appreciable value, does not change with a change in the electrode potential. In this connection the behavior of a silicon electrode is closer to the "ideal" than that of a germanium electrode. As the conductivity of the silicon samples used in [61] was far from intrinsic, the "depleted layer" section which is hardly noticeable in the case of the "intrinsic" germanium samples (see Fig. 11), covers a wide range of potentials here.

Analogous results may be obtained by polarizing the electrode with a current [62].

It was found that the condition $\Delta\varphi = \Delta\varphi_1$ is upset if the electrode surface is first oxidized. In this case the curves of "surface conductivity against potential" (as well as "surface recombination rate against potential," see § 19) become flatter and extend toward the potential axis, i.e., $\Delta\varphi_1 < \Delta\varphi$. The additional potential difference produced is not an ohmic potential drop in the oxide layer as it is not related by Ohm's law to the current passing through the electrode. With the formation of thicker oxide layers, the interphase potential difference apparently depends on polarization and under the conditions in [62], the following equation holds:

$$\frac{d\varphi_0}{d\varphi_1} = \text{const.}$$

---

*In Figs. 26-30 the potentials are relative to a saturated calomel electrode.

Until now, the measurement of the differential capacitance has not given any essential data on the distribution of the potential at the silicon—electrolyte boundary. Although the experimental curves of C against φ as well as the theoretical ones have a minimum (Fig. 27 [63]), the capacitance at the minimum is at least an order greater than the calculated capacitance of the space charge. The form of the curve is also quite different from the theoretical one; the minimum is narrower. Finally, the high-frequency dispersion also indicates that here we are evidently dealing with the capacitance of fast surface states and not with that of the space charge. Thus, measurement of the capacitance cannot be used directly for the calculation of φ_1.

However, certain conclusions may be drawn on the magnitude of the interphase potential difference φ_0 from the shift on the curves of C against φ with a change in the solution pH. First of all, this value is not small although, as has been shown above, it may not change on polarization. In actual fact, a change in pH from 0 to 14 produced no change in the value of the capacitance and the form of the curves and there was only a shift of the curves along the potential axis. If it is assumed that the concentration of surface levels and their energy distribution remain approximately the same at all pH values, then this shift indicates a change in the potential difference at the phase boundary that reaches almost 1 V.

As Fig. 27 shows, the relation of φ_0 to the pH is very complex and cannot be described over the whole pH range from 0-14 by a simple dissociation equilibrium of the type (14.3). At different pH values the surface oxides evidently have different compositions. In the pH range of 0-7 the electrode is covered by a stable oxide film. At pH 8-12 the silicon oxides dissolve slowly and measurement of the capacitance becomes unreliable. Finally, in strongly alkaline solutions, in which silicon oxides are relatively soluble, the electrode surface is again stable. The capacitance is much lower here than on a strongly oxidized surface. The transition of a silicon electrode from an active to a passive state and back again, and the change in capacitance associated with it, will be examined in § 39.

When the sample is illuminated, the capacitance increases with the illumination intensity and approaches the capacitance of the Helmholtz layer [64].

§ 16. Distribution of Potential:Semiconductors with a Wide Forbidden Band

With an increase in the width of the forbidden band E_g $= E_c - E_v$ the concentration of free carriers in a semiconductor with intrinsic conductivity falls rapidly [equation (4.5)]. Simultaneously there is also a fall in the magnitude of the intrinsic conductivity. As has already been mentioned in § 2, when $E_g > 1$ eV, the material is an insulator (and in order to impart to it appreciable electrical conductivity it is necessary to introduce into it 10^{14}-10^{18} cm^{-3} donor or acceptor impurities). Therefore, when $E_g \geq 1$ eV, it is possible to use only samples with extrinsic conductivity for electrochemical measurements. In this section we will examine some results obtained with semiconductors doped with donor impurities and in particular zinc oxide ($E_g = 3.2$ eV) with indium added, cadmium sulfide ($E_g = 2.4$ eV) with cadmium added, and gallium arsenide ($E_g = 1.35$ eV) with sulfur or selenium added. The ionization energy of the impurities selected is low, so that their atoms are completely ionized at room temperature. The concentration n^0 of the majority carriers (electrons) practically equals the donor concentration N_D, while the concentration of minority carriers (holes) $p^0 = n_i^2/n^0$ is vanishingly small. For example, in cadmium sulfide, $n_i = 10^3$ cm^{-3}, and when n^0 $\approx N_D = 10^{16}$ cm^{-3}, $p^0 = 10^{-10}$ cm^{-3}. However, in the space charge region the concentration of holes may increase in principle to an appreciable value with a sufficiently high potential drop φ_1 [equation (9.3)]. In order to obtain the complete charge-potential curve (or capacitance-potential curve, etc.), it is necessary to go to the region of potentials approximately equal to the width of the forbidden band. However, as will be shown in § 27, even with slight anode polarization of the electrode, as a result of the passage of a current the equilibrium is disrupted and there is a fall in the concentration of holes in the space charge region and, consequently, their contribution to the space charge and to the capacitance. Therefore, it is not possible to observe enrichment of the surface in holes with semiconductors with a wide forbidden band. *

a. Zinc Oxide. Dewald showed that the relation of the capacitance to the potential on a zinc oxide electrode with anode polarization is described by equation (12.10) [65, 66]. In Fig. 28 this relation is shown as a plot of $1/C^2$ against φ for two samples with different concentrations of free electrons. The measurements were carried out in 1 N KCl solution at pH 8.5 (borate buffer), in which zinc oxide is stable. The theoretical curves [equation (12.11)] for the same samples (the electron concentration was determined by measurement of the Hall effect) are shown in the figure by broken lines. The theoretical and experimental lines practically coincide over a wide range of potentials. This indicates that the space charge is formed by ionized donor atoms (depletion layer). Furthermore, the change in the electrode potential practically equals the change in the potential difference in the space charge layer, i.e., the Helmholtz potential difference does not change on polarization of the electrode. Finally, there are hardly any fast surface states on the electrode. The latter conclusion is confirmed by the absence of dispersion of the capacitance over the region of frequencies from 50 Hz to 100 kHz.† The discrepancy between the experimental slope of the line of $1/C^2$ against φ and that calculated from equation (12.11) does not exceed 2% and this value may be taken as a measure of the accuracy of the conclusions drawn above. In particular, the density of the surface levels does not exceed 10^9 cm^{-2} if they exist at all at the zinc oxide—electrolyte interphase.

The experimental slope of the $(1/C^2) - \varphi$ line was used by Dewald to measure the bulk concentration of electrons in zinc

---

*The same occurs in the case of heavily doped silicon and germanium (see § 35).

†A frequency dependence of the capacitance was observed in samples in which the atoms of the donor impurity (hydrogen) have appreciable mobility at room temperature.

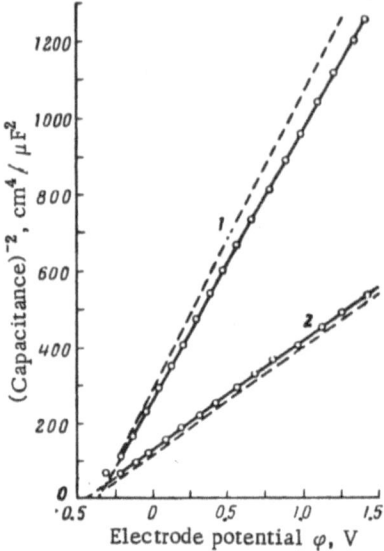

Fig. 28. Relation of the reciprocal square of capacitance C to electrode potential φ of zinc oxide [65]. Specific electrical conductivity: 1) 0.59 $\Omega^{-1} \cdot$ cm^{-1}; 2) 1.79 $\Omega^{-1} \cdot$ cm^{-1}. Broken lines represent theoretical curves calculated from equation (12.11).

oxide. Extrapolation of this line until it intersects the potential axis gives a potential differing from the flat band potential by kT/e (see § 12). The flat band potential depends on the treatment of the electrode surface as it changes the interphase potential difference, and also on the position of the Fermi level in the semiconductor, which determines the galvanic potential at the ohmic contact (see § 8; we should remember that the flat band potential is measured in the system metal – semiconductor – solution – comparison electrode – metal). Both of these effects may be observed in Fig. 29. Here we show the relation of the flat band potential to the concentration of free electrons in samples with different surface treatments.

Let us first look at the concentration region below 0.6 $\cdot 10^{18}$ cm^{-3}. The flat band potential changes by 59 mV with a change in the electron concentration by a factor of 10, the same as the Fermi level. All this change is evidently caused by a change in the galvanic potential at the boundary of the zinc oxide with indium, which was used as an ohmic contact. With a different treatment of the crystal surface (etching in 85% phosphoric acid or in 3 N potassium hydroxide solution) the flat band potential of the sample does not remain constant. Since, at the flat band potential φ_1 always equals 0, this means that the potential difference at the interphase φ_0 changes appreciably. During the etching process there is apparently adsorption of the components of the etching solution on the ZnO surface, and this changes the nature of the surface dipole and the interphase potential difference associated with it.

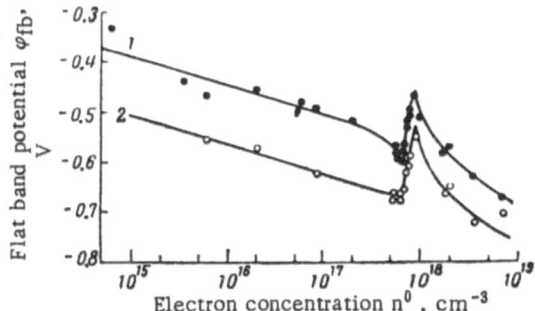

Fig. 29. Relation of flat band potential φ_{fb} of zinc oxide in 1 N KCl solution (pH 8.5) to concentration of free electrons n^0 in the sample [65]. 1) After etching in 85% H_3PO_4 solution; 2) after etching in 3 N KOH solution.

Thus, although only the potential difference in the space charge region changes on polarization of the electrode, the interphase potential difference has an appreciable value, which changes by tenths of a volt. This is apparently connected with the formation of a surface dipole as a result of the adsorption of various ions on the electrode.

The introduction of octyl alcohol and nitrobenzene into the solution does not affect either the form of the $C - \varphi$ curve or the flat band potential.

A break is found on the $\varphi_{fb} - \log n^0$ curve in the region of concentrations n^0 from $0.6 \cdot 10^{18}$ to $0.9 \cdot 10^{18}$ cm^{-3}. Its nature is not quite clear. When $n^0 \approx 10^{18}$ cm^{-3}, anomalous changes are also observed in some bulk properties of ZnO and, in particular, the temperature coefficient of the electrical conductivity. Here there is apparently a transition from semiconductor conductivity to metallic conductivity in ZnO [67].

At electrode potentials more negative than the flat band potential, the surface is enriched in electrons. The capacitance increases exponentially with the potential [equation (12.6)].

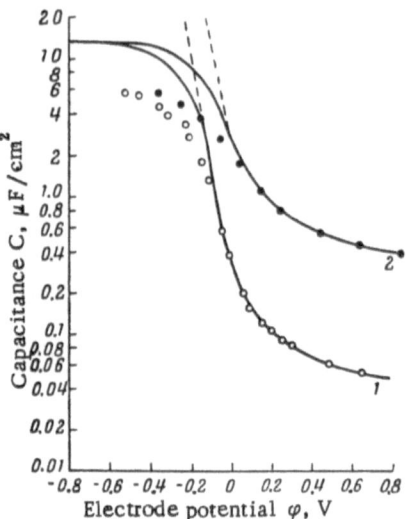

Fig. 30. Relation of capacitance of a zinc oxide electrode to electrode potential φ relative to flat band potential [65]. Specific electrical conductivity: 1) 0.59
$\Omega^{-1} \cdot cm^{-1}$; 2) 9.5 $\Omega^{-1} \cdot cm^{-1}$. O, ●) Experimental
values; solid lines) calculated using Fermi—Dirac sta-
tistics; broken line) calculated from the Poisson—
Boltzmann equation.

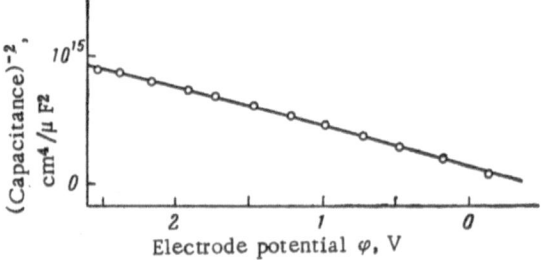

Fig. 31. Relation of the reciprocal square of the capacitance to the electrode potential φ of cadmium sulfide
in 1 N K_2SO_4 solution [68].

With a further shift in the electrode potential in a negative direction the growth of the capacitance slows and then the capacitance reaches a limiting value (about 6 μF/cm^2) and is almost independent of the bulk concentration of electrons (Fig. 30). When $\varphi > \varphi_{fb}$, the experimental data are described well by equation (12.5), and when $\varphi < \varphi_{fb}$, the measured capacitance is less than the calculated value. The reason for this deviation is apparently the fact that at very negative potentials the Fermi level intersects the bottom of the conductivity band. * Thereupon there is degeneration of the carriers at the surface. The surface electron concentration is less than the value calculated by Boltzmann's equation and, consequently, the capacitance is less than that calculated by equation (12.5), which was derived using Boltzmann statistics. Dewald [65] used Fermi−Dirac statistics to calculate the degeneracy of the carriers at the surface (see § 10). The results of calculating the capacitance are shown in Fig. 30 by solid lines, which give a qualitative representation of the C−φ relation in this potential region.

If the sample contains deep donors (i.e., donors with an ionization energy $E_C - E_D \gg kT$) which are incompletely ionized in the bulk of the semiconductor at room temperature, then their degree of ionization in the space charge region increases on anode polarization of the electrode and may reach unity. The free electron concentration in the bulk equals the concentration of ionized donors and the slope of the $(1/C^2) - \varphi$ line is determined by the total concentration of donors and, therefore, is less than the value determined from the bulk electrical conductivity using equation (12.11). Measurement of the capacitance may be used as a method of investigating deep donors in a semiconductor [65] (see also § 57).

b. Cadmium Sulfide. There is much in common in the behavior of electrodes of cadmium sulfide and zinc

---

*In the crystals used in [65], the difference between the Fermi energy and the energy of an electron at the lower boundary of the conductivity band did not exceed 0.25 eV (cf. Fig. 30).

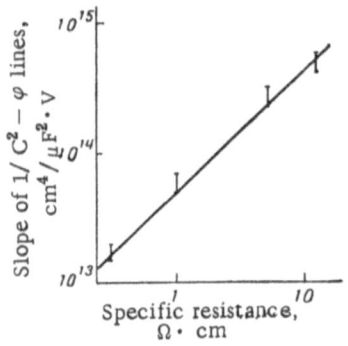

Fig. 32. Relation of the slope of $1/C_2 - \varphi$ lines to the specific resistance of cadmium sulfide (in logarithmic coordinates) [68].

oxide. Figure 31 shows the relation of the square of the reciprocal differential capacitance to the potential for cadmium sulfide in K_2SO_4 solution [68]. The slope of $(1/C^2) - \varphi$ lines for samples with different specific resistances is proportional to the free electron concentration (Fig. 32). The frequency dependence of the capacitance is small (about 20% of the measured value with a change in frequency from 30 Hz to 160 kHz). All this indicates that the measured capacitance is determined by the charge of the depletion layer in the semiconductor, while the density of the surface levels is low. On polarization of the electrode the change in potential is localized in the space charge region, while the potential drop in the Helmholtz region hardly changes.

The flat band potential varies somewhat under the influence of the solution composition and the preliminary etching of the electrode. These changes evidently reflect a change in the surface dipole on the electrode.

On cathode polarization there occurs reduction of the cadmium sulfide. The surface of the electrode is covered with a layer of metallic cadmium, dispersion of the capacitance appears, and hysteresis is observed on $C - \varphi$ curves. Since the electrode decomposes in this region of potentials, the method of measuring the capacitance under steady conditions is inapplicable and the method used consisted of plotting fast galvanostatic charging curves by means of square current pulses [69]. The apparatus for these measurements will be described in §23.

Figure 33 illustrates a cathode charging curve plotted by an oscillographic method. By differentiating it it is possible to obtain a "differential capacitance—potential" relation, which

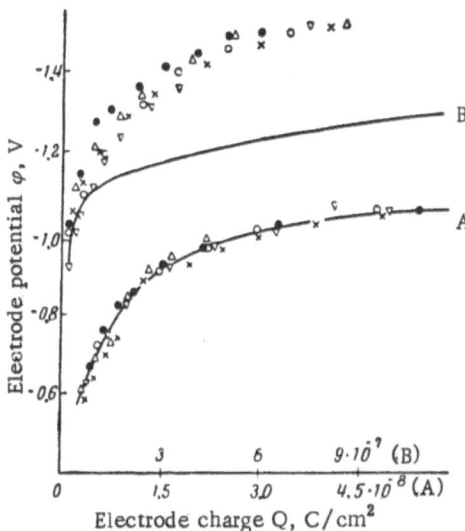

Fig. 33. Cathode charging curve of a cadmium sulfide electrode [69]. Duration of charging pulse (μsec):
●) 5; ○) 10; △) 20; ▽) 50; ×) 100. Solid line, calculated taking into account degeneracy of carriers at
surface.

agrees well with the $C - \varphi$ curve obtained by direct measurement of the impedance with a slow change in the electrode potential. At potentials more negative than −1.1 V, the "charge−
electrode potential" relation measured experimentally deviates
from the $Q_1 - \varphi_1$ relation calculated theoretically [from equation (10.12) and at more negative electrode potentials, from
equations (10.43) and (10.45), taking into account the degeneracy of the free electrons at the CdS surface with strong cathode polarizations]. This discrepancy is evidently connected
with the fact that at an electrode potential more negative than
−1.1 V, the change in the interphase potential difference φ_0 on
polarization is commensurate with the change in the potential
difference φ_1 in the space charge region in CdS. Hence it is
possible to calculate the capacitance of the Helmholtz layer and
this was found to equal 6-8 $\mu F/cm^2$.

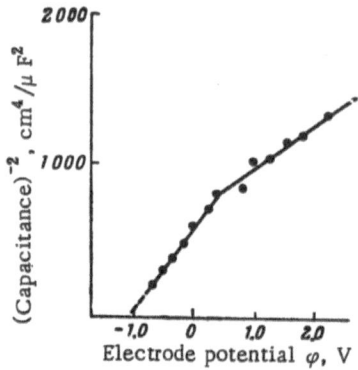

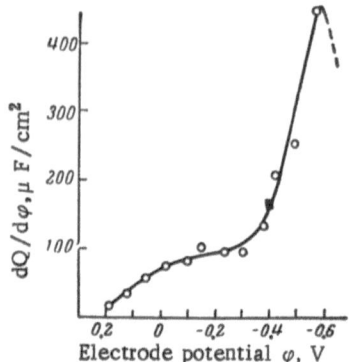

Fig. 34. Relation of reciprocal square of capacitance to electrode potential φ for gallium arsenide (n-type, 0.46 $\Omega \cdot$ cm) in 1 N KOH solution at 20 kHz [70].

Fig. 35. Cathode reduction of oxides on a gallium—arsenide surface (p-type, 0.003 $\Omega \cdot$ cm) in 0.3 N NaOH solution according to data from i — t curves [70].

The form of the $Q-\varphi$ curve is independent of the electrode charging rate: curves plotted with times from 5 to 100 μsec coincide (Fig. 33). Thus, after etching in HCl, the cadmium sulfide surface has no surface states whose relaxation time lies in the region from 5 μsec to tens of minutes. This is an advantage of a cadmium sulfide electrode as compared with germanium and silicon electrodes. As capacitance measurements show, adsorption of sulfide ions occurs on the CdS surface in Na_2S solution [8].

c. Gallium Arsenide. The capacitance of gallium arsenide electrodes was investigated in [70]. A linear relation was found between $1/C^2$ and φ (Fig. 34) in 1 N KOH solution at potentials from −0.6 to 0.4 V. A depletion layer apparently arises in the semiconductor in this case. A break is observed on the $(1/C^2) - \varphi$ line at a potential of 0.4 V. The decrease in the slope on anode polarization is apparently caused by ionization in the space charge region of deep donors which are not ionized in the bulk of the semiconductor and do not contribute to its electrical conductivity (cf. Fig. 96).

With a fall in the pH of the solution the $C-\varphi$ curves are shifted toward negative potentials but do not change their form.

This may be explained by a change in the interphase potential difference which depends on the pH and is probably due to surface oxides. By means of cathode charging curves it was shown [70] that under steady conditions the surface of gallium arsenide is covered by a monolayer of adsorbed oxygen. On anode polarization in KOH solution the amount of this does not increase appreciably, but in H_2SO_4 solution an oxide phase is formed on the surface. When the electrode is kept in solution without a current, the thick oxide layer gradually dissolves. On the other hand, if the surface oxides are reduced by cathode polarization, they are formed again (probably as a result of decomposition of water) in a few minutes. Thus, the reduced surface of gallium arsenide is unstable in aqueous solutions.

The reduction of surface oxides was investigated by determining cathode i—t curves. Gallium arsenide with a high impurity content ($3 \cdot 10^{18}$ cm^{-3}) was used. With such samples the differential capacitance exceeds $1\ \mu F/cm^2$ and the change in potential in the Helmholtz region on polarization apparently constitutes a considerable part of the change in electrode potential. The electrode potential was changed in 50-mV steps by means of a potentiostatic circuit and the amount of electricity passing through the electrode was measured at each step through the i—t curve. The charge Q transmitted to the electrode is used in reduction of the oxides and also in charging the double layer.

The results of these measurements are presented in Fig. 35 in the form of the relation $dQ/d\varphi$ to the electrode potential φ. The value of $dQ/d\varphi$ considerably exceeds the capacitance of the Helmholtz region ($\approx 10\ \mu F/cm^2$) and is apparently determined by reduction of the surface oxides. The bulk of them is reduced at a potential of —0.55 V. The pH dependence of the potential difference in the Helmholtz region may be caused by partial dissociation of the surface oxide as occurs on germanium.

To conclude this section, we should point out the advantage of the "electrochemical" method of changing the surface potential of a semiconductor over the "field effect" method used in semiconductor physics. In the latter case, φ_1 is changed by application of a voltage to a condenser, one of whose plates is the sample investigated. The plates of the condenser are separated by a dielectric spacer with a thickness of the order of 10 μ. The capacitance of such a system is extremely low and the maximum voltage is limited by the breakdown voltage of the condenser (a few hundred volts). Therefore, with this method it is impossible to apply to the surface of the condenser a charge of more than 10^{-8} C/cm^2 and to change the surface potential by more than 0.3-0.35 V.

In the case of the semiconductor—electrolyte system the capacitance of the "dielectric spacer," i.e., the Helmholtz region, is about 10 μF/cm^2 and the capacitance of the system is determined by the space charge region in the semiconductor. Even with slight polarization it is possible to obtain a high applied charge (10^{-8}-10^{-7} C/cm^2) and to change markedly the potential of the semiconductor surface, while with strong polarization degeneration of free carriers at the surface is achieved.

§ 17.   Photopotential

a.   Surface Photo Emf.   Let us consider a semiconductor (as a concrete example, n-type) in contact with a transparent medium. Let the semiconductor surface be illuminated through this medium with the wavelength of the light chosen so that it is adsorbed in the semiconductor in distances comparable with the thickness of the space charge region. As a result of the energy of the absorbed photons, electrons from the valence band pass into the conductivity band. Electron—hole pairs arise as a result of this. These nonequilibrium carriers move into the depth of the semiconductor. They recombine at a distance from the surface of the order of the diffusion length L. * For the most common semiconductor materials, L

---

*The diffusion length is the distance at which the concentration of nonequilibrium current carriers has fallen by a factor "e" when generated locally — see equation (17.13).

is considerably greater than the thickness of the space charge region L_1. The region $L_1 < x < L$ is practically unchanged and, therefore, is called the quasineutral region. The movement of minority carriers in this region is determined by the diffusion process, while the majority carriers move under the influence of the field (more details on the properties of the quasineutral region are given in § 28).

Let the semiconductor—electrolyte boundary be illuminated in such a way that the light intensity increases abruptly from zero to some constant value. Let us examine the behavior of the system such a short time after the light has been switched on that ionic processes are unable to occur and the charge of the surface states has not changed. Then the space charge likewise will not have changed on illumination. *

In accordance with relation (10.12), the space charge is a function of the potential drop in the Debye region $Y = e\varphi_1/kT$ and the concentration of holes at the boundary of the Debye and quasineutral regions. This concentration equals $p_1 = p^0 + \Delta p_1$, where p^0 is the equilibrium (dark) value, and Δp_1 is the increase in the hole concentration on illumination. Thus, on illumination the total charge in the Debye region remains the same as in the dark, while the concentration of carriers at the boundary of the Debye region with the quasineutral region changes by a value Δp_1. Consequently, on illumination there is also a change in the potential drop in the Debye region φ_1, so as to compensate for the change in the concentration of free carriers and maintain the condition $Q_1 = $ const. This change in φ_1 is called the photopotential. †

---

*The initial (dark) value of the space charge is set by slow surface states of the external field.

†As the photopotential and surface recombination (see § 19) are observed under nonequilibrium conditions, they can only be examined provisionally in a chapter devoted to the equilibrium state. However, we should take into account the fact that in some semiconductors (for example, in germanium and silicon), the lifetime of minority carriers is extremely long. Therefore, although there is no thermodynamic equilibrium between electrons and holes, within the limits of each

Following [12, 71], let us examine the relation of the photopotential to Δp_1 and then find the relation between Δp_1 and the hole current produced by illumination. Let us first assume that the semiconductor has no surface levels. The space charge in the semiconductor Q_1 (per unit surface) is related to the electric field at the surface by the relation (12.3)

$$Q_1 = -\frac{\varepsilon_1 E_s}{4\pi}.$$
(17.1)

To calculate the relation of the electric field $E_s(Y)$ to the potential drop Y in the space charge region and the given value of Δp_1, we use equation (10.5). On illumination the concentrations of carriers at the boundary of the space charge region do not equal n^0 and p^0, but $n^0 + \Delta p_1$ and $p^0 + \Delta p_1$, respectively, so that $\Delta p_1 = \Delta n_1$.* For the electric field on illumination we obtain the expression

$$E_s = \sqrt{\frac{8\pi k T n_i}{\varepsilon_1}} \times$$
$$\times \sqrt{\frac{-N_D + N_A}{n_i} Y + \frac{p^0 + \Delta p_1}{n_i}(e^{-Y} - 1) + \frac{n^0 + \Delta n_1}{n_i}(e^Y - 1)} \ .$$
(17.2)

Expression (17.2) may be rearranged in the following way:

___

band a quasiequilibrium state is established. Normal statistics are applicable to electrons and holes in a quasiequilibrium state (see § 3), but instead of the equilibrium Fermi level we introduce the concept of separate quasi-Fermi levels for electrons and holes. Under steady illumination of the semiconductor surface, the quasilevels of electrons and holes do not coincide with the Fermi level, but they may have a constant value in the space charge region. For this two conditions must be fulfilled: 1) the diffusion length of minority carriers L_p is large in comparison with the thickness of the space charge region L_1; 2) the currents of electrons and holes arising with a deviation from equilibrium should not be too great; the same applies to deviations in the concentrations from equilibrium values [12].

*This relation is discussed in § 28.

$$E_s = \sqrt{\frac{8\pi k T n_i}{\varepsilon_1}} \times$$

$$\times \sqrt{\lambda(e^{-Y} - 1) + \lambda^{-1}(e^Y - 1) + (\lambda - \lambda^{-1})Y + \lambda(e^Y + e^{-Y} - 2)\frac{\Delta p_1}{p^0}} .$$

$$(17.3)$$

In the dark the electric field is given by

$$E_s^0 = \sqrt{\frac{8\pi k T n_i}{\varepsilon_1}} \sqrt{\lambda(e^{-Y^0} - 1) + \lambda^{-1}(e^{Y^0} - 1) + (\lambda - \lambda^{-1})Y^0} , \quad (17.4)$$

where Y^0 is the potential in the dark.

As was pointed out previously, the total charge of the semiconductor does not change on illumination and, therefore, in accordance with equation (17.1),

$$E_s^0(Y^0) = E_s(Y, \Delta p_1).$$

By equating expressions (17.3) and (17.4) we find a relation between the potential on illumination Y, the potential in the dark Y^0, and the excess of carriers Δp_1 produced by the light:

$$\frac{\Delta p_1}{p^0} =$$

$$\frac{\lambda(e^{-Y^0} - 1) + \lambda^{-1}(e^{Y^0} - 1) + (\lambda - \lambda^{-1})Y^0 - \lambda(e^{-Y} - 1) - \lambda^{-1}(e^Y - 1) - (\lambda - \lambda^{-1})Y}{(e^Y + e^{-Y} - 2)\lambda}$$

$$(17.5)$$

Equation (17.5) makes it possible to find the potential on illumination Y if we know the potential in the dark Y^0 and the injection level $\Delta p_1/p^0$.

To illustrate the results obtained, we give a graph showing the relation of the photopotential $\Delta Y = Y - Y^0$ for an intrinsic semiconductor to the injection level $\Delta p_1/p^0$ (Fig. 36 [71, 72]). The picture shows that the photopotential increases with an increase in Y with a given value of $\Delta p_1/p^0$.

Let us now find the derivative of the photopotential with respect to Δp_1. For this it is convenient to write equation (17.5) in the form

$$\frac{\lambda \Delta p_1}{p^0}(e^Y + e^{-Y} - 2) = \lambda\,(e^{-Y^0} - 1) + \lambda^{-1}(e^{Y^0} - 1) + (\lambda - \lambda^{-1})Y^0 -$$
$$- \lambda(e^{-Y} - 1) - \lambda^{-1}(e^Y - 1) - (\lambda - \lambda^{-1})\,Y.$$

$$(17.6)$$

By differentiating the left- and right-hand parts of equation (17.6) with respect to Δp_1 at a given value of Y^0, we obtain

$$\frac{\lambda}{p^0}(e^Y + e^{-Y} - 2) + \frac{\lambda \Delta p_1}{p^0}(e^Y - e^{-Y})\frac{dY}{d\Delta p_1} =$$
$$= (\lambda e^{-Y} - \lambda^{-1}e^Y - \lambda + \lambda^{-1})\frac{dY}{d\Delta p_1}.$$

By solving the equation obtained for the derivative $dY/d\Delta p_1$, we find

$$\frac{dY}{d\Delta p_1} = \frac{\lambda}{p^0}\frac{e^Y + e^{-Y} - 2}{-\frac{\Delta p_1}{p^0}\lambda\,(e^Y - e^{-Y}) + \lambda e^{-Y} - \lambda^{-1}e^Y - \lambda + \lambda^{-1}}.$$

$$(17.7)$$

As a concrete example, let us examine in more detail a semiconductor of n-type. Let us express Δp_1 in terms of the current of nonequilibrium carriers from the illuminated surface of the semiconductor into its volume. For this purpose we consider the behavior of holes in the quasineutral region. The condition for conservation of holes is expressed by the following expression:

$$\frac{\partial p\,(x)}{\partial t} = -\frac{p\,(x) - p^0}{\tau_p} - \frac{1}{e}\frac{d}{dx}i_p(x).$$

$$(17.8)$$

Here $i_p(x)$ and $p(x)$ are the hole currents and hole concentration at the point x, respectively, and τ_p is the lifetime of the holes.

Equation (17.8) takes into account the fact that the change in concentration of holes per unit volume $\partial p/\partial t$ occurs as a result of two processes. The first term $(p - p^0)/\tau_p$, in accordance with formula (6.12), describes the recombination-generation of holes process. The second term (the change in the current with the coordinate) denotes that the flow of holes

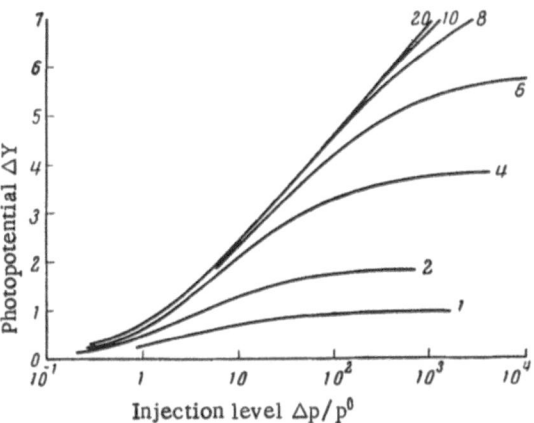

Fig. 36. Theoretical relation of photopotential ΔY to injection level $\Delta p/p^0$ for semiconductor with intrinsic conductivity (in kT/e units) [71]. The parameter is the value of Y^0 (indicated on the curves).

into unit volume of the semiconductor does not equal the flow out of it. The difference in these flows determines the change in the concentration of holes with time $\partial p/\partial t$. Thus, relation (17.8) establishes the balance of holes.

Since the quasineutral region is not charged and the electric field in it is low, the minority carriers move by a diffusion mechanism and for a current i_p we may write

$$i_p = - eD_p \frac{dp}{dx}.$$

$$(17.9)$$

By substituting the expression for the current i_p in relation (17.8), we obtain the equation

$$\frac{\partial p}{\partial t} = -\frac{p - p^0}{\tau_p} + D_p \frac{\partial^2 p}{\partial x^2}.$$

$$(17.10)$$

Taking into account the fact that $p = p^0 + \Delta p$, we find that

$$\frac{1}{D_p} \frac{\partial \Delta p}{\partial t} = -\frac{\Delta p}{D_p \tau_p} + \frac{\partial^2 \Delta p}{\partial x^2}.$$

Under steady conditions, the latter relation assumes the following form:

$$-\frac{\Delta p}{D_p \tau_p} + \frac{d^2 \Delta p}{dx^2} = 0. \tag{17.11}$$

The general solution of equation (17.11) gives the equation

$$\Delta p = Ae^{-ax} + Be^{ax}. \tag{17.12}$$

The integration constant B should be taken as zero since, when $x \to \infty$, Δp must tend to zero.

Thus, we obtain

$$\Delta p = Ae^{-ax}. \tag{17.12a}$$

The constant a may be determined by substituting equation (17.12a) in relation (17.11)

$$a = (D_p \tau_p)^{-1/2} = L_p^{-1}.$$

As pointed out previously, the value L_p is called the diffusion length of the minority carriers.

From the solution of equation (17.11) it is obvious that L_p characterizes the thickness of the region within which the concentration of holes p on illumination of the semiconductor surface differs from the equilibrium value p^0. In germanium with $D_p = 45$ cm^2/sec and $\tau_p = 10^{-4}$ sec, $L_p \approx 0.07$ cm. Thus we obtain

$$p = p^0 + Ae^{-\frac{x}{L_p}}. \tag{17.13}$$

At the boundary of the space charge region and the quasineutral region, i.e., when $x = L_1$,

$$\Delta p = Ae^{-\frac{L_1}{L_p}}.$$

Since $L_1/L_p \ll 1$, then $A = \Delta p(x = L_1) \equiv \Delta p_1$.

By using equation (17.9) we find the hole current at the boundary of the quasineutral and Debye region

$$i_{p_1} = \frac{eD_p \Delta p_1}{L_p} .$$

(17.14)

It is convenient to rearrange the expression found into the form

$$i_{p_1} = \frac{eD_p p^0}{L_p} \frac{\Delta p_1}{p^0} .$$

(17.15)

The value $-eD_p p^0/L_p$ is called the limiting hole current. The meaning of this name will be explained later (see § 27).*

By denoting $-eD_p p^0/L_p$ by $i_p^{\lim}$, we obtain

$$\frac{\Delta p_1}{p^0} = - \frac{i_{p_1}}{i_p^{\lim}} .$$

(17.16)

The latter expression makes it possible to relate the photocurrent i_{p_1} to Δp_1. By substituting $\Delta p_1/p^0$ from equation (17.16) in equation (17.7), we obtain the following expression for the derivative of the photopotential with respect to current:

$$\frac{dY}{di_{p_1}} = - \frac{1}{i_p^{\lim}} \frac{e^Y + e^{-Y} - 2}{\left\{ -\lambda^{-2} \left[\left(1 - \frac{\lambda^2 i_{p_1}}{i_p^{\lim}} \right) e^Y - 1 \right] + \left(1 - \frac{i_{p_1}}{i_p^{\lim}} \right) e^{-Y} - 1 \right\}} .$$

(17.17)

From equation (17.17) it follows directly that

$$\frac{dY}{di_{p_1}} = 0 \quad \text{when} \quad Y = 0.$$

(17.18)

Let us examine some important particular cases of (17.17). If the illumination level is low, so that $i_{p_1}/i_p^{\lim} \ll 1$,

---

* When allowance is made for surface recombination, the expression for the limiting hole current assumes the form

$$- ep^0 (D_p/L_p + s),$$

where s is the rate of surface recombination (see § 19).

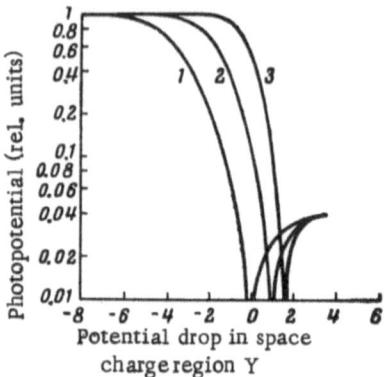

Fig. 37. Theoretical relation of photopotential to the potential drop in space charge layer Y (in kT/e units) for an n-type semiconductor (λ = 0.2) [12]. Concentration of fast surface states: 1) $N_{tD} = N_{tA} = 0$; 2) $N_{tD} = N_{tA} = 8 \cdot 10^8$ cm^2; 3) N_{tD}, $N_{tA} \rightarrow \infty$.

then after rearrangement we obtain

$$\frac{dY}{di_{p1}} = -\frac{1}{i_p^{\lim}} \frac{1 - e^Y}{1 + \lambda^{-2}e^Y} \ . \qquad (17.19)$$

With strong anode and cathode polarization

$$Y \rightarrow -\infty \quad \frac{dY}{di_{p1}} \rightarrow -\frac{1}{i_p^{\lim}},$$

$$Y \rightarrow +\infty \quad \frac{dY}{di_{p1}} \rightarrow \frac{\lambda^2}{i_p^{\lim}}. \qquad (17.20)$$

Equation (17.17) was obtained by Garrett and Brattain [12]. They also calculated the relation of the photopotential to the current in the presence of surface levels, the charge on which varies with the illumination. Figure 37 gives graphs of the relation of the photopotential to the potential drop in the space charge region with various concentrations of donor N_{tD} and acceptor N_{tA} surface levels (the modulus of the photopotential is plotted along the ordinate axis on a logarithmic scale).

Figure 37 shows that curves 1-3 have approximately the same form, but the point at which the photopotential changes sign shifts along the potential scale; when $N_{tD} = N_{tA} = 0$, it coincides with the flat band potential ($Y = 0$).

b. Dember Photo Emf. In addition to the photopotential examined above, which results from a redistribution of the space charge, on illumination of the surface of a semiconductor there arises another bulk photo emf, which is called Dember's electromotive force [73]. It originates from the nonuniform distribution of free carriers in the quasineutral region. Nonequilibrium electrons and holes formed by illumination diffuse from the surface into the depth of the semiconductor. In the general case, the diffusion coefficients of electrons and holes are different (normally $D_n > D_p$). Therefore, the more mobile electrons outstrip the holes and the region more remote from the contact is negative relative to the surface. The resulting field hampers the movement of the electrons and accelerates the movement of the holes, thus compensating for the difference in the diffusion coefficient.

For the mathematical description of the Dember effect we use expressions (7.9) and (7.14), which relate the electrical current to the concentration of free carriers. The total current of electrons and holes is given by

$$i = i_p + i_n = e\left(nU_n + pU_p\right)E + kT\left(U_n \frac{dn}{dx} - U_p \frac{dp}{dx}\right). \quad (17.21)$$

Under steady conditions with illumination the total current equals zero. Then by determining the electric field from equation (17.21) we find that

$$E = \frac{kT}{e} \frac{U_n \dfrac{dn}{dx} - U_p \dfrac{dp}{dx}}{nU_n + pU_p}. \quad (17.22)$$

Since in the quasineutral region $\Delta n = \Delta p$, then $dn/dx = dp/dx$ and relation (17.22) may be written in the form

$$E = \frac{kT}{e} \frac{U_n - U_p}{nU_n + pU_p} \frac{dn}{dx}. \quad (17.23)$$

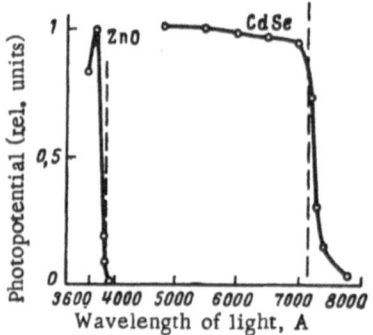

Fig. 38. Spectral distribution of photopotential of ZnO and CdSe electrodes in 0.1 N KCl solution [74]. The broken lines show the main absorption edge.

By introducing the bulk conductivity of the sample through the relation $\sigma_B = e(U_n n + U_p p)$, and differentiating σ_B with respect to the coordinate x, we obtain

$$\frac{d\sigma_B}{dx} = e(U_n + U_p)\frac{dn}{dx}. \quad (17.24)$$

Finally, by eliminating dn/dx from relation (17.23) by means of (17.24), we find for the field

$$E = \frac{kT}{e}\frac{U_n - U_p}{(U_n + U_p)\,\sigma_B}\frac{d\sigma_B}{dx}.$$

The potential difference arising is given by

$$V = \varphi(L_1) - \varphi_B = -\int_{x=\infty}^{x=L_1} E\,dx = -\frac{kT}{e}\frac{U_n - U_p}{U_n + U_p}\ln\frac{\sigma_{ill}}{\sigma_B^0},$$

where $\sigma_B^0 = e(U_n n^0 + U_p p^0)$ is the conductivity of the sample in the dark and $\sigma_{ill} = \sigma_B^0 + \Delta\sigma$ is the conductivity at the point $x = L_1$ under illumination.

Hence, it follows that

$$V = \frac{kT}{e}\frac{U_p - U_n}{U_n + U_p}\ln\left(1 + \frac{\Delta\sigma}{\sigma_B^0}\right).$$

If $\Delta\sigma$ is considerably less than σ_B^0, then

$$V = \frac{kT}{e}\frac{U_p - U_n}{U_n + U_p}\frac{\Delta\sigma}{\sigma_B^0}. \quad (17.25)$$

The expression found for the photo emf is closely connected with the difference in mobilities of the free carriers U_n and U_p. When the mobilities are the same, the Dember photo emf is zero.

§ 18.   Photopotential of Germanium and Silicon Electrodes

Since the photoeffect is produced by the formation of nonequilibrium free carriers on absorption of light by the semiconductor, the relation of the photoeffect to the wavelength of the light normally repeats in general characteristics the spectral distribution of the light absorption coefficient. In particular, the photoeffect reaches its maximum value close to the main absorption edge of the lattice (Fig. 38 [74]).

Below we will examine some results of experimental measurements of the photopotential on the basis of the theory presented in § 17. One of the conditions of applicability of the theory is that the diffusion length of the minority carriers should be much greater than the size of the space charge region and this holds in the case of germanium and silicon. Semiconductors with a wide forbidden band normally do not satisfy this condition and are not considered in the present section (see [75] and § 40 of this book).

Another requirement of the theory is the constancy of the charge on the surface under illumination. Consequently, no electrochemical reactions should occur at the interface. This may be achieved by using short flashes of light, so that the illumination time t is less than the time constant of the electrode $t \ll \tau = RC$ (where R and C are the resistance and the capacitance measured through a parallel circuit). Under these conditions processes involving ions cannot occur and the leakage of charge into solution may be neglected, i.e., we have the so-called fast (instantaneous) photopotential. Below we examine the results of measurements carried out with pulsed illumination (the duration of the flash was of the order of 10^{-5}-10^{-4} sec).*

---

*The results of the calculation given in § 17 may be used in pulsed measurements of the photopotential if the duration of the flash is long in comparison with the relaxation time of the space charge (see p. 40).

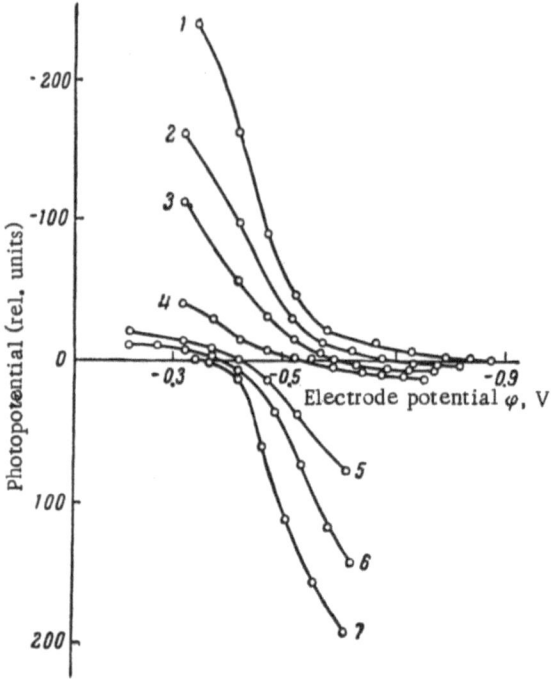

Fig. 39. Photopotential of a germanium electrode in
1 N NaOH solution [40] with pulsed illumination. 1)
n-Type, 0.004 Ω · cm; 2) n-type, 3 Ω · cm; 3) n-
type, 20 Ω · cm; 4) n-type, 40 Ω · cm; 5) p-type, 10
Ω · cm; 6) p-type, 3 Ω · cm; 7) p-type, 0.5 Ω · cm.

The relation of the photopotential to the potential of a germanium electrode in an aqueous solution of KOH is shown in
Fig. 39 [40]. The parameter is the concentration of free electrons in the sample. The form of the curves is similar qualitatively to the theoretical curves (see Fig. 37, curve 1). For
electronic samples with anode polarization the photopotential,
which is shown in Fig. 39 as the change in electrode potential
on illumination, * is large and negative and with cathode pola-

---

*What is referred to as the photopotential of a semiconductor electrode is often
the change in electrode potential on illumination, which is measured experimentally. This value equals the change in the potential difference in the space
charge region on illumination, but has the opposite sign (see footnote on p. 32).

rization it is small and positive. The relation is the reverse for a p-type semiconductor. The potential at which the photo-effect changes sign (the flat band potential) moves into the region of negative values with an increase in the concentration n^0 of free electrons (by approximately 80 mV with a change in n^0 by a factor of 10). The fact that $d\varphi_{fb}/d\log n^0$ was found to equal 80 mV and not 59 mV (cf. § 16) again indicates an appreciable change in the Helmholtz potential difference during slow polarization of the electrode. Hence, it follows that $d\varphi_1/d\varphi \approx 0.7$, and this agrees well with the results of measurements carried out in [32] by the differential capacitance method.

Analogous results were obtained in an investigation of the relation of the photopotential to the type and magnitude of the conductivity of germanium at a steady potential [76].

Boddy and Brattain [77] used measurements of the photopotential to determine the potential drop in the space charge region in a semiconductor φ_1. The electrode and solution were prepared as described on pp. 66 and 67. Equation (17.19) with a correction for the Dember potential in accordance with equation (17.25), was used for calculation of φ_1. It was found that values of φ_1 calculated from the photopotential coincided over a wide range of potentials with the values measured by the differential capacitance method (Fig. 40). Equation (17.19) is valid for the case where fast surface states are absent, i.e., the photoeffect is determined wholly by the space charge in the semiconductor. Therefore, the photopotential measurements confirm the previous conclusion that on a germanium electrode after anode etching in aqueous solutions the concentration of fast surface states close to the center of the forbidden band is low. *

The photopotential method is a very simple experimental procedure for determining φ_1. It is only necessary to en-

---

*If the surface recombination rate s is not small and depends on the potential, it is necessary to know the relation of s to φ_1 for calculating the relation of the photopotential to φ_1 (see footnote on p. 107). The effect of the surface recombination rate on the photopotential was investigated in [54, 78].

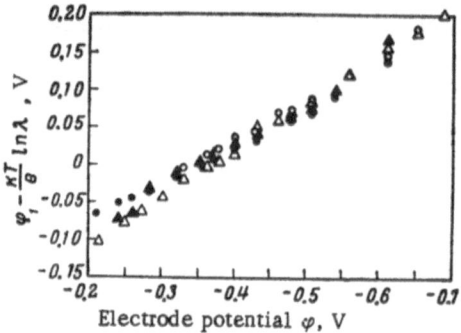

Fig. 40. The relation of $\varphi_1 - (kT \ln \lambda)/e$ to the potential φ of a germanium electrode [O, $\triangle$) p-type, 30Ω · cm; ●, ▲) n-type, 16.5 Ω · cm], obtained by measuring the photopotential with pulsed illumination (O and ●) and by the differential capacitance method ($\triangle$ and ▲) [77].

sure that the injection current is small in comparison with the saturation current of minority carriers $i_p \ll |i_p^{lim}|$ [where equation (17.19) is used] and to introduce a correction for the Dember potential, which is substantial for high-resistance samples. * Under these conditions the change in the sign of the photoeffect occurs at the flat band potential ($\varphi_1 = 0$).

An analogous comparison of the two methods, namely the photopotential and surface conductivity methods, was made for the system germanium—methylformamide [28, 79]. Figure 41 shows the relation of the photopotential (without a correction for the Dember potential) to the potential drop in the space charge region, which was calculated from the surface conductivity. After taking into account the Dember potential, it was found that the photopotential does not change sign at the flat band potential [as is predicted by the theory which does not

---

*To check the Dember potential it is possible to measure the ratio of the photopotentials with strong anode and strong cathode polarization (where the photopotential reaches saturation). When there is no Dember effect the modulus of this ratio is λ^2 [see equation (17.20)].

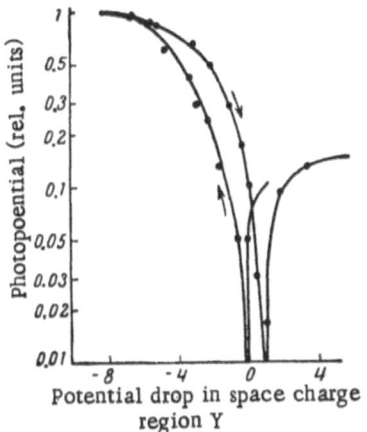

Fig. 41. Photopotential of a germani-um electrode (n-type, 30 Ω · cm) in a solution of KBr in methylformamide with pulsed illumination [82]. Direc-tion of change in potential during the plotting of curves is shown by arrows.

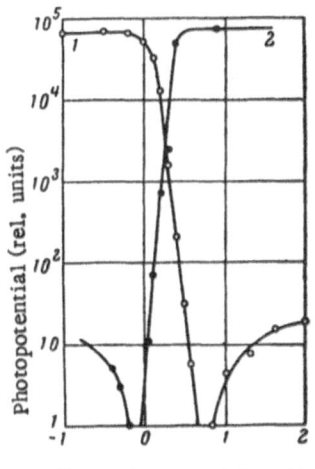

Electrode potential φ, V

Fig. 42. Photopotential of a silicon electrode in 1% H_2SO_4 solution. 1) n-Type; 2) p-type.

take into account surface levels — equation (17.19)]. The shift in the potential of zero photoeffect relative to the flat band po-tential varies, depending on the preliminary treatment of the electrode (Fig. 41) and for different experiments it varied over a range of $2kT/e$ units. On the scale of surface potentials of $\varphi_1 - (kT/e)\ln\lambda$ the photopotential changes sign for samples of n-type (30 Ω · cm) in the range from —0.6 to 1.4 and for p-type (20 Ω · cm) in the range from —0.6 to 1.3 (in kT/e units).

Such a shift in the potential of zero photoeffect relative to the flat band potential may be explained by assuming that the photoeffect is influenced by fast surface states, whose energy levels lie far from the center of the forbidden band. The re-laxation time of these levels does not exceed the duration of the flash during measurement of the photopotential (10^{-5} sec), while their concentration is probably no lower than 10^{11}-10^{12} cm^2. As Johnson [71] showed, with a low concentration the sur-face levels have no influence on the photoeffect.

In the case of silicon, the relation of the photopotential to the potential and the type of conductivity is qualitatively similar to the theoretical picture (Fig. 42) [80]. No quantitative check of the theory has been carried out. However, Muller [81] came to the conclusion that in practice the photopotential depends less on the type of conductivity and the specific resistance and changes more sharply with the light intensity than is predicted by the theory [71].

§ 19.   Surface Recombination Rate

In § 6 we examined the recombination of nonequilibrium carriers at recombination centers in the bulk. The recombination process may also occur at surface levels which, as has already been mentioned, arise as a result of a chemical interaction of the semiconductor material with the surrounding medium, adsorption, an increase in the number of crystal lattice defects at the crystal surface, and for other reasons. The concentration of these levels may be extremely high, so that this recombination process often predominates.

Let the interphase be illuminated. The physical nature of the phenomena occurring was examined in §17. If there are surface levels at the interphase then nonequilibrium electrons and holes produced by the light may be trapped by levels and recombine in them. The rate of surface recombination is defined as the number of carriers recombining on the surface in unit time per unit surface and per unit concentration of excess carriers at the boundary of the space charge region and the quasineutral region

$$s = \frac{R_s}{\Delta p_1}.$$

(19.1)

Before turning to calculations, we give a graph of the relation between the surface recombination rate and the potential drop in the space charge region in the semiconductor φ_1. This relation is represented by a bell-shaped curve (Fig. 43), which is known in the literature as a "Stevenson—Keyes curve" after the authors who first applied the Shockley—Read theory to recombination processes on a surface [82]. Two processes

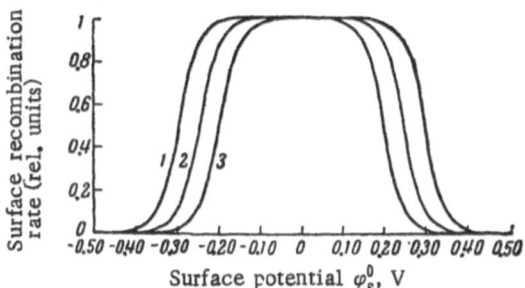

Fig. 43. Theoretical relation of the surface recombination rate s to the surface potential φ_s (when $C_p = C_n$) [82]. Position of recombination level relative to the center of the forbidden band: 1) 0.31 eV; 2) 0.26 eV; 3) 0.21 eV.

occur on recombination of carriers, namely, electrons from the conductivity band and holes from the valence band undergo transition to a surface level and recombine. The flow of electrons to the level is proportional to the number of free sites in it. The flow of holes from the valence band to the level is proportional to the number of electrons in the level. Thus, under steady conditions, the recombination rate is maximal in the case where the recombination levels are only partly filled by electrons, i.e., the Fermi distribution function f_t, which describes the filling of the level by electrons, has an intermediate value between 1 and 0. As is shown by Fig. 4, the change in the function f_t from 1 to 0 occurs over a narrow range close to some potential value, which is determined by the energy of the level. The recombination level is most effective in this range. With a change in the surface potential toward positive or negative values, the filling of the levels by electrons tends toward 1 or 0 and the surface recombination rate falls. Thus, the relation of the surface recombination rate to the surface potential is described by a curve with a maximum.

Let us turn to the calculation of the surface recombination rate [83]. An expression may be written for R_s by using (6.8):

$$R_s = \frac{C_n \cdot C_p \, (p_s \cdot n_s - n_i^2)}{C_n \, (n_s + n') + C_p \, (p_s + p')},$$

$$(19.2)$$

where n_S and p_S are the concentrations of electrons and holes at the semiconductor surface $(x = 0)$.

Assuming that the distribution of carriers in the space charge region is determined by the equilibrium formulas, * we may write

$$n_s \cdot p_s = n_1 \cdot p_1,$$

where n_1 and p_1 are the concentrations of electrons and holes at the boundary of the Debye and quasineutral regions.

We will use the expressions

$$p_1 = p^0 + \Delta p_1,$$
$$n_1 = n^0 + \Delta n_1. \qquad (19.3)$$

From the condition of electrical neutrality of the system it follows that $\Delta n_1 = \Delta p_1$.† By using equation (19.3) with an accuracy up to terms linear with respect to Δp_1, we obtain

$$R_s = \frac{C_n \cdot C_p \, (n^0 + p^0) \, \Delta p_1}{C_n \, (n_s + n') + C_p \, (p_s + p')}. \qquad (19.4)$$

From relations (19.1) and (19.4) we find an expression for the surface recombination rate

$$S = \frac{C_n \cdot C_p \, (p^0 + n^0)}{C_n \, (n_s + n') + C_p \, (p_s + p')}. \qquad (19.5)$$

Then, by using the expressions for the concentrations of free carriers‡

---

*See p. 101.

†See § 28.

‡In the derivation of the expressions for n' and p', it is assumed that the effective masses of an electron and a hole are equal.

$$n_s = n_i e^{\frac{e\varphi_s^0}{kT}} \; ; \quad p_s = n_i e^{-\frac{e\varphi_s^0}{kT}} \; ; \quad n' = n_i e^{\frac{E_t^0 - E_i}{kT}} \; ; \quad p' = n_i e^{\frac{E_i - E_t^0}{kT}} \; ,$$

where $\varphi_s^0 = (kT/e)(Y - \ln\lambda)$ is the surface potential, defined as the difference between the Fermi level and the potential at the middle of the forbidden band at the surface, and E_t^0 is the energy of the surface levels when $\varphi_1 = 0$, we obtain

$$S = \frac{C_n \cdot C_p (p^0 + n^0)}{n_i \left[C_n \exp \dfrac{E_t^0 - E_i}{kT} + C_p \cdot \exp \dfrac{E_i - E_t^0}{kT} + C_n \exp \dfrac{e\varphi_s^0}{kT} + C_p \exp \left(-\dfrac{e\varphi_s^0}{kT} \right) \right]}.$$

It is convenient to introduce the following characteristic value of the surface potential

$$\varphi_a = \frac{kT}{2e} \ln \frac{C_p}{C_n} \; . \tag{19.6}$$

Then the formula for the surface recombination rate assumes the form

$$S = \frac{(C_p \cdot C_n)^{1/2} (n^0 + p^0)}{2n_i \left\{ \mathrm{ch} \dfrac{E_t^0 - E_i - e\varphi_a}{kT} + \mathrm{ch} \dfrac{e\,(\varphi_s^0 - \varphi_a)}{kT} \right\}} \; . \tag{19.7}$$

Expression (19.7) shows that the relation of the surface recombination rate to the potential actually corresponds to Fig. 43.

By comparing the experimental and theoretical $s - \varphi$ curves it is possible to find the ratio of the trapping cross sections for holes and electrons. In actual fact, having determined the position of the maximum surface recombination rate on the scale of surface potentials, i.e., the potential φ_a, the ratio C_p/C_n is readily obtained from equation (19.6).

The surface recombination rate with high injection levels was examined in [84].

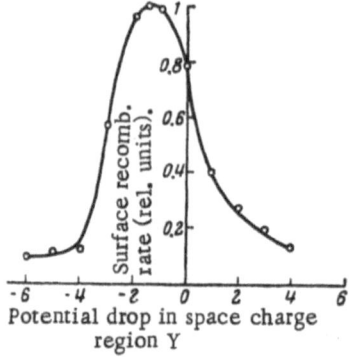

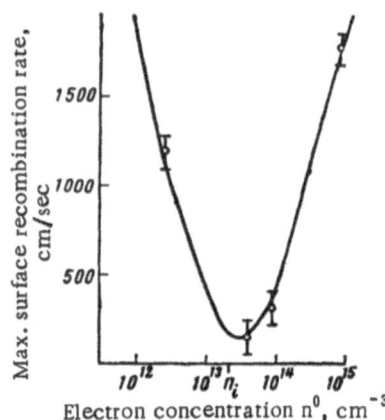

Fig. 44. Surface recombination rate on germanium (n-type, 30 Ω · cm) in a solution of KBr in methylformamide [28].

Fig. 45. Relation of maximum surface recombination rate to free electron concentration n^0 in germanium samples; 1 N KOH solution [85].

§ 20.   Fast Surface Electron Levels.   Surface Recombination Rate on Germanium and Silicon

To determine the surface recombination rate s at the semiconductor—electrolyte boundary we measure the effective lifetime of the minority carriers τ_e by one of the numerous methods.* Knowing the bulk lifetime, s is determined from equation (23.1).

It was found that the potential dependence of τ_e passes through a maximum in a definite range of potentials of a germanium electrode. By measuring simultaneously with τ_e the relation of the potential drop in the space charge region φ_1 to the electrode potential, it is possible to obtain the relation of s to φ_1. Figure 44 shows such a curve for the germanium—methylformamide system [28]. The values of τ_e were determined by the steady photoconductivity method and the values of φ_1 were calculated from the surface conductivity.

---

*Two of them are described in detail in § 23.

The curve shown in Fig. 44 with a characteristic bell shape is similar to the theoretical curve (Fig. 43) and is described qualitatively by equation (19.7) for separate recombination levels. From the shift in the maximum relative to zero surface potential $e\varphi_S^0/kT = Y - \ln\lambda$, it was possible to determine the ratio of the trapping cross sections for holes and electrons $C_p/C_n = 0.21$. Thus, recombination proceeds at donor levels ($C_n > C_p$).

The $s - \varphi_1$ curve for germanium in an aqueous solution of NaOH has an analogous form [85]. In this case it was found that $C_p/C_n = 1800$. It was found that the potential of the maximum surface recombination rate is independent of the type of conductivity and the specific resistance of germanium, while the surface recombination rate increases with a fall in the specific resistance. The latter follows directly from equation (19.7), which may be represented in the following form for a given surface potential:

$$s = \text{const} \cdot (n^0 + p^0). \tag{20.1}$$

Here n^0 and p^0 are the equilibrium concentrations of electrons and holes in the bulk of the semiconductor. The points in Fig. 45 denote the values of s at the maximum for several samples, while the solid curve was calculated from equation (20.1), in which it was assumed that the constant equals $3 \cdot 10^{-12}$ cm^4 per sec. A similar relation for a germanium−gas boundary was first found by Schultz [86].

The dependence of the surface recombination rate on the solution composition in some cases may be reduced to a dependence on the change in the surface potential of the semiconductor under the influence of additives introduced into the solution. Thus, for example, a number of authors have reported a reduction in the surface recombination rate when a solution is saturated with oxygen [76, 87]. At the same time the steady potential of germanium changes and becomes more positive because of acceleration of the spontaneous solution of germanium, whose rate is determined by the concentration of oxygen in the solution (see Chapter IV). It was found that over the

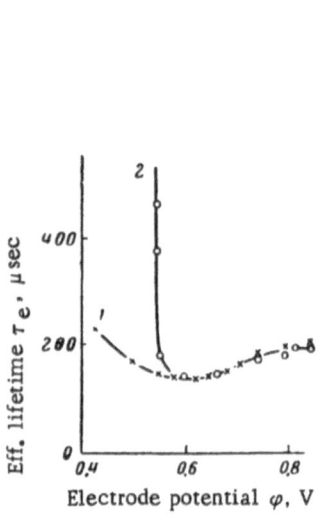

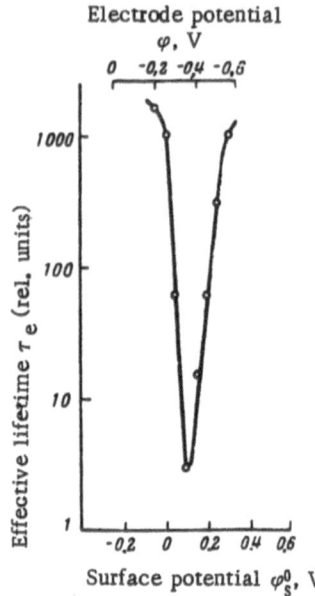

Fig. 46. Relation of effective lifetime τ_e to potential φ of a germanium electrode in 1 N KOH solution [85]. 1) Potential changed by current; 2) potential changed by a change in the oxygen concentration in solution.

Fig. 47. Relation of effective lifetime of minority carriers τ_e to potential φ of a silicon electrode (n-type, 250 Ω · cm) in 1% Na_2SO_4 solution [34].

range of potentials —0.55 to —0.85 V (in 1 N NaOH solution), the effective lifetime is completely independent of the method by which the potential is established, i.e., by the passage of a current or by a change in the oxygen concentration in solution (Fig. 46). Consequently, here the action of oxygen is limited to changing the potential. Only at more positive potentials than —0.55 V is it possible that there is a change in the concentration of recombination centers or the trapping cross section in the presence of oxygen.

The following method was used in [62, 88] to investigate the recombination on the surface of a silicon electrode. The sample consisted of a thin plate, on one side of which was a p—n junction. The other side was in contact with a solution and was illuminated through it. The photopotential at the p—n junction was measured. Its magnitude is determined by the concentration of nonequilibrium minority carriers, which arise on absorption of light in silicon and diffuse across the sample to the p—n junction. The current generated by the light depends on the recombination rate at the silicon—electrolyte boundary. It was shown in [89] that the effects of the silicon—solution boundary on the photoeffect at the p—n junction increases with a decrease in the wavelength of the incident light. This is explained by the fact that thereupon there is a decrease in the effective depth of absorption of light, i.e., the formation of electron—hole pairs occurs closer to the electrode surface. Thus, the photopotential at the p—n junction is proportional to the effective lifetime of the minority carriers in the sample [90]. The relation of the effective lifetime to the potential of a silicon electrode is given in Fig. 47. The minimum in τ_e corresponds to the maximum surface recombination rate s. The absolute value of s may be estimated only approximately on the basis of such measurements, but the position of the maximum of s on the scale of surface potentials of the semiconductor is readily determined. It was found that $C_p/C_n \approx 3000$, i.e., the recombination centers of the acceptor type ($C_p > C_n$).

Let us examine the effect of preliminary polarization of a germanium electrode on the surface recombination rate. As has already been mentioned in § 13, mild anode etching at a current density of the order of 10^{-4} A/cm^2 leads to a sharp fall in the surface recombination rate. The effective lifetime of minority carriers usually reaches the value in the bulk (i.e., $s \approx 0$) and with special purification of the solution remains at this level for an indefinite time, and this procedure was used in [32,40] for accurate measurement of the photopotential and differential capacitance of a germanium electrode. The reason for the decrease in s may be the removal of adsorbed impurities from the surface during etching, the elimination of surface defects in the crystal structure, and also the formation of an oxide layer with a perfect structure.

The behavior of s on cathode polarization was found to be very complex. Experimental data in the literature are often contradictory. Therefore, only cautious conclusions may be drawn at the present time.

In the cathode polarization of germanium in aqueous solutions, two factors should be taken into account. First, the reduction of surface oxides is accompanied by a change in the interphase potential difference φ_0 by several tens of volts [54, 91]. Therefore, at the reduction potential of the oxides (with φ = const) there is also an abrupt change in the potential drop in the space charge region φ_1, so that $\Delta\varphi_1 = -\Delta\varphi_0$. If we measure only the surface recombination rate without checking φ_S° (for example, by the capacitance or photopotential method), we may come to the mistaken conclusion that on the "surface recombination rate—electrode potential" curve there are two maxima corresponding to two types of recombination centers. In actual fact, we should be considering the same maximum which changes its position during the measurements as the $s - \varphi_S^\circ$ curve together with the φ_S° scale shifts abruptly relative to the electrode potential φ scale. This is shown schematically in Fig. 48. If, moreover, the nature and concentration of the surface states changes on cathode polarization (as actually occurs [54]), the relations observed are even more complex. The reduction of the oxides proceeds relatively slowly. This leads to the appearance of hysteresis on curves plotted from anode to cathode potentials and vice versa if the measurements are not carried out very slowly and the electrode is unable to reach the quasiequilibrium state at each potential.

Two such maxima of the surface recombination rate were observed in [92]. Their potentials in 0.1 N sulfuric acid solution were −0.5 and −0.15 V. However, the authors of [92] regarded them as two real maxima and associated the cathode maximum with the adsorption of hydrogen atoms and the anode maximum with the adsorption of hydroxyl groups.

We examined the case where the change in the surface recombination rate measured was produced by a change in the surface potential. With strong cathode polarization there also arise new recombination centers. This is apparently con-

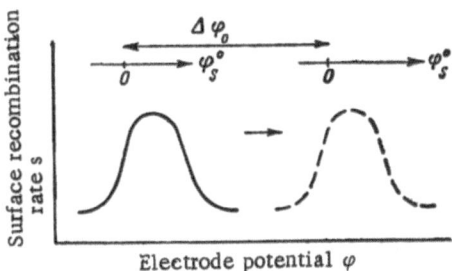

Fig. 48. Shift in the s − φ curve on the electrode
potential scale as a result of a change in the potential difference in the Helmholtz layer of reduction of the surface oxides. $\Delta\varphi_0$ is the shift in the
surface potential scale of the semiconductor and
φ_s^0 is relative to the electrode potential scale φ.

nected with the adsorption and inclusion of atomic hydrogen in
the crystal lattice of germanium.*   After cathode polarization
(at the hydrogen liberation potential) and drying in air, the
surface recombination rate on germanium reaches several
thousand centimeters per second [95, 96].   Analogous results
were obtained in the measurement of s directly in solution in
which hydrogenation was carried out [85, 97, 98].†   When a
current is switched on the surface recombination rate slowly
returns to the original value (after several hours).   The removal of the surface layer of germanium by etching eliminates
the hydrogenation effect.

An investigation of the capacitance [55] and the kinetics
of electrochemical reactions on a hydrogen-treated germanium
electrode (see § 34) showed that such an electrode differs
markedly in properties from one that has not been treated with
hydrogen and is more similar to a metal.   Its capacitance ap-

---

* According to [93], on adsorption of hydrogen on germanium, slow surface states
of the donor type which lie 0.22 V above the center of the forbidden band also
arise, while according to data in [94], acceptor levels arise.

† An increase in the surface recombination rate was also observed when germanium
was treated with atomic hydrogen in a glow discharge [99].

proaches the capacitance of the ionic double layer (more than 10 $\mu F/cm^2$) and many of the peculiarities of electrochemical kinetics which are characteristic of semiconductor electrodes practically disappear. Thus, hydrogen treatment seems to "metallize" the surface of a semiconductor because of the formation of recombination centers and possibly surface levels. The action of hydrogen atoms as generation centers will be examined in connection with the kinetics of cathode liberation of hydrogen in § 38.

The nature of "natural" recombination centers at a germanium—electrolyte interphase has not been elucidated. It is possible that they are associated with defects in the crystal structure of germanium and its oxides or with the adsorption of impurities on the surface. An attempt has been made [100] to compare recombination at germanium—water and germanium—gas interphases.

The increase in the recombination rates on adsorption of heavy metal ions on germanium will be examined in the next section.

§ 21.   Fast Surface Electron Levels.   Effect on Capacitance

If the capture cross sections for electrons and holes of a level differ markedly, then this level is not effective in recombination and is a trapping level [101]. We will examine here fast surface states. The main method of investigating their properties in the electrochemistry of semiconductors is through the measurement of the differential capacitance. The capacitance produced by surface levels was calculated in § 12 [equation (12.15)].

By measuring the capacitance of a germanium electrode, Boddy and Brattain [102, 103] discovered that when gold, silver, or copper salts are added to the solution in small amounts ($5 \cdot 10^{-7}$–$5 \cdot 10^{-6}$ g-equiv/liter) the capacitance increases appreciably in comparison with the capacitance in a pure solution (Fig. 49). Analogous results were obtained in [104]. As these

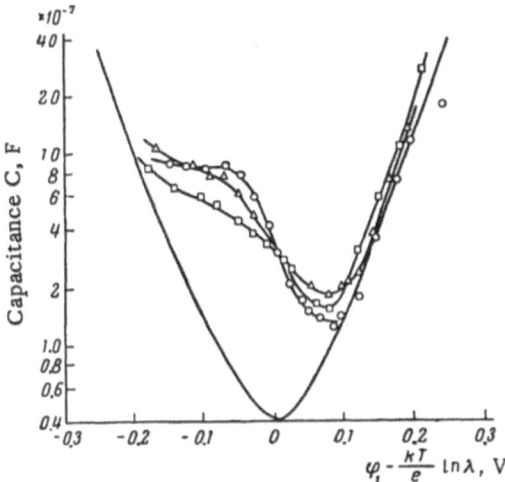

Fig. 49. Increase in capacitance C of a germanium electrode on introduction of silver ions into the solution [103]. O) $5 \cdot 10^{-7}$; □) $2 \cdot 10^{-6}$; △) $5 \cdot 10^{-6}$ g-equiv/liter. Solid curve, pure K_2SO_4 solution.

Table 3.   Characteristics of Levels

	Cu	Ag	Au
$(E_{t_1}^0 - E_i)/kT$	−1.9	−4.1	−0.9
N_{t_1}	10^{11}	$1.9 \cdot 10^{11}$	$1.8 \cdot 10^{11}$
$(E_{t_2}^0 - E_i)/kT$	1.2	−0.55	2.5
N_{t_2}	$1.9 \cdot 10^{10}$	10^{11}	$4.5 \cdot 10^{10}$
σ_p	$6 \cdot 10^{-14}$	$2 \cdot 10^{-14}$	$4.1 \cdot 10^{-15}$
σ_n	$1.5 \cdot 10^{-16}$	$0.74 \cdot 10^{-16}$	$1.4 \cdot 10^{-16}$

metals are more noble than germanium, they are deposited on its surface by an electrochemical replacement mechanism [103, 105] (see also Chapter IV).   The deposition of copper, gold, and silver on the germanium was demonstrated by direct measurements by means of tracers [106-108].

By subtracting the theoretical $C_1 - \varphi_1$ curve of the capacitance of the space charge region from the $C - \varphi_1$ curve obtained experimentally in the presence of traces of heavy metals it is possible to obtain the excess capacitance ΔC due to the deposition of the metals on the semiconductor surface. It is natural to associate it with the formation during this deposition of surface levels with a relaxation time of the order of 1 μsec or less (the current pulse time in the measurement of capacitance in this work was 5 μsec). Figure 50 shows the excess capacitance as a function of the surface potential of the semiconductor $\varphi_s^0 = \varphi_1 - (kT/e)\ln\lambda$. The latter was determined by measurement of the surface conductivity, * since it is not possible to use the differential capacitance method for this purpose in the presence of fast surface states. By using equation (12.15) and selecting the energy of the centers E_t^0 and the concentration N_t, it is possible to represent the experimental $\Delta C - \varphi_s^0$ relation as the sum of the capacitance curves of two discrete surface levels (see Fig. 50). All three of the metals investigated showed two levels, which were effective over a definite range of surface potentials with a ratio of the concentrations N_{t_1}/N_{t_2} from 2 to 5. The characteristics of the levels are given in Table 3. The level with a high concentration is arbitrarily called level 1 and the other level, level 2.

The surface recombination rate was measured simultaneously with the capacitance and the surface conductivity for determining the trapping cross sections σ_p and σ_n. This is close to zero on an uncontaminated surface, but increases on deposition of metals on the electrode to several hundredths of centimeters per second. In the expression for the surface recombination rate (19.7) there are three unknowns, namely, the energy of the recombination center E_t^0 and the trapping con-

---

*In § 13 we pointed out that in the measurement of surface conductivity with an alternating current [32, 102, 103] an error could arise connected with the effect of the low-frequency capacitance of the electrode on the value measured. However, this is not very appreciable in the given case if the low-frequency capacitance changes little on adsorption of heavy metals on the electrode.

stants for holes and electrons $C_{p,n} = N_t \cdot v \cdot \sigma_{p,n}$ (see § 6). The ratio C_p/C_n was determined from the position of the maximum on the $s - \varphi_s^0$ curve. One relation of s to the potential is insufficient for calculating all three unknowns and additional information is required. In the investigation of the "dry" surface of a semiconductor, E_t is usually determined from the temperature dependence of s [83]. Another method was used in [102, 103]. The authors assumed that one of the two surface levels found from the differential capacitance is a recombination center. By substituting the values of $E_{t_1}^0$ and $E_{t_2}^0$ in equation (19.7) they constructed two $s - \varphi_s^0$ curves for each case and one of these was found to agree well with experiment. In all cases level 2 (with a lower concentration) was found to be a recombination level and level 1 was a trapping level.

The microscopic nature of the surface states formed remains unknown. Their concentration does not exceed $2 \cdot 10^{11}$ cm^{-2} though the coverage of the germanium surface by copper and gold is close to a monolayer [106, 107]. Consequently, only an insignificant fraction of the metal atoms gives rise to surface levels. On the other hand, no definite relation was found between the metal content of the solution and the concentration of surface states formed. Finally, although centers 1 and 2 appear simultaneously with the adsorption of an impurity on the germanium, the ratio of their concentrations is arbitrary. An analogous phenomenon was observed on measuring the effect of a field on the "dry" surface of germanium after its contamination by copper from solution [109]. In the opinion of the authors of [102, 103], the surface levels may be associated either directly with metal atoms or with defects in the structure of the oxide layer arising during deposition of the metal atoms and simultaneous oxidation (or solution) of germanium.

On adsorption of copper and gold on a germanium electrode there is also a change in the potential at which the photopotential of the electrode changes sign (Fig. 51) [108]. This change is -0.25 V for copper with a concentration of it on the surface of $3 \cdot 10^{-10}$ g-equiv/cm^2 and it is evidently connected with a change in the interphase potential difference. It may be surmised that in addition to fast levels, there is also the for-

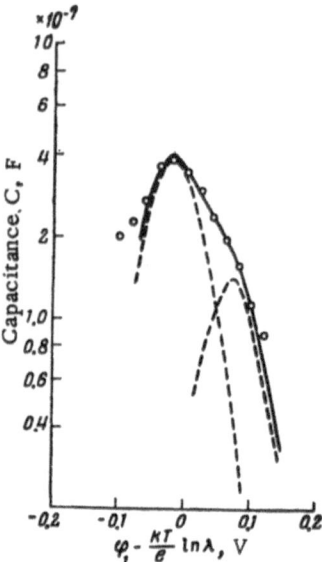

Fig. 50. Curves of excess capacitance due to the adsorption of gold on a germanium electrode [103]. Broken lines show theoretical curves calculated from equation (12.15) with the following values of parameters: $E_{t_1}^0 - E_i = -0.78 kT$, $N_{t_1} = 2.6 \cdot 10^{11}$ cm^{-2}; $E_{t_2}^0 - E_i = 2.73 kT$, $N_{t_2} = 0.93 \cdot 10^{11}$ cm^{-2}. Solid line shows the sum of theoretical curves; circles represent experimental points.

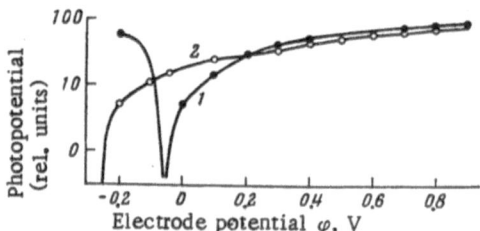

Fig. 51. Effect of adsorption of copper on photopotential of a germanium electrode (n-type, 30 $\Omega \cdot$ cm) in 0.1 N Na$_2$SO$_4$ solution (pH 7.5) [108]. Copper concentration in solutions: 1) 0; 2) 10^{-6} g-equiv/liter.

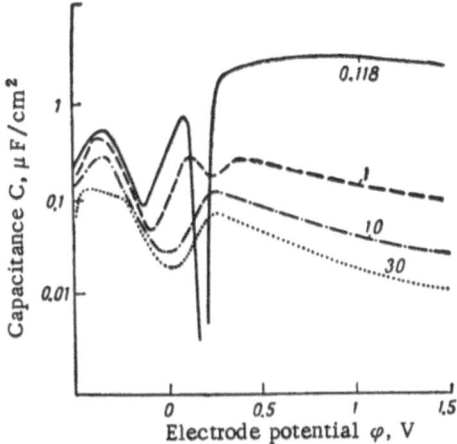

Fig. 52. Relation of capacitance C to potential φ
of a germanium electrode (p-type, 28 $\Omega \cdot$ cm) in
0.1 N H_2SO_4 solution [110]. The frequency (in
kHz) is shown on the curves.

mation of slow surface levels on adsorption, whose number is
close to the number of adsorbed metal atoms and which affects
the potential difference in the Helmholtz region.   Calculation
gives a value of about 0.2 for their degree of ionization.   Moreover, the adsorption of heavy metals may affect the interphase
potential difference through a change in the degree of oxidation
of the surface and the character of the Ge—O bond [108]. This
is all the more probable as the deposition of copper, for
example, by an electrochemical replacement mechanism, is
accompanied by oxidation of the germanium [105].

In addition to levels artificially introduced onto the electrode surface, fast levels arising by a "natural" route during
standard preliminary treatment of the electrode have been observed in some work. It has been reported [48] that in strongly acid and strongly alkaline solutions the capacitance of a germanium electrode is twice as high as at pH 5-11 and no longer
corresponds to the capacitance of the space charge. The excess capacitance apparently may be represented as the capacitance of levels with a concentration of $(1-2) \cdot 10^{11}$ cm^{-2} lying

close to the middle of the forbidden band. The physicochemical nature of these remains unknown.

By using the potential dependence of the capacitance of a germanium electrode in 0.1 N H_2SO_4 solution, Gobrecht, Meinhardt, and Reinicke [110] observed two capacitance minima at -0.15 and -0.5 V (i.e., at approximately the same potentials as the maxima of the surface recombination rate [92]). They treated the capacitance maximum on the $C-\varphi$ curve lying between these two minima as the capacitance of fast surface levels. However, it is possible that in this case they were dealing not with two, but with the same capacitance minimum, which (like the maximum of the surface recombination rate) is shifted along the electrode potential axis during the measurement of the $C-\varphi$ relation due to an abrupt change in the Helmholtz potential drop on reduction of the surface oxides (cf. Fig. 48). In this case, the maximum observed on the experimental $C-\varphi$ curve is fictitious.

The same authors [110, 111] observed an extremely interesting phenomenon which they also associate with an effect of surface levels on the capacitance. Over a very narrow range of potentials the capacitance of p-type germanium at a frequency $\omega \approx 100$ Hz falls sharply and assumes a negative sign. At frequencies from 100 Hz to 1 kHz on the $C-\varphi$ curve in this region there is a minimum, whose depth decreases with an increase in frequency. At $\omega \approx 1$ kHz, the $C-\varphi$ curves have the normal form (Fig. 52). The effect was observed on electrodes of electronic germanium only on illumination. The negative capacitance (or inductance) can be observed only at a pH of about 1.25 and this apparently confirms the relation of this phenomenon to the surface properties of the electrode. Although the explanation of the origin of the inductance put forward in the work of Gobrecht [110] is formal, nonetheless, it may be assumed that with a certain combination of their parameters, the surface levels may be responsible for the inductive properties of the contact. It should be borne in mind that the reason for the appearance of the inductance may also be an electrochemical reaction [112].

§ 22.   Adsorption on Semiconductors from Solutions

Etching and washing are used widely in the production of semiconductor instruments. During the course of these operations impurities present in the etching agents and wash water are deposited on the surface of the semiconductors and change their electrophysical properties. It was mentioned above that after germanium had been kept in a solution of a copper salt and dried there was an increase in the surface recombination rate [109]. The surface conductivity of germanium also changes as a result of the adsorption of metals [113]. This leads to a change in the electrical parameters of the instruments which is difficult to control. We can also expect that the investigation of adsorption will give valuable information on the structure of the electrical double layer on semiconductor electrodes. Therefore, adsorption on semiconductors from solutions and methods of freeing the surface from impurities have long been subjects of investigation.

In analogy with metals it might be expected that both electrostatic and chemical adsorption will be observed on semiconductors. The first is a reversible function of the electrode potential. The charge Q_2 of electrostatically adsorbed ions is determined from relation (14.7).

Since normally $Q_1 \ll Q_t$ (where Q_1 is the space charge and Q_t is the surface charge), then electrostatic adsorption must be determined completely by the surface charge. In very dilute solutions Q_2 approximately equals the charge of the diffuse part of the ionic double layer, which may be calculated if we know the ψ' potential.

Chemical adsorption involves a chemical interaction of the material of the electrode with the adsorbate. It determines the potential drop in the space charge region φ_1. Chemisorption is irreversible to a considerable degree. A particular case of a chemical interaction of an electrode with ions of a solution is the electrochemical replacement of a more noble metal and its conversion to its salts by a less noble metal. It should be noted that the surface charge produced by oxygen, which determines the degree of electrostatic adsorption on a germanium electrode, may also be regarded formally as the result of chemisorption. *

All the investigations of adsorption on semiconductor materials whose results are presented here were carried out with tracers. Reversible and irreversible adsorption of metal ions on gallium arsenide was observed by Larrabee [114]. He measured the radioactivity of a solution containing unstable isotopes of the metals investigated which was in contact with gallium arsenide powder. The noble metals silver, gold, copper, and mercury were firmly adsorbed. Sodium and zinc ions were adsorbed reversibly and were readily washed from the surface. The relation of the adsorption to the concentration of ions in solution followed a Freundlich isotherm.

Harvey, LaFleur, and Gatos [115] observed reversible adsorption of iodine ions on germanium, which reached 10 monolayers in a 10^{-4} M KI solution. The amount of adsorption fell with anode and cathode polarization of the electrode. The amount of adsorption depended on the pH and the degree of oxidation of the surface. Such a large amount of reversible adsorption could not be explained.

The adsorption of metals on germanium and silicon from etching agents and wash water containing ions of the metals investigated has been investigated [106, 116-118]. The adsorption reached 10^{14}-10^{18} atoms/cm^2 with a concentration of the ions investigated in solution of 10^{-5}-10^{-2}%. Heavy metals were adsorbed firmly, but they were desorbed when the samples were washed with solutions of potassium cyanide, Dithizone, and acetonitrile. The degree of adsorption on silicon depended on the concentration of hydrofluoric acid and the solution pH. The adsorption of metals on germanium has also been investigated in [119, 120].

---

*The adsorption of oxygen on germanium and gallium arsenide was examined in §§ 14 and 16 and the adsorption of sulfide ions on cadmium sulfide in § 16.

The pulsed photoeffect method was used for investigating the adsorption of copper and gold on germanium [108]. As was shown in §21, at high coverages, copper and gold change the interphase potential difference and, consequently, the potential drop in the space charge region φ_1 with a constant electrode potential. By checking φ_1 through measurements of the photopotential (with a steady potential, it is possible to follow the kinetics of adsorption and desorption of copper. It was found that even with intensive anode solution of germanium the adsorption is extremely great ($\approx 10^{-10}$ g-equiv/cm^2 from 10^{-6} N solution with an anode current of about 10^{-2} A/cm^2).

In none of the papers listed was there any report of a relation between the degree of adsorption on the one hand and the semiconductor properties of the electrode or the magnitude of the potential drop in the space charge region, on the other. As a rule, irreversible adsorption was measured.

Traces of metals adsorbed on a germanium surface during chemical etching and washing may also be detected by electron diffraction [121]. Their effect on the electrophysical properties on a "dry" surface was investigated by Morrison [122].

§23. Methods of Investigating the Electrophysical Properties of a Semiconductor − Electrolyte Interphase*

a. Surface Conductivity. Thin samples of a semiconductor are used for measuring the surface conductivity so that the thickness and, consequently, the conductivity of the space charge region constitute an appreciable part of the total thickness and conductivity of the sample. In practice, in the case of germanium with a specific resistance of 1 $\Omega \cdot$ cm or above (the thickness of the space charge region is 10^{-4}-10^{-5} cm), the optimal sample thickness is 0.1 mm. Rectangular plates are usually used with a length of 10-20 mm and a width

---

*For methods of investigating the bulk properties of semiconductor materials see, for example, [123-125].

Fig. 53. Germanium electrode for measuring surface conductivity [32]. The light regions marked by num-bers are ohmic contacts. The arms of the bridge for measuring the conductivity are the section 1-5 (thin plate immersed in the electrolyte); 1-2, 2-3 (or 2-4), 4-5.

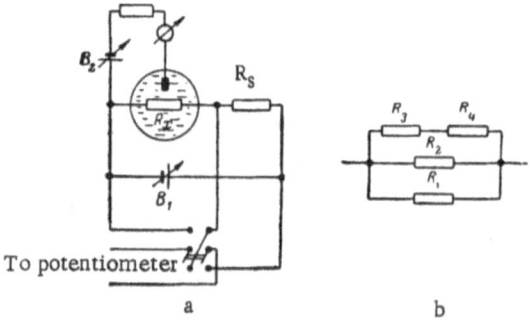

Fig. 54. Circuit of apparatus for measuring surface conductivity (a); equivalent circuit of system for dc measurements (b) [30].

of 3-5 mm.   Two contacts are fitted to the ends of the semiconductor plates and these are usually used simultaneously for
passing the measuring current through the sample, for measuring the potential drop in it, and for changing the electrode
potentials.   After insulation of the leads the sample is placed
in a solution, which contains an auxiliary electrode for polarization and a comparison electrode.   Both of the flat surfaces
of the sample are working surfaces.   In some work the sample
is cemented at the edges into a rectangular opening in the wall
of a cell; in this case no insulation of the contacts is required,
but only one surface is a working surface. *

An original electrode for measuring surface conductivity
was proposed by Brattain and Boddy [32] (Fig. 53).   One part of
it is a thin plate of germanium (like the sample examined
above), which is an arm of a bridge.   The other three arms of
the bridge have a considerable thickness.   In measurements by
the method proposed in [32], the resistance of the thin part,
which is immersed in the electrolyte, is compared with the
bulk resistance of germanium.   This automatically introduces
a correction for a change in temperature (on condition that
there is no temperature gradient inside the electrode itself).

The simplest circuit for dc measurements is given in
Fig. 54a [30].   In series with the resistance of the sample R_x
(which in the case of high-resistance germanium is 1-5 kΩ) is
connected a standard resistance R_s for determining the value
of the measuring current which is supplied by the battery B_1.
The fall in voltage at the resistors R_x and R is measured with
a dc potentiometer fitted with a null instrument with a sensitivity of the order of 10^{-9} A per division, or is amplified with
a dc amplifier and recorded automatically.   The potential of
the electrode investigated is set by the battery B_2.   If there is
no ohmic potential drop along the electrode, its surface is equipotential.   However, in practice, because of the high resistance of a thin sample, even small polarization currents (of the

---

* Instead of a rectangular sample, it is possible to use a thin disk fitted with an annular electrode and a probe at the center of the nonworking surface [93].

order of 10^{-6} A/cm^2) create appreciable ohmic potential drops,
which not only introduce some indeterminacy into the potential
measured, but also add to the potential drop in the sample produced by the measuring current. A correction has to be introduced for this effect.

The constituent resistances of R_X are shown in Fig. 54b;
R_2 is the bulk resistance of germanium, R_3 the differential resistance of the electrode−electrolyte interphase, which is determined from its volt−ampere characteristics, and R_4 the resistance of the solution. Since, in moderately dilute solutions
R_4 is relatively low, to avoid shunting of the sample by the solution it is necessary to carry out the measurements under
conditions such that R_3 is quite large ($R_3 \gg R_1$). In practice in
the case of germanium the polarization current should not substantially exceed 10^{-6} A/cm^2, when the leakage of the measuring current into the solution may be neglected. * Of all the
components of R_X, only the resistance of the space charge region R_1, which is measured along the sample, depends on the
potential, and this makes it possible to separate it from the
constant components.

Measurement of the resistance of a sample by an ac method is less suitable for determining the surface conductivity
of germanium in solutions of high conductivity. This is due to
the fact that the impedance of the germanium−electrolyte interphase is relatively small and the shunting effect of the solution on the alternating current is much more appreciable than
on a direct current. Only with liquids with very low conductivity (carefully purified water and some organic liquids) is it
possible to use ac methods for measuring surface conductivity
[100] as in this case $R_4 \gg R_1$.

---

*With sufficiently high cathode and anode polarization of the electrode an electro-
chemical reaction begins, the differential resistance of the interphase R_3 falls, and
the parallel conductivity of the solution can no longer be neglected [126, 127].
The surface conductivity in the diffuse part of the ionic double layer normally
can be neglected (see pp. 54 and 55). In some cases it is necessary to consider
the possibility of electron conductivity in the layer of oriented molecules adjacent to the electrode surface [128].

b. Surface Recombination Rate. To investigate surface recombination we measure the lifetime of nonequilibrium carriers in a thin sample immersed in an electrolyte. If the sample is sufficiently thin, then recombination occurs mainly on its surface and the effective lifetime τ_e differs markedly from the bulk lifetime τ_B. There is the following relation between these values, the surface recombination rate s, and the thickness of the sample w (see, for
example, [125]):

$$\tau_e^{-1} = \tau_B^{-1} + 2s/w, \qquad (23.1)$$

which holds for a thin plate of infinite width and length.

For measurements using the fall in photoconductivity, the
electrode described in paragraph a is placed in solution and
illuminated with short flashes of light. When the illumination
is cut off, the concentration of nonequilibrium carriers n falls
with time according to the equation

$$n = n_{t=0}e^{-t/\tau_e}, \qquad (23.2)$$

where $n_{t=0}$ is the concentration at t = 0 (at the moment that the
light is cut off) and τ_e is the effective lifetime.

The measurement is reduced to determining the change
in the electrical conductivity of the sample (which is directly
proportional to n) with time after the light has been cut off. For
this purpose a small potential difference ($\approx$20 mV) is applied to
the ends of the sample and the current passing along the electrode is measured. The apparatus includes a source of pulsed
radiation (a lamp with a mechanical light interrupter or a
pulsed lamp), a wide-band amplifier, and a pulsed oscillograph. As in the measurement of surface conductivity, it
must be borne in mind that on polarization of an electrode an
additional ohmic potential drop arises in it. A correction is
introduced for this effect (with a sufficient degree of accuracy)
by maintaining a constant potential difference between the ends
of the sample (with any polarization current). This is achieved
by an appropriate change in the measuring current.

In the steady photoconductivity method, the electrical conductivity of a thin sample is measured while the sample is illuminated with light of constant intensity. The concentration of nonequilibrium carriers (and, consequently, the excess conductivity in comparison with the dark value) is proportional to the lifetime:

$$n = K\tau_e. \tag{23.3}$$

The constant K is determined by calibration, for example, by the fall in photoconductivity at a steady potential.

The effective lifetime in a germanium electrode has also been measured by a photomagnetic method [129] and by measuring the photopotential of a p–n junction on the back of a thin germanium electrode [54, 62, 88] (pp. 121-123).

c.   Photoeffect.   As a rule, the properties of a semiconductor–electrolyte contact are sensitive to illumination (with sufficient radiation energy absorbed by the material of the electrode). It is usual to measure either the photopotential (i.e., the change in the electrode potential on illumination at constant current) or the photocurrent (at constant potential). It is important that in the first case the polarization circuit should be galvanostatic and in the second, potentiostatic.

In principle the steady photoeffect may be measured with a normal polarization circuit with visual determination of the potential and current. For recording long-term changes in the photoeffect, it is convenient to use a recorder or a dc oscillograph with a sufficiently large (in comparison with the impedance of the electrode) input resistor.

Measurements of the short-term component of the photoeffect is of greatest interest. The electrode is illuminated with sufficiently brief flashes of light* by means of a pulsed

---

*See p. 111 on the lower limit of light-flash duration.

lamp or an incandescent lamp with a disk interrupter. The signal (electrode potential and in the measurement of the photocurrent, the fall in potential across a resistance connected in series with the cell in the polarization circuit) is amplified with a wide-band amplifier and displayed on the screen of a pulsed oscillograph. In these measurements it is possible to use the circuit described in the previous section.

It should be noted that the magnitude of the photoeffect depends on the light absorption coefficient of the semiconductor. It is usually maximal close to the natural absorption edge of the semiconductor lattice. An investigation of the spectral distribution of the photoeffect, which is of independent interest, may be carried out with a monochromator or using other methods of selecting monochromatic light (for example, with a set of interference filters).

d. Differential Capacitance and Charging Curves. In principle, the measurement of the differential capacitance of a semiconductor−electrolyte interphase does not differ from the measurement of the capacitance of metal electrodes. It may be carried out with an ac bridge or by a pulse method. However, we should point out some peculiarities of semiconductor electrodes. Their capacitance is usually several orders lower than the capacitance of metal electrodes (and may be, for example, $0.001\ \mu F/cm^2$). The differential resistance is often high, namely, up to $10^8\ \Omega \cdot cm^2$. Moreover, semiconductor electrodes have a considerable ohmic resistance (depending on the specific resistance and the geometry of the sample, the series resistance varied from a fraction of an ohm to hundreds or thousands of ohms). The dispersion of the capacitance of many semiconductor electrodes with frequency remains right up to extremely high frequencies (of the order of 1 MHz). It should be noted that the most valuable information on a semiconductor electrode may be obtained only with simultaneous measurements of several of its characteristics (for example, the differential capacitance and the photoeffect).

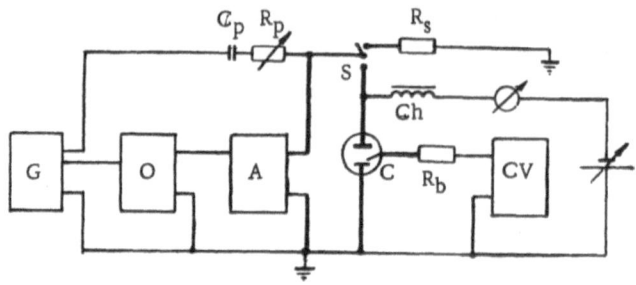

Fig. 55. Circuit of apparatus for pulsed measurements [130].
G) Oscillator; O) oscillograph; A) amplifier; C) cell; CV)
cathode voltmeter.

As many properties of semiconductor electrodes are investigated conveniently by pulse methods, we will briefly describe a pulse apparatus for measuring differential capacitance
and for plotting fast charging curves, which is also suitable for
measuring the pulsed photoeffect and the surface recombination
rate [130].

The principle of the pulse method of measuring capacitance (see, for example, [131]) is as follows. By means of a
galvanostatic circuit the electrode is charged with a square
current pulse. The change in the electrode potential during the
time of the pulse, which is normally several millivolts, is recorded oscillographically. From the time dependence of the
potential we calculate the elements of the circuit equivalent to
the electrode, namely, the differential capacitance and the
series resistance (sample and solution). The double pulse method [132] is used for measuring the parallel resistance, which
is determined by the rate of the electrochemical reaction.
Finally, if large amplitude current pulses are used (such that
a change in the electrode potential of the order of 1 V is at-
tained), then the "potential—time" curves obtained are none
other than the charging curves of the electrode with a current
of constant strength.

The source of square current pulses of constant strength
is GIS-2M voltage pulse generator with a large ballast resistor

R_p connected in series (Fig. 55). The generator will give two square pulses, whose duration, amplitude, and polarity are regulated independently. The resistor R_p, which is connected in parallel with the cell, is considerably greater than the impedance of the latter and is used to set the current. The change in the electrode potential during the pulse is amplified with a USh-2 wide-band amplifier and observed (or photographed) on the screen of an S1-8 oscillograph, which is triggered by the generator.* The current in a pulse is measured through the potential drop across a resistor R_s, which is small in comparison with R_p, and which is connected into the circuit instead of the cell with the switch S. The condenser C_p, whose capacitance is large in comparison with the capacitance of the electrode investigated, protects the cell from dc shunting by the output resistor of the generator.

The polarization circuit is connected to the cell through a high-inductance chock (Ch). The instrument for measuring the potential, which has a high output resistor, is connected through a ballast resistor R_b. The system described is simple to align and gives satisfactory parameters with a pulse of 1-μsec duration (which is equivalent to a frequency of 1 MHz).

In combination with a source of pulsed radiation, this apparatus may be used for the measurements described in the two previous sections. An analogous apparatus combined with a source of sawtooth voltage for polarization was used to obtain $C-\varphi$ curves with a fast change in polarization [44].

e. Some Experimental Problems. (1) Preparation of Ohmic Contacts. As a rule, the attachment of leads to the samples investigated does not involve difficulties in the case of metals, but is a serious problem in the electrochemistry of semiconductor electrodes. Most measurements require contacts which do not rectify an electric current, do not inject minority carriers into the bulk of the semiconductor, and

---

*The signal amplitude is large (several tens of volts) in the plotting of charging curves and the amplifier is not used.

have a moderate contact resistance. Such contacts are called ohmic contacts (since the relation between the current and the potential drop at the contact obeys Ohm's law). *

At a metal—semiconductor contact there arises an electrical double layer, which results from the difference in the work functions of the two phases (see, for example, [134]). If the electronic work function of the semiconductor is greater than that of the metal, then electrons are passed from the metal into the semiconductor until the electric field produced equals the difference in the chemical potentials of electrons in the phases separately. The region of the semiconductor adjacent to the contact is enriched in electrons (see § 9). With the reverse relation between the work functions, the layer adjacent to the contact is impoverished in electrons. If the concentration of majority carriers in the contact region is less than in the bulk of the semiconductor, then this layer has rectifying properties and is called a barrier layer. If the contact is enriched in majority carriers (antibarrier layer), then it does not rectify. Thus, to produce an ohmic contact, it is necessary to select a metal with a suitable work function in each actual case.

In practice, the sample is subjected to thermal treatment after deposition of a metal contact. Thereupon the metal diffuses into the depth of the semiconductor. If a trace of this metal has donor or acceptor properties, the concentration of free carriers in the contact layer is determined by the concentration and energy position of the impurity atoms. If the contact region has the same type, but higher conductivity than the bulk of the semiconductor (so-called n^+-n and p^+-p junctions), then there is no rectification at the contact. If the types of conductivity in the contact layer and in the bulk are different ($p-n$ junction), then the contact shows rectifying properties.

Finally, to eliminate injection of minority carriers from the contact into the volume of the semiconductor, the surface

---

*See [133] for more details on the properties of ohmic contacts.

recombination rate is increased by preliminary mechanical treatment of the surface, during which a large number of structural defects are created, and these act as recombination centers (for example, rough polishing).

Let us examine some methods of depositing contacts on the semiconductors used most. Ohmic contacts on germanium may be obtained by alloying with tin with the addition of small amounts of donor metals (for example, antimony) for n-type samples and acceptors (for example, indium) for p-type samples. The alloying is carried out in a vacuum or in a hydrogen atmosphere, but in the case of low-resistance samples, it may be replaced successfully by soldering in air.

Electrodeposition of palladium or rhodium is usually used for silicon. The following is an example of a bath for palladium plating [135]: $PdCl_2 \cdot 2H_2O$ (16 g/liter), NH_4Cl (30 g/liter), NH_4OH (30 g/liter), temperature 20°C, current density 5-10 mA/cm^2. Rhodium is electroplated from the following solution: $Rh_2(SO_4)_3$ (5 g/liter), H_2SO_4 (50 g/liter) at 50°C and a current density of 5-10 mA/cm^2. It is also possible to use chemical nickel plating with subsequent thermal treatment [136].

Contacts with gallium arsenide are usually made by alloying with tin [137] or electrodeposition of copper with subsequent heating. Indium and gallium contacts are used for cadmium sulfide.

The examples given naturally do not exhaust all the possible methods of preparing ohmic contacts. Some examples of contacts obtained by electrodeposition will be examined in Chapter V.

To check for the absence of rectification, two identical contacts are applied to a sample and the volt−ampere characteristics of the system determined (with dc or ac). From these it is possible to assess the quality of the contacts and also to determine the contact resistance. Then one of the contacts is removed.

Another method of checking is to plot the volt−ampere characteristics of the system in which one of the contacts on the sample is known to be rectifying (in the case of germanium, a tungsten point pressed against an etched surface of the sample), while the second contact is the one investigated. The characteristics are plotted with an alternating current (50 Hz) with an oscillograph. A sharp break on the oscillogram indicates a rectifying contact.

(2). Etching of Electrode Surfaces.* The surface of semiconductor electrodes has special properties in comparison with the bulk of the electrode. This is caused both by a fundamental difference in the energy state of the atoms of the semiconductor at the surface and in the crystal lattice and by the actual preliminary treatment of the sample surface.

The most common semiconductor materials − germanium, silicon, and intermetallic compounds − are extremely brittle and are normally cut and polished with abrasive materials. As a result, a layer with a disrupted crystal structure and with a large number of defects and dislocations is formed close to the surface. Chemical or electrochemical etching is used to remove the layer with a disrupted structure.

In the case of germanium, two etching agents are most commonly used. One of them is a 30% solution of hydrogen peroxide with 1-2 drops of alkali solution added. Etching is carried out at the decomposition point of the hydrogen peroxide. The polishing etching agent SR-4 consists of a mixture of concentrated nitric, hydrofluoric, and acetic acids (in proportions of $5:3:3$) with 1-2 drops of bromine added. The etching time is 1-2 min. Many different etching agents have been described in the literature. Most of them contain nitric, hydrofluoric, and acetic acids and hydrogen peroxides in various combinations with small amounts of iodine, bromine, and some metals added [138].

---

*For more details on chemical etching, see Chapter IV and on electrochemical etching, Chapter V.

Silicon is etched in a mixture of nitric and hydrofluoric acids (3:1). Gallium arsenide is etched with a mixture of nitric and hydrofluoric acids with water (3:1:2), which has a polishing effect [139], or a 10% solution of potassium hydroxide with hydrogen peroxide added (10:1) with heating [140]. Cadmium sulfide is etched with hydrochloric acid [141].

Careful washing follows etching. Since the properties of semiconductor electrodes change markedly with the influence of even very slight adsorption, an investigation must be carried out under particularly clean conditions. Specially pure reagents and doubly distilled water are used for preparing etching agents and solutions.

Brattain and Boddy [32] propose the following method for freeing water and nonoxidizing solutions from traces of heavy metals, which might be deposited on a germanium surface as a result of electrochemical replacement. Before being introduced into the cell, the solution is stirred for a long time over finely divided germanium powder of high purity. As a result, traces of metals more noble than germanium are deposited on the surface of the powder particle (see § 13).

The surface of a semiconductor obtained as a result of etching is not ideal. Adsorbed impurities, oxidation products, etc., are always present on it. Nonetheless, the properties of the surface are determined to a considerable extent by the preliminary treatment and with adequate standardization of all the processes of this treatment it is possible to obtain a surface with parameters that are reproducible within certain limits.

(3). Construction of Electrochemical Cell and Arrangement of Electrode. The electrodes usually consist of parallelopipeds or plane-parallel plates, one side of which is wetted by the electrolyte; on the other side the ohmic contact is deposited. The thickness of the electrode must be several times the diffusion length of the minority carriers in order to eliminate the effects of their injection from the contact (if it occurs) on the properties of the electrode—electrolyte boundary. Purified paraffin wax may be used for insulation of the contact and, in some

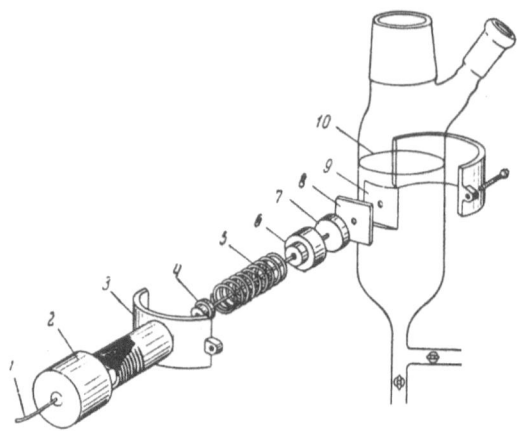

Fig. 56. Cell for electrochemical measurements on semiconductor electrodes [142]. 1) Lead; 2) nut; 3) collar; 4) Teflon gasket; 5) spring; 6) Teflon holder; 7) sample investigated; 8) Teflon gasket; 9) flat polished surface on wall of glass cell; 10) solution level.

cases, epoxide resins or polyfluoroethylene coatings are used.* Another arrangement is proposed in [142]. Into the flat wall of a glass cell is drilled a hole, against which is clamped a flat germanium sample by means of a collar and a spring (Fig. 56). A Teflon gasket may be used to prevent seepage of the solution between the glass and the germanium. An analogous method was used in [94] with a polyethylene cell.

In their other features cells for studying semiconductor electrodes, with the exception of some special cases, do not differ fundamentally from those used normally in electrochemical investigations. The design of a cell for radiochemi-

---

*The use of resins for insulation of electrodes is not permissible in precise electrochemical investigations with metal electrodes. However, in the electrochemistry of semiconductors this method is used sometimes as no effect of organic substances on the electrochemical properties of semiconductor electrodes has been observed up to now.

cal measurements on germanium was described in [115]. Some special methods for electrochemical treatment of semiconductor materials are examined in [143, 144].

Literature Cited

1. M. Green, J. Chem. Phys., 31:200 (1959).
2. B. Lovreček and J. O'M. Bockris, J. Phys. Chem., 63:1368 (1959).
3. J. I. Carasso and M. M. Faktor, Electrochemistry of Semiconductors (ed. P. J. Holmes), Academic Press, London-New York (1962), p. 205.
4. N. de. Zoubov, E. Deltombe, and M. Pourbaix, Centre Belge Étude Corrosion, Rapp. Techn., No. 27 (1955).
5. M. Pourbaix, Atlas d'Équilibres Électrochimiques a 25°C, Gauthier-Villars, Paris (1963), p. 464.
6. J. Besson and W. Kunz, Ann. Univ. Saraviensis, Sci., 7:163 (1958); Chemiker Zg., 13388 (1959).
7. J. F. Dewald, Surface Chemistry of Metals and Semiconductors (ed. H. C. Gatos), Wiley, New York (1960), p. 205.
8. V. A. Tyagai, Dissertation, Institute of Electrochemistry Akad. Nauk SSSR (1964); Elektrokhimiya, 1:387 (1965).
9. L. I. Boguslavskii, Zh. Fiz. Khim., 39:263 (1965).
10. G. Gouy, Collection: Coagulation of Colloids (ed. A. I. Rabinovich and P. S. Vasil'ev), ONTI, Moscow (1936), p. 99.
11. J. Rise, J. Phys. Chem., 30:1501 (1926).
12. C. G. B. Garrett and W. H. Brattain, Phys. Rev. 99: 376 (1955); in: Problems in the Physics of Semiconductors, IL, Moscow (1957), p. 345.
13. Physics of Semiconductors Surfaces (ed. G. E. Pikus) [Russian translation], IL, Moscow (1959), p. 359.
14. R. H. Kingston, and S. F. Neustadter, J. Appl. Phys., 26:718 (1955).
15. C. E. Young, J. Appl. Phys., 32:329 (1961); D. R. Frankl, J. Appl. Phys., 31:1752 (1960).
16. V. G. Litovchenko, Ukr. Fiz. Zh., 7:630 (1962).

17. B. B. Damaskin, Usp. Khim., 30:220 (1961).
18. R. Seiwatz and M. Green, J. Appl. Phys., 29:1034 (1958).
19. M. Green, New Problems in Contemporary Electro-chemistry (ed. J. Bockris) [Russian translation]. IL, Moscow (1962), p. 377.
20. J. McDougall, and E. C. Stoner, Phil. Trans. Roy. Soc. London, A237:67 (1938).
21. J. F. Dewald, Ann. N. Y. Acad. Sci., 101:872 (1963).
22. J. R. Schrieffer, Phys. Rev., 97:641 (1955).
23. N. B. Grover, Y. Goldstein, A. Many, and R. F. Green, J. Appl. Phys., 32:2538, 2540 (1961).
24. V. A. Myamlin, Zh. Fiz. Khim., 35:2166 (1961).
25. J. R. McDonald, Solid-State Electronics, 5:11 (1962).
26. V. A. Myamlin and Yu. V. Pleskov, Usp. Khim., 32:470 (1963).
27. A. N. Frumkin, V. S. Bagotskii, Z. A. Iofa, and B. N. Kabanov, Kinetics of Electrode Processes, Izd. Moscow State Univ., Moscow (1952).
28. M. D. Krotova, and Yu. V. Pleskov, Solid State Phys., 3:2119 (1963).
29. B. V. Ershler, Usp. Khim., 21:237 (1952).
30. M. D. Krotova and Yu. V. Pleskov, Solid State Phys., 2:411 (1962).
31. H. -U. Harten and R. Memming, Phys. Letters, 3:95 (1962).
32. W. H. Brattain and P. J. Boddy, J. Electrochem. Soc., 109:574 (1962).
33. S. M. Repinskii, Vestn. Leningr. Univ. Ser. Fiz. Khim. No. 10:103 (1962); No. 22:134 (1962).
34. S. M. Repinskii, Collection: Surface and Contact Phenomena in Semiconductors, Izd. Tomsk Univ., Tomsk (1964), p. 122.
35. Yu. V. Pleskov, V. A. Tyagai, and M. D. Krotova, Collection: Electron-Hole Transitions in Semiconductors, Izd. Akad. Nauk UzSSR, Tashkent (1962), p. 249.
36. W. H. Brattain and J. Bardeen, Bell System Techn. J., 32:1 (1953).

37. J. T. Law, J. Phys. Chem., 59:543 (1955).

38. A. M. Goodman, J. Appl. Phys., 34:329 (1962).

39. K. Bohnenkamp and H. -J. Engell, Z. Elektrochem., 61:1184 (1957).

40. Yu. V. Pleskov and V. A. Tyagai, Dokl. Akad. Nauk SSSR, 141:1135 (1961).

41. G. V. Smirnov, Yu. M. Polukarov, and V. A. Arsambekov, Collection: Surface Properties of Semiconductors, Izd. Akad. Nauk SSSR, Moscow (1962), p. 93.

42. M. Hofmann-Perez and H. Gerischer, Z. Elektrochem., 65:771 (1961).

43. W. H. Brattain, Surface Chemistry of Metals and Semiconductors (ed. H. C. Gatos), Wiley, New York (1960), p. 9.

44. P. J. Boddy and W. J. Sundburg, J. Electrochem. Soc., 110:1170 (1963).

45. H. -J. Engell and K. Bohnenkamp, Surface Chemistry of Metals and Semiconductors (ed. H. C. Gatos), New York, Wiley (1960), p. 225.

46. P. T. Wrotenbergy, and A. W. Nolle, J. Electrochem. Soc., 109:534 (1962).

47. W. H. Brattain and P. J. Boddy, Proc. Nat. Acad. Sci., 48:2005 (1962); P. J. Boddy, J. Electrochem. Soc., 111:1136 (1964).

48. P. J. Boddy and W. H. Brattain, J. Electrochem. Soc., 110:570 (1963).

49. D. R. Turner, J. Electrochem. Soc., 103:252 (1956).

50. J. Bardeleben, Z. Physik. Chem., Frankfurt, 17:39 (1958).

51. W. W. Harvey, S. Sheff, and H. C. Gatos, J. Electrochem. Soc., 107:560 (1960).

52. E. A. Efimov and I. G. Erusalimchik, Dokl. Akad. Nauk SSSR, 134:1387 (1960).

53. E. A. Efimov and I. G. Erusalimchik, Zh. Fiz. Khim., 36:98 (1962).

54. Yu. V. Pleskov, Elektrokhimiya, 1:4 (1965).

55. E. A. Efimov and I. G. Erusalimchik, Zh. Fiz. Khim., 33:441 (1959).

56.   H. Kruyt, Colloid Science, Vol. 1, Van Nostrand, New York (1949).

57.   W. Eriksen and R. Caines, J. Phys. Chem. Solids, 14:87 (1960).

58.   M. J. Sparnaay, Rec. Trav. Chim., 79:950 (1960).

59.   C. G. B. Garrett, Electrochemistry of Semiconductors (ed. P. J. Holmes), Academic Press, London–New York (1962), p. 141; M. J. Sparnaay, Surface Sci., 1:213 (1964).

60.   E. A. Efimov, I. G. Erusalimchik, and G. P. Sokolova, Zh. Fiz. Khim., 36:1219 (1962); V. I. Strikha and S. S. Kil'chitskaya, Pribory i Tekh. Eksperim., No. 3: 177 (1964).

61.   H. -U. Harten, Z. Naturforsch., 16a:1401 (1961).

62.   H. -U. Harten, Z. Naturforsch., 16a:459 (1961).

63.   R. M. Hurd and P. T. Wrotenbery, Ann. N. Y. Acad. Sci., 101:876 (1963).

64.   M. Seipt, Z. Naturforsch., 14a:926 (1959).

65.   J. F. Dewald, Bell System. Techn. J., 39:615 (1960).

66.   J. F. Dewald, J. Phys. Chem. Solids, 14:155 (1960).

67.   J. F. Dewald, J. Phys. Chem. Solids, 17:344 (1961).

68.   V. A. Tyagai, Izv. Akad. Nauk SSSR, Ser. Khim., 1964:34.

69.   V. A. Tyagai, Elektrokhimiya, 1:377 (1965).

70.   T. P. Birintseva and Yu. V. Pleskov, Izv. Akad. Nauk SSSR, Ser. Khim., 1965:251.

71.   E. O. Johnson, Phys. Rev., 111:153 (1958).

72.   E. O. Johnson, RCA Rev., 18:525, 556 (1957).

73.   W. van Roosbroeck, Phys. Rev., 91:282 (1953).

74.   R. Williams, J. Chem. Phys., 32:1505 (1960).

75.   V. A. Tyagai, Solid State Physics, 6:1602 (1964).

76.   W. W. Harvey and H. C. Gatos, J. Appl. Phys., 29:1267 (1958).

77.   P. J. Boddy and W. H. Brattain, Ann. N. Y. Acad. Sci., 101:683 (1963).

78.   E. A. Efimov, I. G. Erusalimchik, and E. I. Gorgoraki, Zh. Fiz. Khim., 38:1271 (1964).

79. Yu. V. Pleskov and M. D. Krotova, Report of International Conference on the Physics of Semiconductors, Exeter (1962), p. 807; Collection: Surface and Contact Phenomena in Semiconductors, Izd. Tomsk. Univ. (1964), p. 110.

80. H. -U. Harten, Proc. Inst. Electr. Engrs., Suppl. No. 17, 106B:906 (1959).

81. R. S. Muller, J. Electrochem. Soc., 109:1195 (1962).

82. A. Stevenson and R. Keyes, Physica, 20:1041 (1954).

83. A. Many, E. Harnik, and Y. Margoninski, Semiconductor Surface Physics (ed. R. Kingston), Philadelphia (1957), p. 85.

84. R. Memming, Surface Sci., 1:88 (1964); 3:95, 97 (1965).

85. V. A. Tyagai and Yu. V. Pleskov, Fiz. Tverd. Tela, 4:343 (1962).

86. B. Schultz, Physica, 20:1031 (1954).

87. W. W. Harvey, J. Phys. Chem. Solids, 14:82 (1960).

88. H. -U. Harten, J. Phys. Chem. Solids, 14:220 (1960).

89. H. -U. Harten and D. Polder, Philips Res. Repts., 17:125 (1962).

90. H. -U. Harten and W. Schultz, Z. Phys., 141:319 (1955).

91. H. Gobrecht and O. Meinhardt, Phys. Letters, 11:103 (1964).

92. H. Gobrecht, O. Meinhardt, and M. Lerche, Ber. Bunsenges Physik Chem., 67:486 (1963).

93. M. Green, V. Endrašić, and J. McBreen, J. Phys. Chem. Solids, 24:701 (1963).

94. R. M. Lazorenko-Manevich, Dissertation, L. Ya. Karpov Physicochemical Inst., Moscow (1963).

95. S. G. Ellis, J. Appl. Phys., 28:1262 (1957).

96. E. A. Efimov, I. G. Erusalimchik, and E. I. Gorgoraki, Zh. Fiz. Khim., 36:1578 (1962).

97. Yu. V. Pleskov, Dokl. Akad. Nauk SSSR, 126:111 (1959).

98. H. Gerischer, Anal. Real. Soc. Espan. Fis. Quim. (Madrid), B56:535 (1960).

99. N. Holonyak and H. Letaw, J. Appl. Phys., 26:355 (1955).

100. A. V. Rzhanov and I. I. Arkhipova, Fiz. Tverd. Tela, 4:1274 (1962).

101. S. M. Ryvkin, Photoelectric Phenomena in Semiconductors, Fizmatgiz, Moscow (1963).

102. P. J. Boddy and W. H. Brattain, J. Electrochem. Soc., 109:812 (1962).

103. W. H. Brattain and P. J. Boddy, Report of International Conference on the Physics of Semiconductors, Exeter, England (1962), p. 797.

104. R. Memming, Phys. Letters, 7:89 (1963); Surface Sci., 2:436 (1964).

105. M. J. Sparnaay, Surface Sci., 1:102 (1964).

106. V. S. Sotnikov and A. S. Belanovskii, Zh. Fiz. Khim., 35:509 (1961).

107. N. A. Balashova, V. V. Eletskii, and V. V. Medyntsev, Elektrokhimiya, 1:274 (1965).

108. V. V. Eletskii and Yu. V. Pleskov, Elektrokhimiya, 1:194 (1965).

109. D. R. Frankl, J. Electrochem. Soc., 109:238 (1962).

110. H. Gobrecht, O. Meinhardt, and B. Reinicke, Ber Bunsenges. Physik. Chem., 67:493 (1963).

111. H. Gobrecht and O. Meinhardt, Ber. Bunsenges. Physik. Chem., 67:151 (1963).

112. H. Gerischer and W. Mehl, Z. Ekektrochem., 59:1049 (1955).

113. V. I. Fistul' and D. G. Andrianov, Dokl. Akad. Nauk SSSR, 130:374 (1959).

114. G. B. Larrabee, J. Electrochem. Soc., 108:1130 (1961).

115. W. W. Harvey, W. J. LaFleur, and H. C. Gatos, J. Electrochem. Soc., 109:155 (1962).

116. V. S. Sotnikov and A. S. Belanovskii, Zh. Fiz. Khim., 34:2110 (1960); Dokl. Akad. Nauk SSSR, 137:1162 (1961).

117. D. Keiler, Solid-State Electronics, 6:605 (1963).

118. V. S. Sotnikov, A. S. Belanovskii, and F. B. Nikishova, Radiokhimiya, 4:725 (1962).

119. I. M. Kuleshov and A. F. Naumova, Zh. Fiz. Khim., 32:62 (1958).

120. S. P. Wolsky and P. M. Rodriguez, J. Electrochem. Soc., 103:606 (1956).

121. P. J. Holmes and R. C. Newman, Proc. Inst. Electr. Engrs., B106 (Suppl. 15):287 (1959).

122. S. R. Morrison, J. Phys. Chem. Solids, 14:214 (1960).

123. R. Smith, Semiconductors, Cambridge Univ. Press, New York (1959).

124. G. Bush and U. Winkler, Determination of the Characteristic Parameters of Semiconductors [Russian translation], IL, Moscow (1959).

125. Special Laboratory Manual on Semiconductors and Semiconductor Instruments (ed. K. V. Shalimov), Gosenergoizdat, Moscow (1962).

126. W. W. Harvey, J. Electrochem. Soc., 109:638 (1962).

127. W. W. Harvey, Ann. N. Y. Acad. Sci., 101:904 (1963).

128. W. T. Eriksen, J. Appl. Phys., 29:730 (1958).

129. P. P. Konorov and O. V. Ramanov, Fiz. Tverd. Tela, 4:1655 (1962).

130. V. A. Tyagai and Yu. V. Pleskov, Zh. Fiz. Khim., 38:2111 (1964).

131. I. S. Riney, G. M. Schmid, and N. Hackerman, Rev. Scient. Instrum., 32:588 (1961).

132. H. Gerischer and M. Krause, Z. Phys. Chem., Frankfurt, 10:264 (1957).

133. F. Stöckmann, Halbleiterprobleme (ed. F. Sauter), Vol. 6, Braunschweig (1960), p. 279.

134. G. E. Pikus, Collection: Semiconductors in Science and Technology, Vol. 1, Izd. Akad. Nauk SSSR, Moscow-Leningrad, (1957), p. 148.

135. E. A. Efimov and I. G. Erusalimchik, Electrochemistry of Germanium and Silicon, Goskhimizdat, Moscow, (1963), p. 177.

136. M. V. Sullivan, and J. H. Eigler, J. Electrochem. Soc., 104:226 (1957).

137. W. M. Sharpless, Bell System Techn. J., 38:259 (1959).

138. P. J. Holmes, Electrochemistry of Semiconductors, Academic Press, London-New York (1962), p. 329.

139. J. L. Richards and A. J. Crocker, J. Appl. Phys., 31:611 (1960).
140. D. N. Nasledov, A. Ya. Patrakova, and B. V. Tsarenkov, Zh. Tekh. Fiz., 28:779 (1958).
141. J. Nichimura, J. Phys. Soc. Japan, 15:732 (1960).
142. S. Sheff, H. C. Gatos, and S. Zwedling, Rev. Scient. Instrum., 29:531 (1958).
143. E. A. Benjamin and E. F. Duffek, Rev. Scient. Instrum., 35:237 (1964).
144. J. G. Gibson, J. Scient. Instrum., 41:395 (1964).

Additional Literature

Boddy, P. J., The structure of the semiconductor— electrolyte interface, J. Electroanal. Chem., 10:199 (1965).

Boddy, P. J.,and W. H. Brattain, The surface conductance of germanium in aqueous electrolytes, Surface Sci., 3:348 (1965).

Boddy, P. J., and W. H. Brattain, Residual surface recombination on germanium anodes, J. Electrochem. Soc., 112: 1053 (1965).

Eletskii, V. V.,and Yu. V. Pleskov, Effect of adsorption on the surface properties of germanium in an electrolyte solution, Elektrokhimiya, 1:1485 (1965).

Galinker, É. V.,and E. N. Paleolog, Surface recombination at a germanium— electrolyte interphase, Elektrokhimiya, 1:1311 (1965).

Gobrecht, H., J. Bender, R. Blaser, F. Hein, M. Schaldach, and H. -G. Wagemann, Infrarot emission an der pasengrenze germanium— elektrolyt, Physics Letters, 16:232 (1965).

Hall, R.,and J. P. White, Surface capacitance of oxide-coated semiconductors, Solid State Electronics, 8:211 (1965).

Haneman, D., Reaction zone in the germanium—aqueous electrolyte system, Proceedings of the First Australian Conference on Electrochemistry (ed. J. A. Friend and F. Gutmann), Pergamon Press (1965), p. 35.

Krischer, C. C.,and R. A. Osteryound, The effect of cathodic prepolarization on capacity curves of germanium electrodes, J. Electrochem. Soc., 112:938 (1965).

Krotova, M. D.,and Yu. V. Pleskov, On the measurements of the surface conductance of germanium in aqueous solutions, Surface Sci., 3:500 (1965).

Krotova, M. D and Yu. V. Pleskov, Investigation of a germanium – methylformanide interphase by the differential capacitance method, Elektrokhimiya, 2(2) (1966).

Krotova, M. D.,and Yu. V. Pleskov, Investigation of a germanium – methylformamide interphase by the "fast" charging curve method, Elektrokhimiya, 2 (in press).

Marcus, P. M., Calculation of the capacitance of a semiconductor surface with application to silicon, IBM J., 8:496 (1964).

Nemtsov, V. D., Photoelectrochemical investigation of the behavior of p-type silicon in chloride solutions, Zh. Fiz. Khim., 39:2617 (1965).

Nikol'skii, B. P.,M. S. Zakhar'evskii, and T. I. L'vova, The Stannate Semiconductor Electrode and Its Application in Redox Measurements. Contemporary Analysis Methods, Izd. Nauka, Moscow (1965), p. 209.

Ritter, J. S., M. N. Robinson, B. J. Faraday, and J. I. Hoover, Room-temperature oxidation of silicon during and after etching, J. Phys. Chem. Solids, 26:721 (1965).

Romanov, O. V.,and P. P. Konorov, The effect of the field and surface states on the interphase of germanium and electrolytes, Fiz. Tverd. Tela, 8:13 (1966).

Ruetschi, P.,and R. F. Amlie, The electrode potential of the semiconductor CdS in solutions of copper ions and sulphide ions, J. Electrochem. Soc., 112:665 (1965).

Sotnikov, V. S.,and A. S. Belanovskii, Adsorption of some metal ions from electrolytes during etching of germanium, silicon, and quartz., Dokl. Akad. Nauk SSSR, 162:1105 (1965); Izv. Akad. Nauk LatSSR, ser. khim., 443, 449, 659 (1964); 67 (1965); Conprecipitation and Adsorption of Radioactive Elements, Izd. Nauka, Moscow (1965), pp. 149, 154.

Toshima, S., and I. Uchida, Studies of the surface states on a germanium electrode–aqueous electrolyte interface, Denki Kagaku, 32:903 (1964); J. Electrochem. Soc. Japan, 32:247 (1964).

Chapter II

Kinetics of Electrode Reactions

§ 24. Kinetics with a Slow Electrochemical Stage (General Equations)

One of the main peculiarities of a semiconductor electrode is the fact that the rate of an electrochemical reaction depends substantially on the electron structure of the semiconductor. In particular, this appears in the fact that electrons of the conductivity band and valence electrons may participate in the electrochemical reaction. Let us examine the electrode oxidation−reduction reaction

$$A^+ + e^- \rightleftarrows A \qquad (24.1)$$

for the case where conductivity electrons participate in it. Let us denote the currents of the oxidation and reduction reactions by i_n^c and i_n^a, respectively. In analogy with metal electrodes (see, for example [1]) it is possible to write that the current i_n^c is proportional to the number of the electrons at the contact n_s and to the number of ions in the electrolyte c_{A^+} and also to the probability of the transfer of an electron from the conductivity band to an ion [2-4]:

$$i_n^c = k_1 c_{A^+} n_s e^{\frac{\alpha F \varphi_0}{RT}}, \qquad (24.2)$$

where k_1 is the reaction rate constant, α is the transfer coefficient, φ_0 is the potential drop in the Helmholtz layer,* and F is a Faraday.

*See the footnote on p. 32 on the selection of the sign of the potential. Since the

159

Expression (24.2) differs from the kinetic equation for a reaction on a metal electrode in the factor n_S. In the case of metals the concentration of free electrons at the surface is large and independent of the current and, therefore, it is included in the reaction rate constant k_1. By means of analogous considerations it is possible to write an expression for the oxidation current i_n^a:

$$i_n^a = k_2 e^{-\frac{\beta F \varphi_0}{RT}},$$
(24.3)

where k_2 and β are the rate constant and the transfer coefficient of the reverse reaction.

Thus, the total electron current passing through the conductivity band is given by

$$i_n = k_1 c_{A+} n_s e^{\frac{\alpha F \varphi_0}{RT}} - k_2 e^{-\frac{\beta F \varphi_0}{RT}}.$$
(24.4)

Relation (24.4) was obtained on the assumption that the exchange of charges occurs between ions of the solution and the semiconductor surface. This assumption apparently becomes inaccurate with a high potential drop in the space charge region when the concentration of free electrons close to the surface changes sharply at distances of the order of atomic dimensions.* Under these conditions there is the possibility of electrons tunneling into solution at some distance from the surface and the magnitude of the current is determined by the integral of the product of the concentration of free electrons and the probability

---

x axis is directed into the semiconductor, the positive direction of the current corresponds to the transfer of positive charges from the solution to the electrode (cathode current).

* The question of the applicability of the usual concepts of energy bands for the case of a strong field at the surface of a semiconductor is discussed in [5]. It should be remembered that with high electrode charges the thickness of the space charge region may become comparable with the wavelength of an electron in a semiconductor (10^{-6} cm) and with the mean distance between charges on the surface.

of this tunneling (each of these values is a function of the coordinate).

It is assumed below that the electron concentration and the field at the contact and, consequently, the current density, have the same values at all points on the surface. In the development of a stricter theory it is necessary to take into account the discreteness of the charges both in the ionic part of the electrical double layer* and in the space charge region in the semiconductor. Thus, for example, because of the discreteness of the charges of the surface states (in particular, adsorbed atoms) and ionized impurities in the crystal the space charge density does not remain constant over the whole of the interphase.

Let us examine the relation of the current to the applied voltage. As was shown in § 9, in the simplest case the potential drop in the diffuse part of the double layer in the electrolyte may be neglected. Therefore, c_{A^+} may be regarded as independent of the externally applied potential. It is then convenient to introduce the exchange current of free electrons i_n^0 through the relation

$$i_n^0 = k_1 c_{A^+} n_s^0 e^{\frac{\alpha F \varphi_0^0}{RT}} = k_2 e^{-\frac{\beta F \varphi_0^0}{RT}},$$

where n_s^0 is the concentration of free electrons at the surface at equilibrium and φ_0^0 is the potential drop in the Helmholtz layer at equilibrium.

By defining the value η by the relation

$$\eta = \varphi_0^0 - \varphi_0,$$

we may rewrite expression (24.4) in the form

_____

*This is also a problem in the electrochemistry of metal electrodes.

$$i_n = i_n^0\left(\frac{n_s}{n_s^0}e^{-\frac{\alpha F\eta}{RT}} - e^{\frac{\beta F\eta}{RT}}\right).$$

(24.5)

Reaction (24.1) may also proceed with the participation of electrons from the valence band. In this case the expressions for the reduction and oxidation current have the form

$$i_p^c = k_1' c_{A^+} e^{\frac{\alpha F\varphi_0}{RT}}; \quad i_p^a = k_2' p_s e^{-\frac{\beta F\varphi_0}{RT}}.$$

(24.6)

In writing equations (24.6) we take into account the fact that the number of electrons in the valence band is very great and hardly changes when an external field is applied. Therefore, their concentration is included in the coefficient k_1'. On oxidation of the ion A^+ an electron passes into the valence band. It is assumed that the current is proportional to the number of free spaces in this band. This is taken into account in the second equation of (24.6) by the introduction of the concentration of holes at the contact p_s.

Thus, the total current of electrons passing through the valence band is given by

$$i_p = k_1' c_{A^+} e^{\frac{\alpha F\varphi_0}{RT}} - k_2' p_s e^{-\frac{\beta F\varphi_0}{RT}}.$$

(24.7)

Expression (24.7) is conveniently written by introducing the exchange current i_p^0 for the reaction which proceeds through the valence band:

$$i_p^0 = k_1 c_{A^+} e^{\frac{\alpha F\varphi_0^0}{RT}} = k_2' p_s^0 e^{-\frac{\beta F\varphi_0^0}{RT}}.$$

(24.8)

Then the expression for the total current through the valence band assumes the form

$$i_p = i_p^0\left(e^{-\frac{\alpha F\eta}{RT}} - \frac{p_s}{p_s^0}e^{\frac{\beta F\eta}{RT}}\right),$$

(24.9)

where p_s and p_s^0 are the concentrations of holes at the surface during the passage of a current and in equilibrium, respectively.

The total current through the interphase is given by $i = i_p + i_n$. By using (24.5) and (24.9) we obtain

$$i = i_n^0 \left(\frac{n_s}{n_s^0} e^{-\frac{\alpha F \eta}{RT}} - e^{\frac{\beta F \eta}{RT}} \right) + i_p^0 \left(e^{-\frac{\alpha F \eta}{RT}} - \frac{p_s}{p_s^0} e^{\frac{\beta F \eta}{RT}} \right).$$

(24.10)

It should be emphasized that the current is determined not only by the potential drop in the Helmholtz layer, as occurs in the electrochemistry of metals, but also by the potential drop in the space charge region in the semiconductor. In actual fact, the relation of the current to the potential drop in the Debye region of the semiconductor is contained in the concentrations n_s and p_s. Let us express in a clear form the relation between the total current, the potential drop in the Helmholtz layer, and the potential drop in the semiconductor, assuming that the concentrations of electrons and holes follow a Boltzmann distribution. By using equations (9.3) and substituting formula (24.10) in them, we find

$$i = i_n^0 \left[e^{\frac{F}{RT} (-\alpha \eta + \varphi_1 - \varphi_1^0)} - e^{\frac{\beta F \eta}{RT}} \right] + i_p^0 \left[e^{-\frac{\alpha F \eta}{RT}} - e^{\frac{F}{RT} (\beta \eta - \varphi_1 + \varphi_1^0)} \right],$$

(24.11)

where φ_1^0 is the potential drop in the space charge region in the semiconductor at equilibrium.

Let us examine the case where the current i_p^0 is so small in comparison with i_n^0 that it is possible to neglect reactions involving the valence band. In this case, the total current in accordance with relation (24.11) is determined by the expression

$$i = i_n^0 \left[e^{\frac{F}{RT} (-\alpha \eta + \varphi_1 - \varphi_1^0)} - e^{\frac{\beta F \eta}{RT}} \right].$$

(24.12)

The form of the polarization curve is discussed in § 26.

It was pointed out in § 9 that, depending on the polarization conditions, two limiting cases may exist. With a large number of surface levels, the whole potential drop occurs in the Helm-

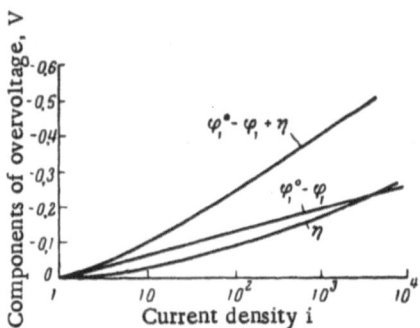

Fig. 57. Theoretical curves of the relation
of $\varphi_1^0 - \varphi_1$, η and the total overvoltage
$\varphi_1^0 - \varphi_1 + \eta$ to the current density (in units
of the current density in the bulk i^0) [6].

holtz layer. In this case, $\varphi_1 \approx \varphi_1^0$, and for the current we obtain the expression

$$i = i_n^0 \left(e^{-\frac{\alpha F \eta}{RT}} - e^{\frac{\beta F \eta}{RT}} \right),$$

(24.13)

which is also obtained in the electrochemistry of metals [1].

With high galvanic potentials the fall in the overvoltage in
the Helmholtz layer is also high. A numerical calculation for
the relation between the current and the components of the overvoltage was carried out by Green [6] and is illustrated in Fig.57.
With low overvoltages the $\varphi - \ln i$ curve has a constant slope and
at high overvoltages, $\partial \varphi / \partial \ln i \approx 5RT/2F$.

In the other limiting case the whole potential drop occurs
in the Debye region in the semiconductor. Assuming that $\eta = 0$,
we obtain from formula (24.12)

$$i = i_n^0 [e^{\frac{F}{RT}(\varphi_1 - \varphi_1^0)} - 1].$$

(24.14)

On comparing expression (24.14) with the kinetic equation
for a reaction on a metal electrode [for example (24.13)], it is
obvious that the transport coefficients α and β on a semiconductor electrode formally equal 1 and 0, respectively.

By differentiating equation (24.14) with respect to potential we find that when $\varphi_1 - \varphi_1^0 \gg RT/F$,

$$\frac{\partial \ln i}{\partial \varphi} = \frac{F}{RT}.$$

(24.15)

Thus, on semiconductor electrodes, the value of $\partial \ln i/\partial \varphi$ is approximately the same as on metals, though the derivation of this coefficient has a different character (see also [7]).

It should be noted that the assumption that a Boltzmann distribution applies is valid only with small currents (see § 7).

The limits of applicability of this assumption should be established in each separate case. For example, let the surface of n-type germanium in contact with a solution be enriched in holes (inversion layer) with a potential drop in the space charge region φ_1 of 0.1 V. Then with a thickness of the space charge region $L_1 = 10^{-5}$ cm the field at the contact is given by $E \approx \varphi_1/L_1 = 10^4$ V/cm. With a specific resistance of germanium of $\approx 1 \ \Omega \cdot$ · cm, the concentration of holes in the bulk of the semiconductor $p^0 = 3 \cdot 10^{11}$ cm^{-3}. The drift current of holes in the space charge region is given by

$$eU_p E p^0 \approx 1 \ \text{A/cm}^2.$$

As follows from § 7, as long as the current of the electrochemical reaction remains much less than this value there is a Boltzmann distribution of holes.

The Boltzmann distribution may not apply to carriers whose concentration at the surface is considerably less than in the bulk of the semiconductor. Then the term $eU_n nE$ will not be great enough for the inequality (7.15) to hold. In § 26 we examine deviations from the equilibrium Boltzmann distribution.*

---

*The effect of the current on the concentration of free carriers is considered in the calculation of the volt-ampere characteristics of a semiconductor—metal contact in the diffusion theory of rectification.

§ 25.   Kinetics with a Slow Electrochemical Stage (Quantum Mechanical Theory)

In a number of theoretical papers attempts were made to use quantum mechanical methods to obtain the basic equations of the theory of slow discharge for oxidation-reduction reactions on metal and semiconductor electrodes [8-14]. We present in a qualitative form the theory developed by Dogonadze, Chizmadzhev, and Kuznetsov [13,14] as applied to a semiconductor electrode.

Electron transitions from an ion to a semiconductor and back are examined with the following assumptions.

1. The ions are discharged independently of each other.

2. The solvent is a continuous medium which is characterized by the specific polarization.

3. The vibrations of the solvent molecules are considered using a harmonic approximation.

4. For simplicity the solvent is characterized by one frequency of the optical vibrations ω_e (libration vibrations of water molecules).

5. Changes in the first solvation envelope of the ion are ignored.

6. It is assumed that there is no formation or rupture of any chemical bonds during the reaction and, in particular, processes associated with the adsorption of the reagents on the electrode are ignored.

7. The calculations were carried out for an adiabatic approximation, i.e., two subsystems are examined, namely, a slow subsystem — the solvent — and a fast subsystem — an electron. The fast subsystem interacts strongly with the solvent and follows it adiabatically. This means that the energy of an electron depends on the coordinate of the solvent.

In accordance with the assumption made, the expression for the potential energy of the solvent has the form

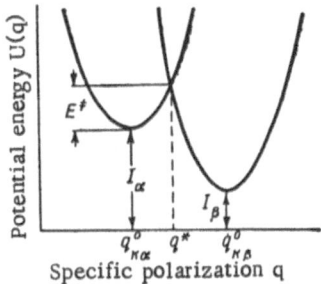

Fig. 58. Energy terms for the system ion A^2 + electron (in electrode) $\rightleftharpoons$ ion A^+ (+ electrode).

$$U = \frac{1}{2} \hbar\omega_e \sum_k (q_{k\alpha} - q_{k\alpha}^0)^2 + I_\alpha,$$

where $q_{k\alpha}^0$ is the equilibrium coordinate of the solvent in the state "α"; $q_{k\alpha} - q_{k\alpha}^0$ is the deviation of the coordinate of the solvent from the equilibrium state, and I_α is the equilibrium energy of the system in the state "α."

Figure 58 gives the potential energy of the system in relation to the coordinate in two states of the system. (For simplicity, the potential energy is represented as a function of only one coordinate). The state "β" corresponds to the electron being on the ion in solution. In the state of the system "α" the electron is inside the electrode. The transition of the electron from one state to the other in the high-temperature approximation is achieved at the point of intersection of the electronic terms (Franck–Condon principle). At this point the coordinate of the solvent equals q* (see Fig. 58). In [13, 14] it was shown that the probability $W_{\alpha\beta}$ of a transition from the state "α" to the state "β" is given by

$$W_{\alpha\beta} = Ae^{-\frac{E^{\neq}}{kT}}, \tag{25.1}$$

where the value $E^{\neq}$ is the activation energy.

Calculation gives the following expression for $E^{\neq}$:

$$E^{\neq} = \frac{(I_\beta - I_\alpha + E_n)^2}{4E_n}, \tag{25.2}$$

where I_β is the equilibrium energy of the system in the state "β," and E_n is the energy of overpolarization of the solvent, which is given by

$$E_n = \frac{1}{2} \hbar\omega_e \sum_k (q_{k\alpha}^0 - q_{k\beta}^0)^2. \tag{25.3}$$

The preexponent A may be calculated exactly:

$$A = |L|^2 \left(\frac{\pi}{h^2 kTE_n} \right)^{1/2},$$

where L is the matrix element of the transition of the electron from state "α" to state "β" (exchange integral).

Let us now turn to the calculation of the electric currents corresponding to a reaction of the type

$$A^{2+} + e^- \rightleftarrows A^+.$$

Let some potential difference be applied between the semiconductor and the electrolyte. Then the equilibrium energies I_β and I_α may be written in the form

$$I_\beta = -E^* + e\varphi(x) - a_1, \tag{25.4}$$

$$I_\alpha = E_1 - U_0 - e\varphi_B + 2e\varphi(x) - a_2, \tag{25.5}$$

where E^* is the ionization energy of the ion A^+ in vacuum, α_1 and α_2 are the solvation energies of the ions A^+ and A^{2+}, respectively, E_1 is the energy level of the electron in the semiconductor, U_0 is the depth of the potential well in the semiconductor, φ_B is the potential in the depth of the semiconductor, and $\varphi(x)$ is the potential at the point at which the discharging ion lies.

For the density of the electron current from the semiconductor into the solution we may write the expression

$$i = e \int_0^\infty c_{A^{2+}}(x)\,dx \int W_{\alpha\beta} f(E)\, \rho(E)\, dE, \tag{25.6}$$

where $\rho(E)$ is the density of electron states in the semiconductor corresponding to an electron energy of E, and $f(E)$ is the Fermi distribution function.

In actual fact the electrical current is proportional to the number of electrons with an energy lying in the range from E to $E + dE$, which equals $f(E)\rho(E)dE$, and also the number of ions A^{2+} to which the electrons are transferred. In order to find the total current it is necessary to sum for all the ions (integration with respect to the coordinate x) and for all energies of the electrons in both the valence band and in the conductivity band.

To calculate the current from formula (25.6), simplifying assumptions were made. As the probability of an electronic transition W(x) falls sharply with an increase in the distance between the discharging ion and the electrode, it is convenient to introduce the concept of a reaction region with an effective thickness l_{eff} in which the discharge occurs.

In this approximation formula (25.6) may be rewritten in the form:

$$i = e\Gamma \int W_{\alpha\beta}(\delta) f(E) \rho(E) dE, \tag{25.7}$$

where $\Gamma = l_{eff} c_{A^{2+}}(\delta)$. The value δ represents the most probable distance from which electrons pass to the electrode. The value δ can evidently be identified with the thickness of the Helmholtz layer (the closest approach of ions to the electrode) and the value Γ should be regarded as the surface concentration of the ions A^{2+} in the outer plane of the Helmholtz layer. By substituting the value of the transition probability $W_{\alpha\beta}$ from equation (25.1) in formula (25.7) and assuming that the preexponential factor in equation (25.1) depends little on the electron energy E, we obtain

$$i = B \int f(E) \exp\left\{-\frac{(I_\beta - I_\alpha + E_n)^2}{4E_n kT}\right\} \rho(E) \, dE, \tag{25.8}$$

where the constant B is given by

$$B = e\Gamma |L|^2 \left(\frac{\pi}{h^2 kTE_n}\right)^{1/2}. \tag{25.9}$$

Let us first find the current of electrons from the valence band. For this purpose it is necessary to integrate equation (25.8) with respect to the energies of electrons of the valence band. In the integration it is necessary to substitute expressions (25.4) and (25.5) for I_β and I_α and to take into account the fact that the function $f(E)$ may be replaced by unity for electrons of the valence band.

Omitting the calculations we give the final result:

$$i_p = \frac{eN_v l_{\mathrm{eff}}}{h} |L|^2 \sqrt{\frac{4\pi kT}{E_n}} \sqrt{c_{A^+} \cdot c_{A^{2+}}} \times$$

$$\exp\left\{ - \frac{[E_n + E_F - E_v - e\,(\varphi_B - \varphi_s^0) + e\eta]^3}{4E_n kT} \right\},$$

(25.10)

where N_V is the density of states in the valence band [equation (3.9)].

Rough calculations show that the following relations usually hold in semiconductors:

$$E_n \gg E_F - E_v - e\,(\varphi_B - \varphi_s^0); \quad |E_n| \gg e\eta.$$

(25.11)

Taking into account the condition (25.11), the expression (25.10) may be simplified:

$$i_p = i_p^0 e^{-\frac{1}{2}\frac{e\eta}{kT}},$$

(25.12)

where the exchange current i_p^0 is given by

$$i_p^0 = \frac{eN_v l_{\mathrm{eff}}}{h} |L|^2 \sqrt{\frac{4\pi kT}{E_n}} \sqrt{c_{A^+} c_{A^{2+}}} \times$$

$$\exp\left\{ - \frac{[E_n + E_F - E_v - e\,(\varphi_B - \varphi_s^0)]^2}{4E_n kT} \right\}.$$

(25.13)

Comparison of equation (25.12) and the first term in relation (24.9) readily shows that they coincide with a transport coefficient α of $\frac{1}{2}$ (allowing for the fact that $F/RT = e/kT$). It should be emphasized that in the theory examined the transport coefficient may differ from $\frac{1}{2}$ if the inequalities (25.11) do not hold.

Integrating equation (25.8) with respect to the energies of the electrons of the conductivity band gives the following expression for the current of electrons from the conductivity band into the solution:

$$i_n = i_n^0 \frac{n_s}{n_s^0} e^{-\frac{1}{2}\frac{e\eta}{kT}}.$$

(25.14)

This expression coincides with the first term in formula (24.10) when $\alpha = \frac{1}{2}$. Transitions of electrons from ions into the semiconductor are regarded analogously. This gives the macroscopic expression for the current written in §24.

In the developed theory it is shown that electrons with energies equal to the Fermi energy pass into solution and back with the highest probability. However, in contrast to metals, in semiconductors the Fermi level lies in the forbidden band so that only levels in the valence or free band may participate in electron exchange. The probability of a transition falls rapidly as the distance from the Fermi level increases and, therefore, electron and hole exchange currents on semiconductor electrodes are many orders less than the exchange current on a metal and fall with an increase in the width of the forbidden band.* The ratio of the electron and hole exchange currents is determined by the position of the Fermi level in the semiconductor relative to the edges of the energy bands at the surface

$$\frac{i_n^0}{i_p^0} = \exp\left\{-\left(1 + \frac{E_g}{2E_n}\right)\frac{e(\varphi_B - \varphi_s^0) + E_l - E_F}{kT}\right\}$$

(25.15)

In [15] the concepts developed are applied to the system metal—thin semiconductor film—electrolyte.

---

*It may be assumed that the decrease in the exchange current (and consequently the degree of reversibility) of electrode reactions on metal electrodes when their surface is oxidized is connected with the fact that the oxide layer formed is a semiconductor.

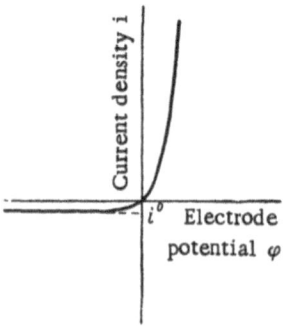

Fig. 59. Volt-ampere curve of the system semiconductor—electrolyte [according to equation (26.2)].

§ 26.   Rectification at a Semiconductor—Electrolyte Interphase

The semiconductor—electrolyte interphase has rectification properties. * To obtain a qualitative picture of the rectification effect we will use formula (24.11) and assume that the participation of minority carriers in the total current may be neglected. This may be done (in a certain potential region) if the exchange currents differ strongly, for example, $i_n^0 \gg i_p^0$, as usually occurs for semiconductors with a wide forbidden band.   Then equation (24.12) assumes the form

$$i_n = i_n^0 \left[e^{\frac{F}{RT}(-\alpha\eta + \varphi_1 - \varphi_1^0)} - e^{\frac{\beta F\eta}{RT}} \right]. \qquad (26.1)$$

[Formula (26.1) was obtained on the assumption that a Boltzmann distribution may be written for the electrons and, generally speaking, this is only true for small currents. The deviation from a Boltzmann distribution will be calculated below.] If we neglect the change in potential in the Helmholtz layer, i.e., assume that $\eta = 0$, then in this case equation (26.1) may be rewritten in the form

$$i_n = i_n^0 \left[e^{\frac{e(\varphi_1 - \varphi_1^0)}{kT}} - 1 \right]. \qquad (26.2)$$

---

* A rectification effect is also observed at a metal—electrolyte interphase, for example, when the electrochemical overvoltages of the anode and cathode reactions differ markedly or when the rate of one of these reactions is limited by concentration polarization to a much greater extent than the rate of the other. The same phenomena may be observed on semiconductor electrodes. However, here we are examining the rectifying effect which is connected with the specific characteristics of the semiconductor—electrolyte interphase.

The relation of the current to the potential obtained (Fig. 59) is reminiscent of the volt-ampere characteristics of a semiconductor p-n junction [16]. It should be noted that electrochemical rectifiers may have advantages over normal p-n junctions as the back current in electrochemical systems equals the exchange current of the electrochemical reaction i^0 and may be much less than the limiting diffusion current of minority carriers, whose magnitude determines the back current of a p-n junction.

As will be shown later, when the change in potential in the Helmholtz layer η is taken into account, the back current increases slowly with potential according to the law

$$|i_n| \sim i_n^0 \exp\left(\frac{d_0}{L_1}\sqrt{-\frac{2e\varphi_1}{kT}}\right),$$

$$(26.3)$$

where L_1 is the Debye length.

With high potential drops in the space charge region, close to the width of the forbidden band, minority carriers begin to play a part in contact phenomena and the rectifying properties of the contact are disrupted [17].

Let us turn to a detailed calculation of the volt-ampere characteristics. First we will examine a semiconductor (as a definite example, n-type) with such a wide forbidden band that the effect of minority carriers (holes) on the field adjacent to the contact may be neglected over the whole range of potentials considered. We assume that in the semiconductor there are only completely ionized donor levels. We will then consider that the layer adjacent to the surface is impoverished in electrons, so that $n_s \ll N_D$. In the bulk of the semiconductor $n^0 \approx N_D$ (from the condition of electrical neutrality). In accordance with the assumptions made, the electron concentration $n(x)$ falls on approaching the surface and, as will be evident below, this fall is quite rapid. Therefore, in the Debye region it is possible to neglect the electron concentration n in comparison with the concentration of donors N_D. We arrive at the conclusion that the space charge region (layer impoverished in electrons) is described by the following equation in accordance with equation (9.4):

$$\frac{dE}{dx} = \frac{4\pi e}{\varepsilon_1} N_D = \frac{kT}{eL_1^2} .$$

(26.4)

The potential of the semiconductor at the point x, $\varphi(x)$, is related to the electric field $E(x)$ by the equation

$$\varphi(x) = - \int\limits_{x=\infty}^{x} E\,dx = - \int\limits_{E_B}^{E(x)} E\,\frac{dx}{dE}\,dE,$$

(26.5)

where E_B is the field in the bulk of the semiconductor.

By substituting dE/dx from the relation (26.4) and integrating, we obtain a relation between the electric field $E(x)$ and the potential close to the surface*:

$$\varphi(x) = - \frac{eL_1^2 E^2(x)}{2kT} .$$

(26.6)

In the absence of surface levels the potential drop in the Helmholtz layer is given by $\varphi_0 = E_s d_0$. By using equation (26.6) we find that

$$\varphi_0 = - \frac{kT}{e}\frac{d_0}{L_1}\sqrt{-\frac{2e\varphi_1}{kT}} .$$

(26.7)

Relation (26.7) shows that the potential drop in the Helmholtz layer φ_0 is considerably less than the potential drop in the semiconductor φ_1 because of the presence of the coefficient d_0/L_1.

Let us turn to the calculation of the relation of the concentration n_s to the potential drop φ_1. For this purpose we rewrite relation (7.14) in the form

$$\frac{dn}{dx} + \frac{eEn}{kT} = \frac{i_n}{U_n kT} .$$

(26.8)

---

*In obtaining relation (26.6) we neglect the weak field E_B in comparison with the field close to the surface $E(x)$.

Equation (26.8) may be written in the following way:

$$\frac{dn}{dE}\frac{dE}{dx} + \frac{eEn}{kT} = \frac{i_n}{U_n kT}.$$

(26.9)

By using equation (26.4), we obtain

$$\frac{dn}{dE} + \frac{e^2 E n L_1^2}{(kT)^2} = \frac{e i_n L_1^2}{U_n (kT)^2}.$$

(26.10)

Equation (26.10) is a linear equation with respect to n. The integration of this equation with the condition that $n \to N_D$ when $E \to E_B$ leads to the expression

$$n_s = N_D e^{\frac{e\varphi_1}{kT}} \left(1 - \frac{\sqrt{2}\, i_n}{i_n^{\lim}} \int_0^{\sqrt{-e\varphi_1/kT}} e^{z^2}\, dz \right),$$

(26.11)

where z is the integration variable and $i_n^{\lim}$ is determined by the relation

$$i_n^{\lim} = \frac{kT U_n N_D}{L_1} = \frac{e D_n N_D}{L_1}.$$

(26.12)

The second term in brackets in (26.11) determines the deviation from the equilibrium distribution of free electrons.

By substituting expression (26.11) in formula (24.5) and solving the equation obtained for the current i_n, we find

$$i_n = i_n^0 \frac{e^{\frac{e(\varphi_1 - \varphi_1^0)}{kT}} - e^{\frac{\beta e(\varphi_0^0 - \varphi_0)}{kT}}}{1 + \sqrt{2}\, \frac{i_n^0}{i_n^{\lim}} e^{\frac{e(\varphi_1 - \varphi_1^0)}{kT}} \int_0^{\sqrt{-e\varphi_1/kT}} e^{z^2} dz}.$$

(26.13)

By using (26.7) we obtain the volt-ampere characteristics:

$$i_n = i_n^0 \frac{e^{\frac{e(\varphi_1 - \varphi_1^0)}{kT}} - e^{\frac{\beta e \varphi_0^0}{kT} + \frac{\beta d_0}{L_1} \sqrt{-2e\varphi_1/kT}}}{1 + \sqrt{2}\, \frac{i_n^0}{i_n^{\lim}} e^{\frac{e(\varphi_1 - \varphi_1^0)}{kT}} \int_0^{\sqrt{-e\varphi_1/kT}} e^{z^2} dz}.$$

(26.14)

We then find the conditions under which it is possible to neglect the integral in the denominator of formula (26.14):

$$\frac{i_n^0}{i_n^{\lim}} e^{\frac{e(\varphi_1-\varphi_1^0)}{kT}} \int_0^{\sqrt{-e\varphi_1/kT}} e^{z^2}dz < \frac{i_n^0}{i_n^{\lim}} e^{-\frac{e\varphi_1}{kT}} \sqrt{-\frac{e\varphi_1}{kT}} e^{\frac{e(\varphi_1-\varphi_1^0)}{kT}} =$$

$$= \frac{i_n^0}{i_n^{\lim}} \sqrt{-\frac{e\varphi_1}{kT}} e^{-\frac{e\varphi_1^0}{kT}} .$$

In estimating the integral we strongly overestimated its value, by taking the maximum value of the subintegral expression $e^{z^2} = e^{-e\varphi_1/kT}$. When the following relation holds:

$$\frac{i_n^0}{i_n^{\lim}} \sqrt{-\frac{e\varphi_1}{kT}} e^{-\frac{e\varphi_1^0}{kT}} \ll 1 ,$$

the expression for the current i_n has the form

$$i_n = i_n^0 \left(e^{\frac{e(\varphi_1-\varphi_1^0)}{kT}} - e^{\frac{\beta e\varphi_0^0}{kT}} \cdot e^{\frac{\beta d_0}{L_1} \sqrt{-\frac{2e\varphi_1}{kT}}} \right) .$$

In the barrier direction the relation of the current to the potential has the form

$$i_n = - i_n^0 e^{\frac{\beta e\varphi_0^0}{kT}} \cdot e^{\frac{\beta d_0}{L_1} \sqrt{-\frac{2e\varphi_1}{kT}}} .$$

(26.15)

The latter expression coincides with formula (26.3).

For later use we need to estimate the thickness of the depletion layer. This may be done in the following way:

$$L_{eff} = \int_0^{L_{eff}} dx = \int_{\varphi_s}^{\varphi_B} \frac{dx}{d\varphi} d\varphi = - \int_{\varphi_s}^{\varphi_B} \frac{d\varphi}{E} .$$

(26.16)

By using relation (26.6) and taking into account the fact that $|\varphi_s|$ is considerably greater than $|\varphi_B|$, we obtain

$$L_{\text{eff}} = \sqrt{2} L_1 \sqrt{-\frac{e\varphi_1}{kT}} \,.$$

$$(26.17)$$

We see that the thickness of the depletion layer increases in proportion to the square root of the potential drop in the space charge region.

The volt-ampere curve of a system with weak ionization of the donor levels was considered in [3]. It was shown that there is also rectification in this case. Its efficiency is characterized by the ratio $\Delta\varphi(+i)/\Delta\varphi(-i)$, where $\Delta\varphi(+i)$ and $\Delta\varphi(-i)$ are the changes in electrode potential when positive and negative currents of density i, respectively, are passed.

Calculations give for this value

$$\frac{\Delta\varphi(+i)}{\Delta\varphi(-i)} = \frac{1}{\beta} \frac{L_1^2}{d_0 l} \ln \frac{|i_n|}{i_n^0} \,,$$

$$(26.18)$$

where l is the thickness of the semiconductor sample.

When $l = 10^{-3}$ cm, $L_1 = 10^{-4}$ cm and $\ln(i_n/i_n^0) = 10$, we obtain

$$\frac{\Delta\varphi(+i)}{\Delta\varphi(-i)} = 200,$$

which corresponds to considerable rectification.

§ 27.   Kinetics with Slow Diffusion of Minority Carriers

In § § 24 and 26 we examined the relations found when the slow stage of the process is the electrochemical stage and the supply of free carriers to the electrode surface does not limit the rate of the process. However, if the reaction involves minority carriers, whose concentration in the depth of the semiconductor is low, then beginning at some potential the current density will be determined by the rate of supply of the minority carriers to the electrode surface. An example of such a reaction is the anode solution of n-type germanium. As a concrete example we will examine precisely this reaction [18, 19]. The results obtained are equally valid for other anode and cathode processes in which minority carriers participate.

As will be shown below (§§ 31 and 32), holes must be supplied to the interphase for the dissolution of germanium. Let us assume that the current of germanium dissolution is proportional to the concentration of holes at the contact:

$$i = - k p_s e^{-\frac{\beta F(\varphi_0 - \varphi_0^0)}{RT}} = - i^0 \frac{p_s}{p_s^0} e^{-\frac{\beta F(\varphi_0 - \varphi_0^0)}{RT}},$$

(27.1)

where $\varphi_0 - \varphi_0^0$ is the overvoltage in the Helmholtz region, k is the reaction rate constant, i^0 is the exchange current, and p_s^0 is the concentration of holes at the surface in equilibrium.

In an experimental investigation of this reaction it was observed that in the anodic dissolution of germanium the current through the interphase is carried by free electrons as well as holes. Therefore, in the space charge region there arise two currents, namely, a current of electrons in the conductivity band i_n and a current of holes in the valence band i_p. From a comparison of the theoretical hole current with the total current of the electrochemical reaction measured experimentally ($i_p + i_n$), it was found that the ratio

$$\frac{i_p (x = 0)}{i_n (x = 0)} = \frac{r}{m},$$

(27.2)

which is determined by the peculiarities and conditions of the electrode reaction, is independent of the potential.* Before turning to the calculation, let us examine the qualitative considerations which make it possible to understand the form of the polarization curve of anodic dissolution of germanium. As will be shown in §29, this curve is described by the formula

---

*As will be shown below, this does not always hold. However, at the present time there is hardly any information on the nature of the value r/m and as a first approximation we will assume that r/m does not change with potential.

$$\varphi = \varphi^0 + \frac{RT}{\beta_1 F} \ln \frac{|i|}{i^0} - \frac{RT}{F} \ln \left(1 - \frac{i}{i^{\lim}}\right).$$
(27.3)

Here φ^0 is the equilibrium electrode potential, β_1 is a constant, and $i^{\lim}$ is the limiting current determined by the relation

$$i^{\lim} = -\left(1 + \frac{m}{r}\right) n_i^2 D_p e^2 U_n \frac{\rho}{L_p},$$
(27.4)

where ρ is the specific resistance of germanium.

At currents far from the limiting value $(i \ll i_p^{\lim})$ it is possible to neglect the third term in formula (27.3):

$$\varphi = \varphi^0 + \frac{RT}{\beta_1 F} \ln \frac{|i|}{i^0},$$
(27.5)

i.e., Tafel's law applies. The coefficient β_1 may have various values. To explain this, we write formula (27.1) in the form

$$i = -i^0 e^{-\frac{F}{RT}\left[\left(\varphi_1 - \varphi_1^0\right) + \beta\left(\varphi_0 - \varphi_0^0\right)\right]}.$$
(27.6)

Here, $\varphi_1 - \varphi_1^0$ is the overvoltage in the space charge region in germanium. The "effective" value of the transport coefficient β_1 depends on the ratio of the components of the total overvoltage. In actual fact, from comparison of formulas (27.5) and (27.6) it follows that

$$\beta_1 = 1 - (1 - \beta)\frac{\varphi_0 - \varphi_0^0}{(\varphi_0 - \varphi_0^0) + (\varphi_1 - \varphi_1^0)}.$$

In cases where the rate of the process is limited by the supply of minority carriers, holes, to the surface there arises concentration polarization with respect to holes and at currents close to the limiting current, the current dependence of the potential is determined by the third term of formula (27.3):

$$\varphi \approx \frac{RT}{F} \ln\left(1 - \frac{i}{i^{\lim}}\right).$$
(27.7)

To elucidate the nature of the limiting current it is necessary to examine the distribution of the hole concentration in a

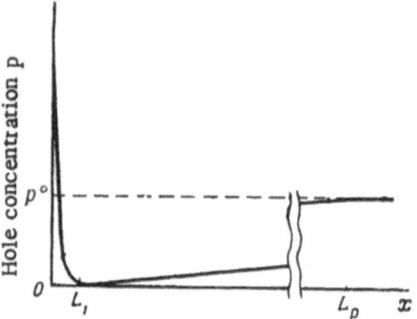

Fig. 60. Distribution of hole concentration
close to the surface of an electrode of n-type
germanium during anode solution at the
limiting current.

semiconductor electrode, at whose surface anode solution is occurring. The distribution of holes at currents close to the limiting value is illustrated in Fig. 60. As a result of the effect of
the electric field, the hole concentration is high close to the
electrode surface. Beyond the limits of the Debye region ($x >$
L_1) there is no field. Here the hole concentration is less than
equilibrium ($p < p^0$), as holes are carried away through the interphase by the electrochemical reaction. Under the conditions
of the limiting current, the concentration of holes at the boundary of the Debye and quasineutral regions falls practically to
zero. The concentration of holes increases with distance from
this boundary and at a distance of the order of diffusion length
L_p it reaches the equilibrium value p^0. It should be noted that
the same form of distribution is shown, for example, by cations
in an electrolyte during their discharge at a negatively charged
electrode surface under conditions of concentration polarization
(L_1 is equivalent to the thickness of the diffuse part of the electrical double layer and L_p is the thickness of the diffusion layer).
The magnitude of the limiting current may be estimated by means
of the following obvious considerations. Since the whole region
$L_1 < x < L_p$ lies beyond the limits of the space charge, the supply of holes to the contact occurs mainly through diffusion. The
limiting diffusion current of holes is determined by the expression

$$i_p^{\lim} = -eD_p \frac{dp}{dx}\bigg|_{\lim}, \tag{27.8}$$

and the order of magnitude of the gradient $(dp/dx)|_{\lim}$ is given by

$$\frac{dp}{dx}\bigg|_{\lim} = \frac{p^0}{L_p}. \tag{27.9}$$

By using formula (27.2) we find the total limiting current

$$i^{\lim} = -\left(1 + \frac{m}{r}\right)eD_p\frac{p^0}{L_p}. \tag{27.10}$$

Taking into account the fact that the specific resistance of the semiconductor

$$\rho = \frac{1}{en^0U_n},$$

and $n^0p^0 = n_i^2$, we immediately obtain formula (27.4). *

Until now we have not taken into account the recombination and generation of holes in the Debye region and at the surface. The generation of holes leads to an increase in the limiting current on the solution of germanium. The actual mechanism of generation was examined in [18, 19], in which it was assumed that the generation of holes occurs at surface centers. The generation current i_s is proportional to the deviation of the concentration of holes at the boundary of the quasineutral and Debye regions p_1 from the equilibrium value p^0:

$$i_s = -es(p_1 - p^0), \tag{27.11}$$

---

*Formula (27.4) is analogous to the expression for the limiting current in a p-n junction (see, for example, [16]) and differs only in the coefficient

$$\left(1 + \frac{m}{r}\right).$$

where s is the surface recombination rate.* Under conditions of the limiting current $p_1 = 0$, and we obtain

$$i_s = - esp^0.$$

(27.12)

Thus, the limiting current of holes is given by

$$i_p^{\lim} = - eD_p \frac{p^0}{L_p} - esp^0.$$

(27.13)

The diffusion length of the holes L_p is related to the lifetime of the holes by the expression $L_p = \sqrt{D_p \tau_p}$. Consequently

$$i_p^{\lim} = - ep^0 \left(\sqrt{\frac{D_p}{\tau_p}} + s \right).$$

(27.14)

Naturally, in the anode solution of p-type germanium, there is no limiting current as there is no diffusion hindrance to the supply of holes to the contact.

§ 28.   Quasineutral or Diffusion Region

In §§ 17 and 27 we pointed out that in examining processes involving the flow of current through a semiconductor electrode it is convenient to separate out two regions in the semiconductor,

---

*In using relation (27.11) it is necessary to bear in mind the fact that it is usually assumed that the hole generation current i_s is proportional to the deviation of the hole concentration from the equilibrium value $p_1 - p^0$, i.e., $i_s = f(p_1 - p^0)$, where f is some unknown function of $p_1 - p^0$. With a small deviation from equilibrium ($p_1 - p^0 \ll p^0$), the function f may be expanded into a series with respect to the small parameter $p_1 - p^0$ and for the generation current we obtain the expression

$$i_s = \frac{\partial f}{\partial p}\bigg|_{p=p^0} (p_1 - p^0) = - es(p_1 - p^0),$$

where s is taken at the equilibrium point and is independent of p_1. With large deviations from equilibrium ($p_1 - p^0 \approx p^0$) the expansion is groundless but sometimes the use of formula (27.11) is continued. In this case, s is an unknown function of p_1 and is not determined by relation (19.7), which was derived for conditions of slight deviation from equilibrium.

namely the space charge region and the quasineutral region. Here we will consider in detail phenomena occurring in the quasineutral region. Our task includes the determination of the limits of applicability of two assumptions made previously in calculations:

   1) the quasineutral region is not charged, i.e., the deviations of the concentrations of electrons and holes from the equilibrium values equal each other ($\Delta n = \Delta p$);

   2) in semiconductors with conductivity that is far from intrinsic, the current of minority carriers is determined solely by diffusion, while the drift current of minority carriers may be neglected.

   We will also calculate the relation of the concentration of minority carriers, the electric field, and the currents in the quasineutral region to the coordinate .

   In accordance with § 17, the distribution of holes in an n-type sample is described by equation (17.11), whose solution with the limiting condition $\Delta p = 0$ when $x \to \infty$ has the form

$$p(x) - p^0 = \Delta p(x) = \Delta p_1 e^{\frac{-x}{L_p}}.$$
(28.1)

Here $L_p = \sqrt{D_p \tau_p}$ is the diffusion length and Δp_1 is the change in the hole concentration at the point $x = L_1$.

   Assuming that the current i_{p_1} is known, by means of the relation $i_p = -eD_p(dp/dx)$ we find $\Delta p_1 = p_1 - p^0$:

$$\Delta p_1 = \frac{i_{p1} L_p}{eD_p}.$$
(28.2)

   The hole current is part of the total current; according to equation (27.2),

$$i_{p1} = \frac{r}{m+r} i.$$
(28.3)

By using equations (28.2) and (28.3) we can express Δp_1 in terms of the total current:

$$\Delta p_1 = \frac{L_p \, ir}{eD_p \, (m+r)} \, .$$

(28.4)

The hole current at any point x may be found by means of formula (28.1)

$$i_p(x) = \frac{r}{m+r} \, ie^{-\frac{x}{L_p}} \, .$$

(28.5)

Let us find the distribution of the electron current along the x axis. If the relation $\Delta p = \Delta n$ holds, then the diffusion current of electrons equals $eD_n(\partial \Delta p / \partial x)$. By using the symbols $D_n/D_p = U_n/U_p = b$ and the expression given above, $i_p = -eD_p \cdot (dp/dx)$, we obtain for the diffusion current of electrons the value $-bi_p$. In accordance with relation (28.5), the diffusion current of electrons equals $-b(r/r + m) \, i \exp(-x/L_p)$.

The drift current of electrons equals approximately en^0U_nE. This approximation is good if in the semiconductor $p^0 \ll n^0$, and, at the same time, we will consider those steady states where $\Delta p \ll n^0$. The total electron current, which equals the sum of the drift and diffusion currents, may be written in the form

$$i_n = en^0U_nE - \frac{r}{m+r} \, bie^{-\frac{x}{L_p}} \, .$$

(28.6)

Since the electron current equals $i_n = i - i_p$, then, by using equation (28.5), we rewrite relation (28.6) in the form

$$i \left(1 - \frac{r}{m+r} e^{-\frac{x}{L_p}}\right) = en^0U_nE - \frac{r}{m+r} \, bie^{-\frac{x}{L_p}} \, .$$

The latter relation makes it possible to determine the electric field E:

$$E = E_B \left[1 + (b-1)\frac{r}{m+r} e^{-\frac{x}{L_p}}\right],$$

(28.7)

where $E_B = i/en^0 U_n$ is the electric field when $x \to \infty$. Now it is possible to find the limits of applicability of the assumptions made previously. We assume that the drift current of holes is much less than the diffusion current, i.e., that

$$e(p^0 + \Delta p) U_p E \ll \frac{r}{m+r} \, ie^{-\frac{x}{L_p}}$$
(28.8)

For the estimation we substitute in condition (28.8) the maximum values of the field (at the point $x = L_1$) and the deviation of the concentration of holes from the equilibrium value $(\Delta p \approx p^0)$

$$\frac{2ep^0 U_p i \left[1 + (b - 1) \frac{r}{m+r} \right]}{en^0 U_n} \ll \frac{r}{m+r} \, i.$$
(28.9)

From the inequality (28.9) there follow the conditions of validity of the assumption made:

$$\frac{p^0}{n^0} \ll \frac{b \dfrac{r}{m+r}}{2 + (b-1) \dfrac{2r}{m+r}}.$$
(28.10)

For a semiconductor which is far from intrinsic (i.e., when $p^0 \ll n^0$), this inequality normally holds.

Let us now examine the accuracy with which the relation $\Delta n = \Delta p$ holds. For this we will use Poisson's equation

$$\varepsilon_1 \frac{dE}{dx} = 4\pi \rho (x),$$
(28.11)

where $\rho (x)$ is the charge density at the point x in the semiconductor.

By substituting in equation (28.11) the electric field from (28.7) and differentiating, we arrive at the conclusion that the charge density in the quasineutral region is given by

$$\rho(x) = -\frac{E_B(b-1)\varepsilon_1 r}{4\pi L_p(m+r)} e^{-\frac{x}{L_p}}.$$

(28.12)

The approximation $\Delta p = \Delta n$ is sufficiently accurate if the following condition is fulfilled:

$$|\Delta p| \gg \left|\frac{\rho}{e}\right| = \left|\frac{i(b-1)\varepsilon_1 r}{4\pi L_p e^2 n^0 U_n (m+r)}\right| e^{-\frac{x}{L_p}}.$$

(28.13)

By using for Δp the expression $\Delta p = \Delta p_1 e^{x/L_p}$ in combination with expression (28.4), we can rearrange condition (28.13) into the form

$$L_p^2 \gg \left|\frac{\varepsilon_1(b-1)kT}{4\pi e^2 n^0}\right|.$$

(28.14)

The latter is the equivalent of the relation

$$\frac{L_p^2}{L_1^2} \gg \left|\frac{b-1}{b}\right|,$$

(28.15)

which holds for a large class of semiconductors.

We should then note that the boundary between the quasineutral and Debye regions may be regarded as the point at which the diffusion current of minority carriers becomes equal to the drift current. It may be shown that the field at this point approximately equals $E(L_1) = kT/eL_p$.

§ 29.   Kinetics with Slow Diffusion of Minority Carriers (Calculation) [18, 19]

Let us turn to phenomena occurring in the space charge region. By solving equation (28.2) with respect to p_1 we find the concentration of holes at the boundary of the quasineutral and space charge regions:

$$p_1 = p^0\left(1 - \frac{i_{p1}}{i_p^{\lim}}\right),$$

(29.1)

where $i_p^{\lim} = -eD_p p^0/L_p$.

Estimates show (§24) that with enrichment of the contact in holes, Boltzmann's distribution holds for them. Then for the concentration of holes at the contact we obtain the expression

$$p_s(\varphi_1) = p_1 e^{-\frac{e\varphi_1}{kT}} = p^0 \left(1 - \frac{i_{p1}}{i_p^{\lim}}\right) e^{-\frac{e\varphi_1}{kT}}.$$

(29.2)

By rearranging equation (29.2) we obtain

$$-\frac{e\varphi_1}{kT} = \ln \frac{p_s}{p^0} - \ln\left(1 - \frac{i_{p1}}{i_p^{\lim}}\right).$$

(29.3)

By assuming that the total current is related to the hole current by relation (28.3) and taking into account relation (27.10) we can express the potential drop in the space charge layer φ_1 in terms of the total current i

$$-\frac{e\varphi_1}{kT} = \ln \frac{p_s}{p^0} - \ln\left(1 - \frac{i}{i^{\lim}}\right).$$

(29.4)

The change in the electrode potential equals the sum of the changes in the potential in the Helmholtz layer and in the space charge layer with opposite signs: $\Delta\varphi = -(\varphi_1 - \varphi_1^0) - (\varphi_0 - \varphi_0^0)$. Let us now find the potential drop in the Helmholtz layer and the electric field at the electrode surface. For this purpose we write the Poisson equation which, with expression (29.2) taken into account, has the form

$$\frac{dE}{dx} = \frac{4\pi e}{\varepsilon_1}\left[N_D + p_1 e^{-\frac{e\varphi}{kT}} - n_1 e^{\frac{e\varphi}{kT}}\right].$$

(29.5)

In the anodic dissolution of germanium it is possible to neglect the concentration of electrons in the whole Debye region in comparison with the value of N_D. By integrating equation (29.5), as was done in § 10, we obtain for the electric field E_s the expression

$$E_s = -\sqrt{\frac{8\pi e}{\varepsilon_1}\left[-N_D \varphi_1 + \frac{kT}{e}p_s\right]}.$$

(29.6)

The potential drop in the Helmholtz layer may be found by means of relation (10.14)

$$\varphi_0 = - \frac{4\pi Q_t(\varphi_1)d_0}{\varepsilon_0} + \frac{\varepsilon_1 E_s(\varphi_1)d_0}{\varepsilon_0}.$$

$$(29.7)$$

Here Q_t is the charge of the surface levels and the electric field E_s is found from formula (29.6).

By substituting the value of p_s in formula (27.1) we obtain

$$i = -i^0\left(1 - \frac{i}{i \lim}\right) e^{-\frac{F}{RT}\left(\varphi_1 - \varphi_1^0\right)} e^{-\frac{\beta F}{RT}\left(\varphi_0 - \varphi_0^0\right)}.$$

$$(29.8)$$

Equation (29.8) is a transcendental equation. It defines the relation of the current i to the potential φ_1. This equation cannot be solved in an analytical form. We will examine two limiting cases. We should note that the first exponential term in equation (29.8) is determined by the change in the potential drop in the Debye region of the semiconductor, while the second depends on the overvoltage in the Helmholtz layer.

We will first assume that the change in potential in the Helmholtz layer may be neglected. Then relation (29.8) may be written in the form

$$i = -i^0\left(1 - \frac{i}{i \lim}\right) e^{-\frac{F}{RT}\left(\varphi_1 - \varphi_1^0\right)}.$$

$$(29.9)$$

By taking logarithms we find the volt-ampere characteristic for this case:

$$\varphi \approx \text{const} + \frac{RT}{F} \ln \frac{-i}{i^0} - \frac{RT}{F} \ln\left(1 - \frac{i}{i \lim}\right).$$

$$(29.10)$$

On the other hand, if the change in the potential in the semiconductor may be neglected in comparison with the overvoltage in the Helmholtz region, then from equation (29.8) it follows that

$$i = -i^0\left(1 - \frac{i}{i \lim}\right) e^{-\frac{\beta F}{RT}\left(\varphi_0 - \varphi_0^0\right)}$$

$$(29.11)$$

and the volt-ampere characteristic has the form

$$\varphi \approx \text{const} + \frac{RT}{\beta F} \ln \frac{-i}{i^0} - \frac{RT}{F} \ln \left(1 - \frac{i}{i^{\lim}} \right). \tag{29.12}$$

On comparing relations (29.10) and (29.12) we see that at currents far from the limiting value the slope of the overvoltage curve $\partial \varphi / \partial \ln i$ is independent of the current and with a change from one limiting case to the other it changes from RT/F to $RT/\beta F$.

§ 30. Anodic Dissolution of n-Type Germanium of Finite Size

Let us examine the dissolution of a germanium electrode, whose thickness is comparable with the diffusion layer of holes or less than it [20]. In this case it is impossible to ignore processes occurring at the ohmic contact (in contrast to §29, where we examine the solution of an electrode of infinitely great thickness).

Let an ohmic contact of the semiconductor with a metal lie at a distance x = w from the germanium−electrolyte interface. We will characterize the contact by the rate of surface recombination s. In other words, we will assume that the hole current passing through the ohmic contact is determined by the relation

$$i_p(w) = -e\,s\,[p(w) - p^0], \tag{30.1}$$

where p(w) is the concentration of holes in the plane of the ohmic contact. Let us find the limiting current of anodic dissolution. For this purpose it is sufficient to examine the movement of holes in the quasineutral region. The solution of the continuity equation for holes may be written in the form (see §17)

$$p = p^0 + Ae^{-\frac{x}{L_p}} + Be^{\frac{x}{L_p}}. \tag{30.2}$$

To determine the constants A and B it is necessary to formulate two limiting conditions. The first limiting condition is provided by relation (30.1). The second is the condition

$$p(x = L_1) = 0. \tag{30.3}$$

In actual fact, with the limiting currents the hole concentration gradient must be maximal and this will be the case if condition (30.3) is fulfilled. By calculating i_p from formula (17.9) and using (30.1), we obtain the relation

$$e\frac{D_p}{L_p}\left[-Ae^{-\frac{w}{L_p}} + Be^{\frac{w}{L_p}}\right] = es[p(w) - p^0]. \tag{30.4}$$

By means of formula (30.2) we write equation (30.3) in the form

$$p^0 + A + B = 0. \tag{30.5}$$

By solving the system of equations (30.4) and (30.5) let us find A and B:

$$A = -p^0 - \frac{p^0}{2} \frac{se^{-\frac{w}{L_p}} - D_p L_p^{-1} e^{-\frac{w}{L_p}}}{D_p L_p^{-1} \operatorname{ch}\frac{w}{L_p} + s\operatorname{sh}\frac{w}{L_p}},$$

$$B = \frac{p^0}{2} \frac{se^{\frac{w}{L_p}} - D_p L_p^{-1} e^{-\frac{w}{L_p}}}{D_p L_p^{-1} \operatorname{ch}\frac{w}{L_p} + s\operatorname{sh}\frac{w}{L_p}}. \tag{30.6}$$

The limiting current is given by

$$i_p^{\lim} = -eD_p \frac{dp}{dx}\bigg|_{x=0} = -\frac{eD_p}{L_p}(-A + B). \tag{30.7}$$

By means of expression (30.6) we obtain

$$|i_p^{\lim}| = \frac{eD_p p^0}{L_p} \frac{D_p \operatorname{sh}\frac{w}{L_p} + sL_p \operatorname{ch}\frac{w}{L_p}}{D_p \operatorname{ch}\frac{w}{L_p} + sL_p \operatorname{sh}\frac{w}{L_p}}. \tag{30.8}$$

The total anodic current is the sum of the hole and electron currents. By assuming that the currents are related by relation (27.2), we find the expression for the limiting current of anode solution

$$|i^{\lim}| = \left(1 + \frac{m}{r}\right) \frac{eD_p p^0}{L_p} \frac{D_p \, \text{sh} \, \frac{w}{L_p} + sL_p \, \text{ch} \, \frac{w}{L_p}}{D_p \, \text{ch} \, \frac{w}{L_p} + sL_p \, \text{sh} \, \frac{w}{L_p}}. \tag{30.9}$$

Let us examine particular cases of formula (30.9). If $w/L_p \gg 1$, then sh (w/L_p) and ch (w/L_p) may be written with sufficient accuracy in the form ch (w/L_p) = sh (w/L_p) = $e^{w/L_p}/2$. By substituting the values found in formula (30.9) we arrive at expression (27.10) for an infinitely thick electrode in which the recombination rate at the ohmic contact plays no part.

In the opposite limiting case where $w/L_p \ll 1$, the hyperbolic cosines and sines may be expanded into series:

$$\text{sh} \, \frac{w}{L_p} \approx \frac{w}{L_p}, \quad \text{ch} \, \frac{w}{L_p} \approx 1.$$

Then for the limiting current when $sw/D_p \ll 1$, we obtain the relation

$$i^{\lim} = \left(1 + \frac{m}{r}\right) \frac{eD_p p^0}{L_p} \left(\frac{w}{L_p} + \frac{sL_p}{D_p} - \frac{s^2 L_p w}{D_p^2}\right). \tag{30.10}$$

When $s \to \infty$, the limiting current [equation (30.9)] tends to the value

$$i^{\lim} = \left(1 + \frac{m}{r}\right) \frac{eD_p p^0}{L_p} \, \text{cth} \, \frac{w}{L_p}. \tag{30.11}$$

When $(sL_p/D_p) \cdot \text{th} \, (w/L_p) \ll 1$, the fraction in formula (30.9) may be expanded into a series and we obtain

$$i^{\lim} = \left(1 + \frac{m}{r}\right) \frac{eD_p p^0}{L_p} \left(\text{th} \, \frac{w}{L_p} + \frac{sL_p}{D_p \, \text{ch}^2 \, \frac{w}{L_p}}\right). \tag{30.12}$$

The theory examined in this section has not been checked quantitatively. However, the results of investigation of the anodic dissolution of n-type germanium, which were obtained with thin electrodes ($w = 0.3$ mm $< L_p$) indicate that the relation of the limiting current of anodic dissolution to the thickness of the electrode predicted by relation (30.9) is found qualitatively [21, 22].

§ 31. Kinetics with Slow Electrochemical Stage (Experimental)

In this section we will examine reactions whose kinetics are determined by the electrochemical stage itself. This condition is usually satisfied by processes in which the majority carriers participate and also reactions involving the minority carriers if the current in the system remains much less than the limiting currents of the minority carriers. In this case the concentration of free carriers in the quasineutral region hardly differs from the equilibrium value.

a. Reduction of $K_3F(CN)_6$ at Cadmium Sulfide and Zinc Oxide Electrodes.

The reduction of the ferricyanide ion at a CdS electrode involves free electrons [23]. Over a wide range of potentials the cathode overvoltage curves are described by the Tafel equation and the slope b is not constant. Depending on the conditions of preliminary etching of its surface in hydrochloric acid, with the same electrode it is possible to obtain values of b from 0.3 to 0.9 (in units of RT/F).

It was found that the slope is greater, the more positive the flat band potential of the CdS after etching. * On the basis of differential capacitance measurements it may be surmised that in those cases where the flat-band potential is shifted toward positive values, the surface of a cadmium sulfide electrode does not bear appreciable amounts of specifically adsorbed anions. Under these conditions, $b \approx RT/F$, in agreement with the theory developed in § 24.

With an increase in the specific adsorption of anions the flat-band potential becomes more negative, while the coefficient b decreases.

In the system $K_3Fe(CN)_6/K_4Fe(CN)_6$ the slope of the cathode overvoltage curves on a ZnO electrode is close to RT/F as a rule (Fig. 61). However, some samples give higher slopes

---

*A similar principle was found with a CdS—gold contact if the gold was deposited on the crystal after its surface had been etched with hydrochloric acid [24].

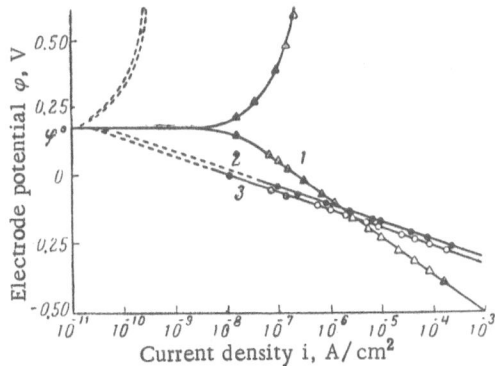

Fig. 61. Overvoltage curves in the system $Fe(CN)_6^{3-}$ —
$Fe(CN)_6^{4-}$ on a zinc oxide electrode [25]. Specific
electrical conductivity: 1) 0.025 $\Omega^{-1} \cdot cm^{-1}$; 2) 0.59
$\Omega^{-1} \cdot cm^{-1}$; 3) 1.7 $\Omega^{-1} \cdot cm^{-1}$.

close to 2RT/F [25]. The author explains the anomalies observed by the presence in the semiconductor of deep donors,
ionized in the space charge region, in addition to the main doping additive. It should be noted that there are very few experimental data on this system.

b. Anodic Dissolution of Germanium. The an-
odic dissolution of germanium is the reaction which has been investigated most fully experimentally. When one germanium atom
passes into solution with the formation of a tetravalent ion, four
negative charges are liberated. A specific characteristic of
this reaction is the fact that charges pass simultaneously into
the conductivity band and into the valence band. The latter process is only possible if there are holes in the valence band. The
ratio of the hole and electron currents has been determined by
direct experiment (see §32). The overall equation for the reaction in an alkaline medium may be written in the form

$$Ge + 2.4e^+ + 6OH^- \rightarrow GeO_3^{2-} + 1.6\,e^- + 3H_2O. \qquad (31.1)$$

Here, e^+ and e^- denote a hole and a free electron, respectively.

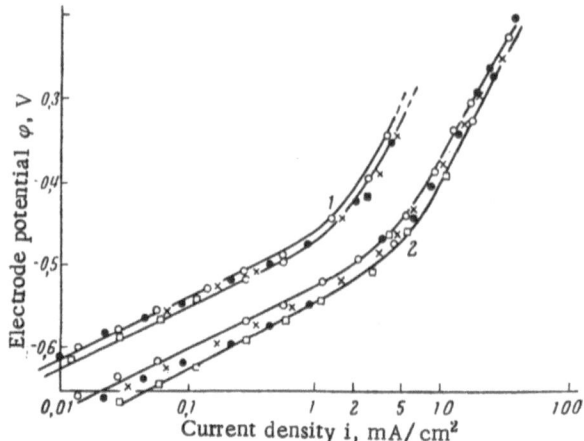

Fig. 62. Overvoltage curves of the anode solution of p-type germanium in alkaline solutions [31]. ○) 0.3 Ω · cm; ×) 1 Ω · cm; ●) 11 Ω · cm; □) 27 Ω · cm. 1) 0.01 N NaOH + 0.9 N NaClO$_4$; 2) 0.1 N NaOH + 0.9 N NaClO$_4$. The potential is relative to a saturated calomel electrode.

At low and medium current densities the current yield calculated on tetravalent germanium practically equals 100% [26, 27]. Germanates are formed as a result of the reaction and at a very high current density the reaction products are liberated at the anode as white germanium dioxide GeO$_2$ and the orange oxide GeO. It is believed that the germanium oxide arises as a result of the secondary reaction

$$GeO_2 + Ge \rightleftarrows 2\,GeO. \qquad (31.2)$$

With an increase in the current density there is an increase in the amount of GeO$_2$ at the electrode and the equilibrium (31.2) is shifted to the right [28]. *

---

*No compounds of divalent germanium have been detected in solution during anode dissolution of germanium, though they are apparently formed in the reverse process, i.e., the cathode reduction of tetravalent germanium ions [29,30].

The relation between the potential and the current density for p-type germanium (and also for n-type when i < 10^{-4} A/cm²) is expressed by a Tafel relation (Fig. 62).

From theory [equations (29.10) and (29.12)] it follows that the coefficient b in the Tafel relation may have values from RT/F (first-order reaction with respect to holes and the overvoltage falls completely in the space charge region) to RT/βF (overvoltage concentrated in Helmholtz region). In the first case the slope of the overvoltage curves in the coordinates φ—log i equals 0.06 V and in the second case, when β = 0.5, it equals 0.12 V. In practice, the slope has intermediate values: 0.07-0.08 V in solutions of NaOH and KOH [31, 32] and 0.1 V in solutions of H_2SO_4 [33] and NaOH [34]. This may be explained by assuming that the values of $\Delta\varphi_0$ and $\Delta\varphi_1$ are comparable (the overvoltage is distributed approximately equally between the space charge region and the Helmholtz layer) and, consequently, a case intermediate between the two limiting cases examined above is observed. According to [35], with an increase in the overvoltage the slope changes from 0.13 to 0.08 V. Here there may be a gradual increase in the fraction of the total overvoltage appearing in the space charge region. This is all the more probable since close to the steady potential, $\Delta\varphi_0 \approx \Delta\varphi_1$ (see §13), while with considerable overvoltages the potential difference in the Helmholtz region is practically unchanged (at least in n-type samples) and the whole of the potential drop is concentrated in the space charge region (see §32). High slopes close to 0.12 V have been observed in acid solutions [21, 26].

It must be stated that the nature of the slow stage in the anodic dissolution of germanium remains unknown. Therefore, the theoretical form of the kinetic equation for the reaction also remains unknown. We assumed that the rate of anodic dissolution was proportional to the concentration of holes at the surface [equation (27.1)]. However, some experimental data can be explained only by assuming that the concentration of holes appears in the kinetic equation to the power 0.5. It should be noted that in the case of the anodic dissolution of metals the value of the coefficient b cannot always be explained by a simple theory (see, for example, [36]).

The anode overvoltage on germanium depends on the crystallographic orientation of the surface [12, 37]. It is maximal
on the (111) plane and minimal on the (100) plane. In the latter
case a surface germanium atom is attached to the crystal by
only two covalent bonds and therefore it passes into solution
relatively readily.

The overvoltage of the anode solution of germanium is
practically independent of the specific resistance of germanium
(with the conditions given on p. 192 fulfilled) [31]. This is readily explained, since, according to relation (27.1), the current
is determined by the surface concentration of holes which depends little on the electrode potential with slight deviations from
equilibrium and is the same for samples with different concentrations of holes in the bulk.

The overvoltage depends on the alkali concentration in the
solution (Fig. 62). In dilute NaOH solutions there is a limiting
current from the diffusion of hydroxyl ions in the solution. With
a sufficiently high overvoltage, water molecules participate in
reaction (31.1) instead of hydroxyl ions and the current again increases.

Beck and Gerischer [31] proposed the following scheme
for the process:

$$\text{(31.3)}$$

$$\text{(31.4)}$$

$$\text{(31.5)}$$

$$\text{(31.6)}$$

$$\text{(31.7)}$$

In the equations for reactions (31.5) and (31.6) the sign $\oplus$ denotes a hole localized at a Ge$-$Ge bond and the sign $\bullet$ denotes an unpaired electron. Stage (31.6) is the slow stage and determines the rate of the process. The coefficients m' and m" together give the total number of holes per germanium atom passing into solution [2.4 according to equation (31.1)].

The Peltier heat liberated at the germanium$-$electrolyte contact during the anode reaction has been measured [38].

c. Cathodic Liberation of Hydrogen. The cathodic liberation of hydrogen involves free electrons (see also §36). This process is complicated by the adsorption of atomic hydrogen on the electrode surface. As has already been mentioned, hydrogen atoms are recombination centers and, apparently, trapping centers. Therefore, in discussing the mechanism of this reaction, it is necessary to take into account the possibility of the production of ionized surface states at the in-

terface. When their density is sufficiently great, the overvoltage is localized in the Helmholtz region and the kinetics of
the reaction at a semiconductor electrode are the same as at a
metal electrode. The kinetic parameters, namely, the coeffi-
cient $b = \partial \varphi / \partial \ln i$ in the Tafel equation and the stoichiometric
number* ν for different courses of this reaction were calcu-
lated by Green [6] and are given in Table 4.

The most reliable data on the liberation of hydrogen were
apparently obtained in [39, 40]. In 1 N $HClO_4$ solution the slope
of the overvoltage curves equals 0.1-0.11 V, $\nu = 1$, and the exchange current $i^0 = 1.4 \cdot 10^{-8}-4.7 \cdot 10^{-6}$ A/cm^2 [6, 39]. In 0.2 N
KCl and HCl solutions, the slope equals 0.1 V, $\nu = 0.9$, $i^0 = 1.2$
$\cdot 10^{-6}$ A/cm^2, and the transport coefficient $\alpha = 0.59$ [40].

By comparing the experimental values of b and ν with
those given in Table 4, and assuming that during the liberation
of hydrogen the electrode is covered with a monolayer of adsorbed hydrogen, Green [6] came to the conclusion that for the
liberation of hydrogen on germanium we should select mechanism "D" (fast discharge and slow electrochemical desorption
with the limiting coverage of the surface).

We will give the results of some other investigations. Depending on the conditions, the slope of the overvoltage curves
varies from 0.11-0.13 [41] to 0.23 V (on polycrystalline samples)

---

*The stoichiometric number is the number of times that the slow stage must oc-
cur for the formation of one molecule of the reaction product (H_2). It may be
determined from the slope of the polarization curve close to the equilibrium potential

$$\nu = i^0 \frac{nF}{RT} \left(\frac{\partial \varphi}{\partial i} \right)_{\eta \to 0}$$

(n is the number of electrons participating in the reaction).

In the case of germanium, the corrosion process usually interferes with its
determination. If the germanium corrosion current is higher than the exchange
current of the reaction $H^+ + e^- \rightleftharpoons \frac{1}{2}H_2$, then the reversible hydrogen potential is
not established (the steady electrode potential is the corrosion potential) and it is
not possible to determine ν by this method.

Table 4. Kinetic Parameters of the Cathode Liberation of Hydrogen [6]

Mechanism	Stoichiometric number	$\partial\varphi/\partial \ln i$ when $\alpha=1/2$		
		Metal electrode	Semiconductor electrode	
			Nondegenerate; without surface states	With high density of ionized surface states
A $\quad$ H$_3$O$^+$ + e$^-$ $\xrightarrow{\text{slow}}$ MH + H$_2$O MH + MH $\xrightarrow{\text{fast}}$ 2M + H$_2$ ($\theta \to 0$)	2	$2RT/F$	RT/F	$2RT/F$
B $\quad$ H$_3$O$^+$ + e$^-$ $\xrightarrow{\text{fast}}$ MH + H$_2$O MH + MH $\xrightarrow{\text{slow}}$ 2M + H$_2$ ($\theta \to 0$)	1	$RT/2F$	$RT/2F$	$RT/2F$
C $\quad$ H$_3$O$^+$ + e$^-$ $\xrightarrow{\text{slow}}$ MH + H$_2$O H$_3$O$^+$ + MH + e$^-$ $\xrightarrow{\text{fast}}$ H$_2$ + H$_2$O + M ($\theta \to 0$)	1	$2RT/F$	RT/F	$2RT/F$
D $\quad$ H$_3$O$^+$ + e$^-$ $\xrightarrow{\text{fast}}$ MH + H$_2$O H$_3$O$^+$ + MH + e$^-$ $\xrightarrow{\text{slow}}$ H$_2$ + H$_2$O + M ($\theta \to 0$)	1	$2RT/3F$	$RT/2F$	$2RT/3F$
E $\quad$ Mechanism D with $\theta \to 1$	1	$2RT/F$	RT/F	$2RT/F$

[42, 43]. The high slopes are probably explained by the degree of oxidation of the electrode and contamination of the solution. In some work the authors obtained the value b ≈ 5RT/F [44-47] which for samples of n type may be explained by degeneracy of free electrons at the electrode surface with high negative potentials (see § 24).* Values from 10^{-7}-10^{-8} A/cm^2 [44] to 10^{-6} A

---

*The high slopes of the overvoltage curves may also be the result of a change in the interphase potential drop because of reduction of the surface oxide during the measurements (see § 14).

per cm^2 [47] have been given for the exchange current and the transport coefficient $\alpha = 0.3$ [47].

In none of the work listed was it possible to detect differences in the behavior of n- and p-type samples at a current density below the limiting current of minority carriers.

As a result of hydrogen treatment, the slope of the overvoltage curves falls from 0.16-0.18 to 0.08-0.11 V [44]. In some cases the overvoltage curve for an electrode that has been subjected to prolonged cathode polarization consists of three linear sections with slopes of 0.12, 0.47, and 0.12 V [48].

d. Reactions of Silicon and Gallium Arsenide Electrodes. The anode overvoltage curve for silicon in hydrofluoric acid solution is described by a Tafel equation with a slope of 0.1 V [49]. For the cathode liberation of hydrogen from HF and H$_2$SO$_4$ solutions on a hydrogen-treated electrode the slope equals 0.17-0.18 V [50, 51]. In alkaline solutions the slope of the Tafel line equals 0.095 V [52]. It should be noted that at high overvoltages, on a silicon electrode, there is retardation of a cathode reaction even when majority carriers participate in it. The nature of this phenomenon remains unclear.

The overvoltage curve of anode solution on gallium arsenide has a slope of 0.1 V and that of the cathode liberation of hydrogen has values from 0.12 V (in H$_2$SO$_4$ solutions) to 0.25 V (in NaOH solutions) [53]. The electrochemical kinetics of an organic semiconductor, namely, thermally treated polyacrylonitrile, is discussed in [54].

§ 32. Anodic Dissolution of n-Type Germanium

The anodic dissolution of n-type germanium was first investigated in detail by Brattain and Garrett in 1954-1955 [55-57]. These papers laid the foundation of the field to which the present book is dedicated, namely, the electrochemistry of semiconductors. The anode process on n-type germanium is still the subject of intensive investigation and this is connected not only with its practical importance, but also with the fact that it

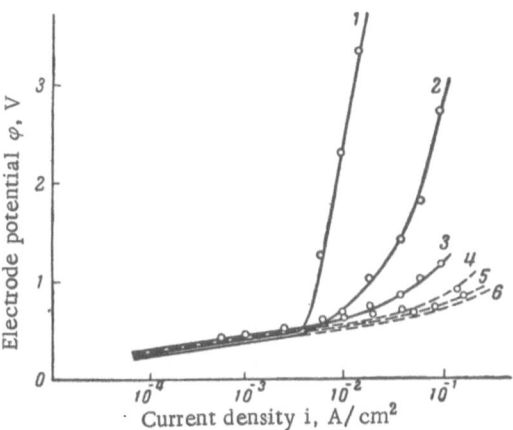

Fig. 63. Polarization curves of the anode solution of germanium in 0.1 N HCl solution [22]. 1) n-Type, 1 Ω · cm; 2) n-type, 6 Ω · cm; 3) n-type, 25 Ω · cm; 4) p-type, 30 Ω · cm; 5) p-type, 6 Ω · cm; 6) p-type, 1 Ω · cm.

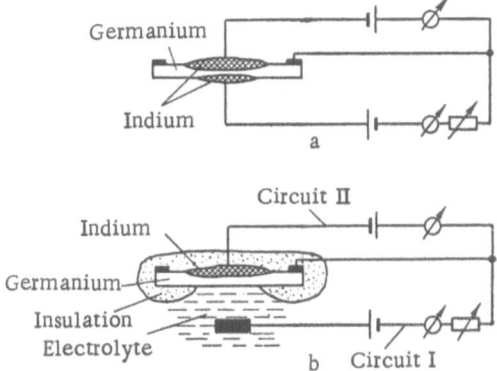

Fig. 64. Cross section of a p-n-p transistor (a) and a thin germanium electrode with a p-n junction for investigating the kinetics of electrode reactions (b).

is an excellent example of the effect of semiconductor properties of an electrode on electrochemical kinetics.

While an anode of p-type germanium does not differ from
a metal in the first approximation, n-type germanium has a
specific anode polarization curve. The comparative properties
of germanium anodes with different types and degrees of conductivity are illustrated in Fig. 63, which is taken from the
work of Efimova and Erusalimchik [22]. When $i \leq 3 \cdot 10^{-3}$ A
per cm^2, the "current−potential" relation is represented by a
Tafel line and was discussed in §31. At higher current densities the reaction is retarded on n-type germanium and this effect is greater, the lower its specific resistance, i.e., the
lower the equilibrium concentration of holes p^0. *

Brattain and Garrett [55] observed that there is the following relation between the potential of the germanium anode
and the concentration of holes beyond the space charge region
p_1:

$$\varphi = -\frac{kT}{e} \ln \frac{p_1}{p^0} + \text{const},$$

(32.1)

where p^0 is the equilibrium concentration.

In addition, a well-expressed limiting current of anode
solution was found in a series of studies of n-type germanium.
In accordance with the concepts presented in §27, the existence
of a limiting current in the anode solution of n-type germanium
and the relation found between the overvoltage and the hole concentration indicate that in the anode solution of germanium the
current is carried through the electrode−electrolyte interphase
by holes. In samples with a low equilibrium concentration of
holes the slow stage of the reaction is the transport of holes to
the interphase.

---

*It is interesting that this effect is also observed in a less-marked form with high-
resistance p-type germanium (25 Ω · cm), in which the concentration of holes is
still not very great. In low-resistance p-type germanium the deviation from the
Tafel relation at currents of the order of 10^{-1} A/cm^2 is produced by the ohmic
potential drop in the sample (Fig. 63).

The participation of holes in the anodic dissolution of germanium was demonstrated by direct experiment with the electrode apparatus which was proposed in [55, 58] and is illustrated in Fig. 64b. The electrode was a disk of n-type germanium 6-8 mm in diameter and with a thickness of the order of 0.1 mm (which is considerably less than the diffusion length of holes). On one side of the disk was an alloyed indium contact giving a p-n junction of large area. An annular ohmic contact was on the same side. On the opposite side of the disk there remained a section of free surface, approximately equal in area to the p-n junction, while the rest of the surface of the germanium and contacts was covered with a lacquer or paraffin wax. The electrode was placed in a solution, in which there were also an auxiliary electrode for polarization and a comparison electrode.

The anodic dissolution curves plotted for such an electrode by means of circuit I do not differ in principle from curves obtained with massive samples. However, the limiting current is lower as the thickness of the electrode and, consequently, of the quasineutral region in which holes are generated are less than the diffusion length of holes L_p. If direct bias is now applied to the p-n junction by means of circuit II, then the density of the limiting anode current is increased. This is explained in the following way. The passage of a current through the p-n junction in a forward direction is accompanied by the injection of holes from the p-type region into the n-type region. The nonequilibrium holes recombine at a distance of the order of the diffusion length from the p-n junction. In the case examined the thickness of the electrode, i.e., the distance between the plane of the p-n junction and the plane of the germanium–electrolyte interface is much less than the diffusion length of holes. Therefore, the nonequilibrium holes diffuse across the disk practically without recombination, * reach the electrode–solution interphase, and participate in the anode reaction. The li-

---

*The slight correction for recombination in the bulk of the electrode and at its side surface was calculated in [55].

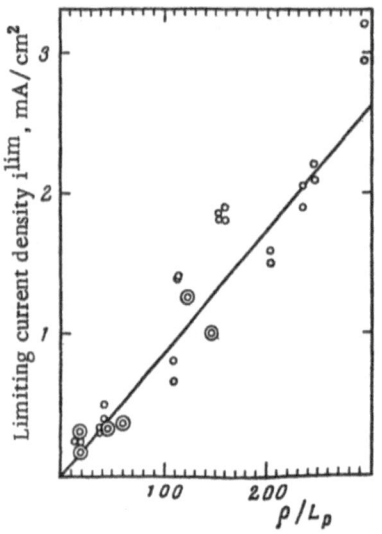

miting anode current, which is determined by the flow of holes to the electrode surface, increases [55, 59]. It is readily seen that the apparatus described is analogous in principle of operation to a junction p-n-p transistor (Fig. 64a). The p-n junction acts as a source of holes (emitter) and the germanium − electrolyte interphase with anode polarization acts as a collector. The limiting anode current i_p^{lim} is equivalent to the saturation current of the collector i_{co}.

Fig. 65. Relation of the limiting current density of anode solution of n-type germanium to the ratio ρ / L_p [60].

For an experimental check of the theory it is possible to use Flynn's data [60] on the relation of the limiting current on n-type germanium to its specific resistance ρ and the diffusion length of holes L_p. It was found that the current is proportional to the ratio ρ / L_p (Fig. 65). In accordance with equation (27.4) it may be concluded that the current is produced practically completely by the bulk generation of holes, * while the contribution to the hole current of generation through surface levels is insignificant.

Thus, the rate of surface recombination − generation at the germanium − electrolyte interphase under conditions of anodic dissolution was found to be insignificant. It should be noted that under conditions of the limiting current of anode solution of a germanium electrode, the actual concept of the rate of surface

---

*The characteristics of generation centers in the bulk may be determined by investigating the temperature dependence of the limiting current and using the Shockley−Read recombination theory [61].

recombination becomes indefinite (see footnote on p. 182). In this case the most direct information on surface recombination may be obtained precisely by measuring the limiting current of holes. The usual methods of measuring the effective lifetime of holes, which were developed for quasiequilibrium conditions, lead to results which are difficult to interpret unequivocally under the essentially nonequilibrium conditions of anode solution of n-type germanium. Thus, for example, according to [62], the effective lifetime of holes falls appreciably during anode polarization of germanium. More careful measurements, which were carried out by the photoconductivity decay method [63], showed that the disappearance of nonequilibrium holes, introduced by pulsed illumination, is described by two exponents. The decay constant of the "slow" exponent is close to the bulk lifetime of holes and this apparently confirms the absence of recombination through surface levels. The decay constant of the "fast" exponent is several times less and apparently characterizes the "extraction" of nonequilibrium holes through the semiconductor—electrolyte interphase as the result of the electrochemical reaction.

For a detailed comparison of the results of an experimental investigation with the calculation (§27) it is necessary to determine the ratio of hole and electron currents in the anode solution of germanium, i.e., r/m. This value may be measured by means of the electrode apparatus described above. We compare the current i_p of nonequilibrium holes injected by the p-n junction (emitter) and the increase in the limiting anode current produced by this.* Generally speaking, the hole current at the germanium—electrolyte interphase does not equal the emitter current because of bulk and surface recombination in the electrode. Moreover, it depends on the injection coefficient of the emitter, i.e., the ratio of the hole current to the total current

---

*In this type of measurement holes may be injected not only by means of the p-n junction, but also by illumination (with a light source of calibrated intensity) [34,55] or by means of an electroreduction reaction involving the valence electrons (see § 36).

of the emitter. The latter value practically equals unity for
markedly unsymmetrical p–n junctions. * With the chosen geometry of the sample (the thickness much less than the diffusion
length of holes and also the diameter of the p–n junction) the recombination of holes in the bulk and at the side surface of the
sample may also be neglected. Therefore, the current of injected holes practically equals the emitter current, i.e., the
current in circuit II. (In actual fact, in triodes of similar configuration the total recombination losses are no more than 3%.)

By measuring the ratio of the currents in circuits I and
II, Brattain and Garrett [55] found that the increase in the anode
solution current exceeds the additional hole current. This phenomenon was called "current multiplication." The ratio of the
limiting solution current to the limiting hole current $\alpha' =
i_a^{lim}/i_p^{lim} = (r + m)/r$ is called the current multiplication fac-
tor. It was found that with small injection currents which do
not exceed the limiting anode current in the absence of injection,
$\alpha' = 1.65$. A similar value of α' was obtained in [59].† This
also made it possible to write the equation for the anode reaction in the form (31.1).

Turner [26] attempted to relate α' to the molecular mechanism of the reaction. He surmised that the anode oxidation
of germanium proceeds in two stages with holes participating in
the first of these and free electrons in the second:

$$Ge + 2e^+ \rightarrow Ge\,(II),$$

$$Ge\,(II) \rightarrow Ge\,(IV) + 2e^-. \qquad (32.2)$$

---

*The concentration of acceptors in the p-type region is made much greater than
the concentration of donors in the n-type region (i.e., in the disk which is itself
the electrode). The current in a forward direction through such a junction is
carried by the majority carriers in the heavily doped region, i.e., by holes.

†In the work mentioned above [60] it was found that $\alpha' \approx 4$ by comparison of the
experimental values of the limiting anode current with values calculated from
equation (27.4). The reason for this discrepancy remains unknown.

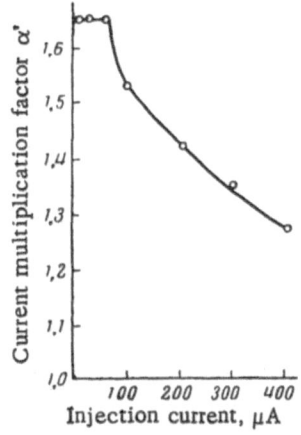

Fig. 66. Relation of the current multiplication factor α' at a germanium anode to the injection current strength [65].

According to this scheme the current multiplication factor should equal exactly 2. In experiments α' is always less than 2 and evidently depends on the crystallographic orientation of the surface and the solution composition [64]. By means of the method described above, it was shown that α' depends on the injection current strength i_p and on the same electrode it may have values from 1.65 (i_p less than the initial value of the limiting anode current) to 1 (i_p much greater than the initial limiting current). The character of the change in α' in relation to i_p is illustrated in Fig. 66 [65]. Analogous results were obtained independently by other authors by means of indirect methods [34, 66].

Thus, the current multiplication factor α' is not constant. Apparently it cannot be related to any definite molecular reaction mechanism; α' is a statistical value and characterizes the degree of participation of the valence band and the conductivity band in anode solution. On the basis of the results obtained in [65], it may be concluded that with an increase in the concentration of holes p_1 beyond the space charge region, α' falls, i.e., the fraction of holes in the anode current increases. The calculation of α' on the basis of detailed analysis of the nature of the slow stage of anodic dissolution is the next problem in the theory.

Let us turn to the polarization curve of anodic dissolution. Three factors should be noted. Firstly, the limiting current measured in a series of experimental studies was found to be several times (sometimes several tens of times) higher than that calculated from equation (27.4). Secondly, the limiting current is often poorly expressed (see, for example, Fig. 63). In actual fact, here the current is a linear function of the voltage. Finally, the magnitude of the limiting current depends on the composition of the solution [41].

The surface recombination rate apparently is not always low. In any case, for samples with a very high bulk lifetime surface recombination-generation cannot be neglected. In particular, this refers to the interpretation of the results of [55], in which samples with $\tau > 4000$ μsec were used. In the case of electrodes with a conductivity close to the intrinsic value, it is necessary to take into account another effect. Since the concentrations of majority and minority carriers here are comparable in magnitude, in the bulk of the semiconductor holes participate in the transfer of current and, consequently, the hole current includes a drift component in addition to a diffusion component. Therefore, the solution rate is found to be higher than that calculated from equation (27.4).

There is another possible source of the increase in the hole current, which is examined in detail in §37 and which is as follows. In addition to the generation of holes in the quasineutral region, there is generation in the space charge region and under certain conditions this effect becomes appreciable. The generation current in the space charge region increases with the potential, and this probably explains the poor "saturation" of the anode solution current of n-type germanium. Another reason for this may be the distribution of surface recombination levels with respect to energy in the forbidden band. Then, with an increase in potential, the rate of surface recombination and, consequently, the generation current, varies in a complex way in relation to the energy and the concentration of recombination centers of various types. Finally, dislocations may be a source of excess current in the anode solution of germanium [67].

With a sufficiently high anode potential, the surface barrier breaks down and the current increases sharply. This phenomenon is analogous to the breakdown of semiconductor rectifiers with a p-n junction. The breakdown voltage falls with a fall in the specific resistance of the germanium [31]. The breakdown mechanism has not been investigated. It is only known that breakdown normally occurs at isolated points where there is a

local increase in the concentration of donors N_D and not over
the whole germanium – electrolyte interphase [59]. As follows
from formulas (9.9a) and (26.17), an increase in N_D is accompanied by a decrease in the thickness of the space charge region. Consequently, at these points the field strength is higher
than over the rest of the surface and with an increase in potential these points are the first to reach the critical field strength
at which there may begin, for example, collision ionization lead-
ing to breakdown. This phenomenon may be used for checking
the uniformity of distribution of impurities in germanium
samples (see §57).

Let us examine the effect of the solution composition on
the limiting current of anode solution. Efimov and Erusalimchik
[68-70] found that the addition of potassium oxalate or iodide to
HCl or H_2SO_4 solution increases, while the addition of potassium
bichromate decreases the limiting anode current on n-type germanium. This effect is expressed more strongly in the case of
massive electrodes than in the case of thin plates (10-20 μ). In
the opinion of the authors of [68-70], anions of hydriodic and
oxalic acids are oxidized at a germanium anode simultaneously
with its solution and these reactions proceed not through the valence band, but through the free band, i.e., are accompanied by
the injection of free electrons into the germanium. As a result
of the injection of electrons and the increase in their concentration at the electrode surface, a strong electric field is produced
and this, in its turn, promotes the appearance at the surface of
a large number of holes largely due to drift. The retarding effect of bichromate was explained by the adsorption of bichromate ions on the germanium surface and the positively charged
chromium nuclei repelling holes from the surface. However,
this explanation is not clear cut. In actual fact, surface processes, namely adsorption and electrodeposition, affect the electric field and the concentration of free carriers only in the Debye
region and not in the quasineutral volume of the semiconductor.
Therefore, they cannot accelerate the generation of holes in the
bulk or their movement from the bulk of the sample to its surface. It is more likely that here there are phenomena of a pure-

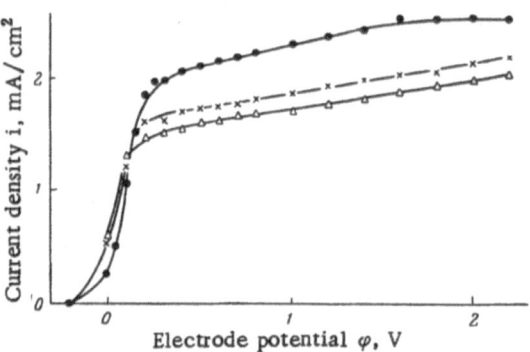

Fig. 67. Effect of the addition of $K_4Fe(CN)_6$ on the limiting current of anode solution of n-type germanium (9.7 $\Omega \cdot$ cm) in 0.1 N H_2SO_4 solution [73]. $K_4Fe(CN)_6$ concentration: $\bullet$) 0; x) 0.085 mole/liter; $\triangle$) 0.14 mole/liter.

ly surface character.* The peculiarities of anode solution of n-type germanium were also investigated in [71, 72].

In addition to anodic dissolution, the slow diffusion of holes occurs in two other anodic processes on a germanium electrode. One of them is the oxidation of $Fe(CN)_6^{4-}$ ions, which occurs at the same potentials as the anodic dissolution of germanium. There is competition between these two processes for holes, whose flow to the electrode surface is limited by the bulk generation rate. It was found that on introduction of $K_4Fe(CN)_6$ into the electrolyte solution the limiting anode current on n-type germanium falls (Fig. 67). Apparently there is no multiplication of current in the oxidation of ferrocyanide ions, i.e., $\alpha' = 1$ and, therefore, the total anode current is found to be less than in the anodic dissolution of germanium [73].

In Ziegler's electrolyte (a mixture of triethylaluminum and sodium fluoride) there occurs decomposition of the complex

---

*It was shown in [63] that the introduction of iodide ions into solution increases the rate of surface recombination of germanium with anode polarization.

Na[Al(C$_2$H$_5$)$_3$]$_2$F on a germanium anode.   The relation of the current to the type of conductivity and the specific resistance of
germanium and also to the illumination are qualitatively the
same as in the anode solution of germanium.   This gives
grounds for assuming that here the anode reaction rate is also
limited by the supply of holes to the surface [74].

There is the widely held opinion that the participation of
holes in the anode solution of germanium (and also silicon and
some other semiconductor materials, see below) is connected
with dislocation of the crystal lattice: it is possible that localization of a hole at a surface germanium atom weakens its covalent bonds with the crystal and facilitates their rupture. However, the data presented above on other oxidation reactions,
which also involve the participation of holes, compel us to approach this statement with care.

§ 33. Special Techniques for Investigating Electrochemical Kinetics at Semiconductor Electrodes

The electrode with a p-n junction described in §32 may be
used for a number of other measurements in addition to the determination of the current multiplication factor [55, 65] and the
qualitative investigation of the effect of injection on anode solution [58, 59, 75].

a. Effect of Bulk Concentration of Holes
on Overvoltage.   The potential drop at the p-n junction in
the absence of a current through it, i.e., the so-called floating
potential V_f, is a measure of the deviation of the concentration
of minority carriers from the equilibrium value. For the concentration of holes in the n-type region close to the junction we
may write

$$p_1 = p_1^0 e^{\frac{eV_f}{kT}} \tag{33.1}$$

(p_1^0 is the equilibrium value of this concentration). If the thickness of the n-type region is less than the diffusion length of

holes, then with a sufficient degree of accuracy it may be assumed that p_1 = const over the whole of the neutral volume, *
and thus the measurement of the floating potential with the p-n
junction on open circuit may be used to determine the concentration of holes p_1 at the boundary of the space charge region at
the germanium−electrolyte interphase. This method was used
by Brattain and Garrett [55] for investigating the relation of the
overvoltage of anode solution of germanium to the concentration
of holes in the sample p_1, which was varied by illumination of
the sample on the electrolyte side.

b. Investigation of the Participation of
Valence Electrons in Cathode Reactions. The
transfer of valence electrons from a semiconductor into solution means the formation of holes in the valence band. Therefore, when valence electrons participate in a reduction reaction, the electrode−electrolyte interphase injects holes into the
germanium. These nonequilibrium holes diffuse across the
sample to the p-n junction (see Fig. 64b) and may be observed
by measuring the floating potential. To allow for the loss in recombination, we make a preliminary measurement of the minimal value of the floating potential V_f^{min} when $p_1 \approx 0$ close to the
germanium−electrolyte contact (i.e., at the limiting rate of
anode solution). The hole current through the germanium−
electrolyte interphase during the cathode reaction is given by
[55]:

$$i_p = i_p^s \frac{e^{eV_f/kT}}{e^{eV_f^{min}/kT} - 1},$$

(33.2)

where i_p^s is the saturation current of the p-n junction in the absence of injection.

---

*This statement becomes inaccurate if $p_1 \ll p^0$ close to the semiconductor−
electrolyte interphase (particularly during anode solution under conditions of
limiting current). In this case it is necessary to introduce a correction taking into account the bulk generation of holes.

Another method of measuring the fraction of valence electrons in the reduction current was developed in [76]. A negative bias is applied to the p-n junction. Then the barrier current through the junction is determined by the rate of generation of holes in the n-type region. Holes injected into the germanium by the cathode reaction diffuse from the electrode surface to the p-n junction and increase its barrier current. Since
the losses in recombination in the bulk are usually insignificant
(see p. 206). then the increase in the barrier current of the
junction Δi_p^s practically equals the hole current i_p through the
germanium−electrolyte interphase. The ratio of the current in
circuit I to the increase in current in circuit II directly gives γ,
i.e., the fraction of valence electrons in the reduction current.
It is only necessary for the rate of surface recombination at the
germanium−electrolyte interphase to be small. It is readily
seen that the system examined is the same in the principle of
operation as a p-n-p transistor with the p-n junction acting as
a hole collector and the germanium−electrolyte acting as a peculiar form of emitter. The value γ is equivalent to the injection coefficient of the emitter of the transistor. The results of
such measurements are discussed in §36.

An analogous electrode was used in [77], but the thickness
of the "base," i.e., the n-type region, was only 25 μ. When a
sufficiently high back voltage was applied to the p-n junction the
thickness of the depletion layer in the p-n junction reached the
thickness of the plate. Then the space charge regions at the
p-n junction and at the germanium−electrolyte interphase
merged. This phenomenon may be used for producing thin
plates of a semiconductor by electrochemical etching (see § 59).

The apparatus illustrated in Fig. 64b has also been used
for determining the surface recombination rate at a semicon-
ductor−solution interphase by measurement of the photopotential (see §20). Such an electrode may also be used apparently
for measuring the potential drop in the Debye region in a germanium electrode by the method described in [78].

   c.   Thin Double-Sided Germanium Electrode.   The form of electrode described above is a thin

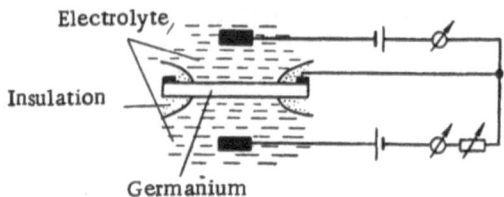

Fig. 68. Thin double-sided germanium electrode
for investigating the kinetics of electrode reactions
(cross section) [79].

double-sided germanium electrode, which was proposed in [79]
and is illustrated in Fig. 68. It consists of a disk of n-type
germanium, 6-8 mm in diameter and about 0.1 mm in thickness,
fitted with an annular lead and placed between two solutions
which are not directly in contact with each other. Each solution contains an auxiliary electrode for polarization and a comparison electrode. One of the sides of the germanium electrode
is the one investigated. The second side is always under an
anode potential under conditions of limiting current and is used
as an indicator of the hole concentration in the bulk of the electrode. It consequently fulfills the function of the p-n junction
in the electrode illustrated in Fig. 64b. Otherwise, the two
electrode systems are identical in principle of operation.

The double-sided electrode is particularly convenient in
cases where it is difficult to apply a p-n junction to the material investigated. It has been used by a number of authors for investigating the injection of holes in cathode processes on a germanium electrode and also for investigating surface recombination at a germanium — solution interface [47, 80, 81]. In actual
fact, the limiting current on the indicator side is a function of
the surface recombination rate on the side investigated (see
footnote on p. 181). Transient processes in the electrode with
pulsed polarization of the side investigated, produced by diffusion of nonequilibrium holes across the germanium plate, were
investigated in [82].

In conclusion, it should be noted that the methods described are applicable in principle to other semiconductor ma-

terials besides germanium if they have a sufficiently great diffusion length, for example, silicon. The use of electrodes of p-type material makes it possible to investigate cathode processes involving free electrons [83].

§ 34. Cathodic Liberation of Hydrogen on p-Type Germanium

Polarization curves of the liberation of hydrogen on a p-type germanium cathode obtained by Brattain and Garrett [55] are mirror images of anode curves plotted for n-type germanium. Only the limiting current is less well expressed and there is no current multiplication (i.e., $\alpha' = 1$). The characteristic form of the $i - \varphi$ curve, the increase in the current on illumination of the electrode, and also the dependence observed in [55] of the electrode potential on the concentration n_1 of free electrons at the boundary of the Debye and quasineutral regions

$$\varphi = \frac{kT}{e} \ln \frac{n_1}{n^0} + \text{const}$$

gave grounds for concluding that free electrons participate in the liberation of hydrogen and the slow stage of the cathode reaction on p-type samples is their transfer to the electrode surface.

However, complications arising in the investigation of the cathode liberation of hydrogen are the reason for the fact that this point of view was considered incorrect for a long time. There appeared papers (see, for example, [48]) in which no differences were detected in the polarization curves for n-type and p-type germanium. At the present time it may be regarded as proved that the reason for these experimental contradictions is a change in the state of the cathode surface during the liberation of hydrogen. If the cathode curve is plotted rapidly on a freshly etched electrode surface, then the limiting current is relatively well expressed and increases on illumination [6, 47]. The current increases with time at constant potential, while the photosensitivity decreases. On a hydrogen-treated electrode the current increases exponentially with potential as on n-type electrodes (Fig. 69).

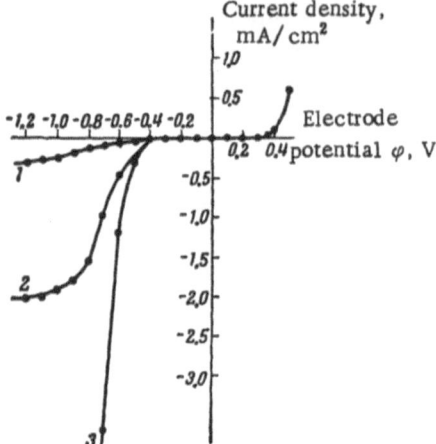

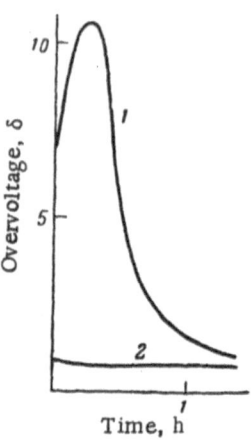

Fig. 69. Polarization curve of the cathode liberation of hydrogen on p-type germanium (1 Ω · cm) in H_2SO_4 solution [47]. 1) In the dark; 2) with illumination; 3) on a hydrogenated electrode.

Fig. 70. Relation of cathode overvoltage (with a current of 10 mA per cm^2) to time in 1 N KOH solution [85]. 1) p-Type germanium, 0.5 Ω · cm; 2) n-type germanium, 1.1 Ω · cm.

The hydrogenation process has been studied very little. Its kinetics are apparently very sensitive to the properties of the sample, the preparation of its surface, and the electrolysis conditions. In some cases the hydrogenation process lasts for a few seconds [84] and in others it is tens of minutes [85]. It was found that with a constant current the cathode overvoltage on p-type germanium passes through a maximum (Fig. 70). The overvoltage is relatively stable on n-type samples. The overvoltage finally becomes the same on electrodes with different types of conductivity and this agrees with the data in [48], in which the electrodes were subjected to prolonged cathode polarization before the measurements.

The fall in the overvoltage in Fig. 70 can apparently be explained by the fact that the absorption of hydrogen on germanium leads to an increase in the surface recombination rate

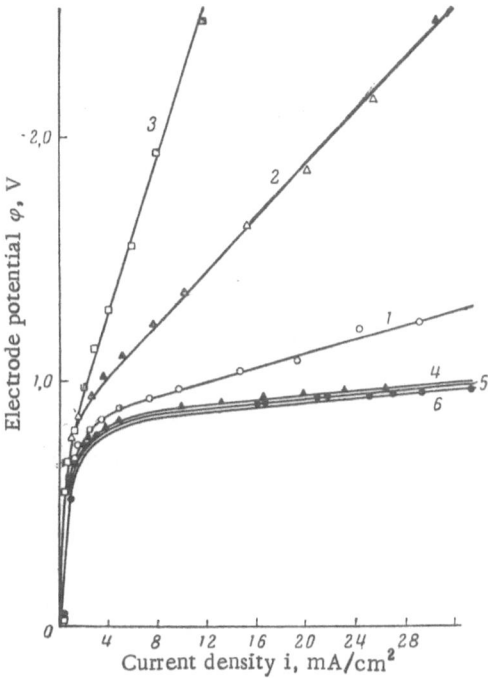

Fig. 71. Cathode polarization curves on a germanium electrode in H_2SO_4 solution (pH 1) [86]. 1) p-Type, 1.3 $\Omega \cdot$ cm; 2) p-type, 12 $\Omega \cdot$ cm; 3) p-type, 20 $\Omega \cdot$ cm; 4) n-type, 1 $\Omega \cdot$ cm; 5) n-type, 10 $\Omega \cdot$ cm; 6) n-type, 20 $\Omega \cdot$ cm.

(see §20).* The latter was demonstrated by direct measurements with a thin double-sided germanium electrode, which was described in §33c [47, 79] and also by the photoconductivity decay method [63]. The increase in the overvoltage in the initial period after the cathode current has been switched on was explained by Lazorenko-Manevich by the appearance of acceptor properties of hydrogen adsorbed on the electrode [44]. The in-

---

*Another possible reason is a change in the Helmholtz potential drop because of cathode reduction of adsorbed oxygen (see § 14).

crease in the number of acceptor levels changes the potential drop in the space charge region and, consequently, the degree of filling by electrons of the recombination centers already present on the surface. Therefore, the rate of generation of electrons falls, while the cathode overvoltage increases. However, as the number of recombination−generation levels (also connected with adsorbed hydrogen) increases, the second effect predominates over the effect of hydrogen as an acceptor and the generation rate again increases and the overvoltage falls.

The relation of the cathode current to the specific resistance of germanium was investigated by Paleolog, Fedotova, and Tomashov [86]. Under the conditions of their experiments the current was linearly related to the overvoltage. With an increase in the specific resistance of p-type germanium the current decreases (Fig. 71), while according to equation (27.4), the limiting diffusion current of free electrons must increase because of the increase in their equilibrium concentration in the bulk of the sample. The relation of the hydrogen liberation rate to the specific resistance found here and the absence of a well-expressed limiting current are the foundation of the hypothesis [84, 86] that it is valence electrons and not free electrons which participate in the liberation of hydrogen on germanium and that the retardation of the reaction on p-type germanium is explained by "the ohmic potential drop in the depletion layer of the semiconductor." A criticism of this point of view will be given in §38 and here we will only point out that the deviation of the experimental cathode $i - \varphi$ curve on p-type germanium from the curve calculated from equation (27.3) may be for the same reasons as in the case of the anode curve on n-type germanium (§32). In our opinion, it is most probable that the additional source of free electrons is generation in the space charge region. This mechanism is examined in detail in §38 where we also discuss the relation of the current to the specific resistance of the semiconductor.

In addition to the liberation of hydrogen, free electrons apparently participate in the reduction of hydrogen peroxide and the persulfate anion (these reactions are also retarded on elec-

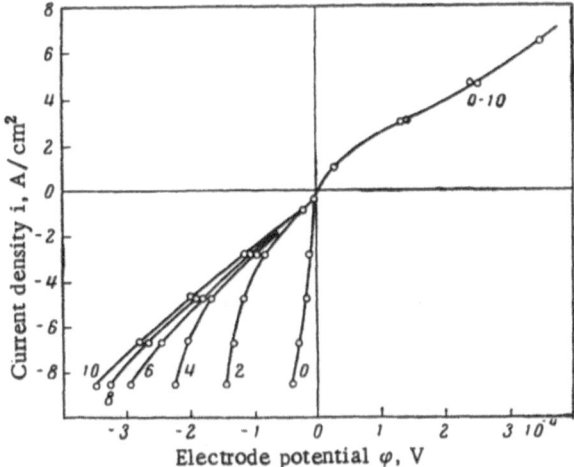

Fig. 72. Polarization curves for p-type selenium in 0.1 N
H_2SO_4 solution on illumination with green light [88]. The
light intensity (in relative units) is given on the curves.

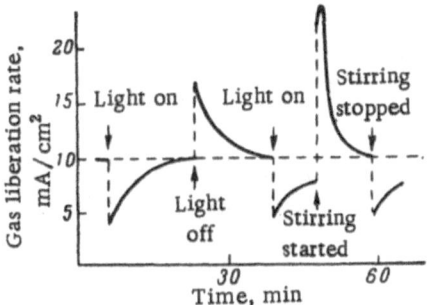

Fig. 73. Effect of illumination on the rate of liberation of
gaseous hydrogen on a germanium cathode (p-type) in KOH
solution [85].

trodes of p-type germanium [83, 86, 87]) and in the cathode process on a selenium electrode [88]. Illumination of the electrode
accelerates the cathode reaction on p-type samples (Fig. 72).
The volt-ampere characteristics on an illuminated selenium
electrode are determined by the ohmic resistance of selenium.

To conclude this section we will examine some phenomena which occur on the surface of a germanium cathode when it is illuminated [44, 85, 89]. They are observed only on p-type germanium and only in alkaline solutions at a current density above the calculated limiting diffusion current of free electrons (consequently, here the current is determined by generation at the surface or in the space charge region).

With a constant cathodic current density illumination decreases the rate of gaseous hydrogen liberation. After the light has been switched off, the gas liberation rate increases and for some time it exceeds the value calculated by Faraday's law from the density of the polarization current. Stirring the solution reduces the effect of light (Fig. 73). It may be surmised that hydrogen formed on the electrode dissolves in the electrolyte more actively in light than in the dark, producing local supersaturation of the solution. When the light has been switched off, it separates as bubbles, increasing the effective gas liberation rate. This is apparently connected with changes in surface tension at the interphase. On illumination there is a change in the potential drop in the Helmholtz layer (by approximately 0.1 V) and the surface tension increases, reducing the possibility of the formation of nuclei for hydrogen bubbles on the electrode surface.

Illumination also produces partial desorption of hydrogen from the surface of a germanium cathode (it was mentioned above that during the liberation of hydrogen the surface of the germanium is covered by a monolayer of hydrogen atoms). The amount desorbed increases in proportion to the intensity and the logarithm of the duration of illumination and is up to 10% of the monolayer. *

---

*The total amount of hydrogen adsorbed on an electrode was measured by means of anode charging curves, which were plotted in the dark immediately after the cathode current and illumination were switched off. On these curves there is a well-expressed delay caused by the oxidation of atoms of adsorbed hydrogen [89].

The photodesorption of hydrogen is apparently also connected with the decrease in the potential drop in the Helmholtz layer on illumination. The following mechanism for this process was put forward in [44, 89]. It was considered that even with a considerable cathode overvoltage, in addition to the discharge and electrochemical adsorption, there is also the reverse (anodic) process of ionization of adsorbed hydrogen. This hypothesis is not improbable in the case of semiconductor electrodes in contrast to metals. In actual fact, the rate of ionization of hydrogen, which involves the participation of only free electrons, in accordance with equation (24.2) may be written in the following form:

$$i = k\theta e^{\alpha F \Delta\varphi_0 / RT} , \qquad (34.1)$$

where θ is the coverage of the surface by hydrogen and $\Delta\varphi_0$ is the overvoltage in the Helmholtz layer.

Under the conditions of the experiment in [89], under cathode polarization there was a change mainly in the potential drop in the space charge region in the semiconductor (so that the value of φ_0 remained practically constant). Since θ increases with an increase in the cathode polarization, the ionization current may even exceed the exchange current at the equilibrium potential.

The discharge and electrochemical desorption currents hardly change under illumination under galvanostatic conditions. The illumination produces an increase in $\Delta\varphi_0$ and as a result of this the ionization current increases [see equation (34.1)] and discharge of the "hydrogen capacitance" of the electrode begins, i.e., θ decreases.

§ 35. Effect of Minority Carrier Current on Differential Capacitance of Electrode

In § 12 we examined the dependence of the differential capacitance of the electrode on the applied potential under equilibrium conditions. When an electric current passes, the distribution of the carriers begins to deviate from equilibrium and,

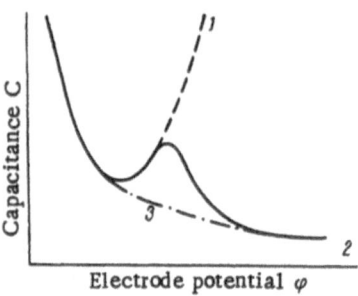

Fig. 74. Relation of the capacitance
C to the electrode potential φ. 1)
Under equilibrium conditions; 2) with
the limiting current of minority carriers passing, frequency $\omega \to 0$; 3) the
same, $\omega \to \infty$.

naturally, with a marked deviation from equilibrium we must expect new effects. In actual fact, as Fig. 74 shows, with an increase in the potential and the corresponding current, the "capacitance—potential curve" deviates from the equilibrium curve
and passes through a maximum. This effect is qualitatively
understandable. With strong anode polarization the electrode
surface is enriched in holes, which also form the space charge
Q_p. In accordance with equation (12.2), the capacitance of the
space charge region C^+ produced by the holes may be written in
the form

$$C^+ = -\frac{dQ_p}{d\varphi_1}.$$

$$(35.1)$$

With an approach to the limiting current the concentration of
holes at the surface tends to its limiting value $p_s^{lim} =$
$[i^{lim}]/i^0(p_s^0)$, which is determined from equation (27.1) if we
neglect in it the change in the potential in the Helmholtz layer.
As Fig. 60 shows, the concentration of holes at the boundary of
the quasineutral and Debye regions p_1 tends to 0. Thus, the
concentrations of holes at both boundaries of the space charge
region assume certain fixed values. Therefore, it is natural to
assume that the charge of holes Q_p at points close to the limit

practically cease to depend on the potential and the derivative $dQ_p/d\varphi_1$ tends to zero. Consequently, the experimentally observed capacitance in the region of the limiting current is caused by the change in the charge of the donor levels with a change in potential.

Let us turn to quantitative calculation. In accordance with formula (12.3), the capacitance of the space charge region of a semiconductor electrode is given by

$$C_1 = -\frac{dQ_1}{d\varphi_1} = \frac{\varepsilon_1}{4\pi}\frac{dE_s}{d\varphi_1}.$$

By using formula (29.6) for E_S and differentiating, we obtain

$$C_1 = -\sqrt{\frac{\varepsilon_1 e}{2\pi}}\;\frac{-N_D + \dfrac{\partial p_s}{\partial \varphi_1}\cdot\dfrac{kT}{e}}{2\sqrt{-N_D\,\varphi_1 + \dfrac{kT}{e}\,p_s}}. \tag{35.2}$$

For calculating the derivative $\partial p_S/\partial\varphi_1$ we will examine the simplest case. We will assume that the relation between the concentration p_S and the current is given by $i = -k_1 p_S$, which is obtained from equation (27.1) if we neglect the potential drop in the Helmholtz layer. Then we arrive at the expression

$$C_1 = \sqrt{\frac{\varepsilon_1 e}{2\pi}}\;\frac{N_D + \dfrac{1}{k_1}\dfrac{\partial i}{\partial \varphi_1}\cdot\dfrac{RT}{F}}{2\sqrt{-N_D\cdot\varphi_1 - \dfrac{RT}{F}\cdot\dfrac{i}{k_1}}}. \tag{35.3}$$

In order to find the derivative $\partial i/\partial\varphi_1$ it is necessary to use expression (29.9)

$$\frac{d\varphi_1}{di} = -\frac{RT}{Fi} - \frac{RT}{F\left(1 - \dfrac{i}{i^{\lim}}\right)i^{\lim}}. \tag{35.4}$$

After elementary rearrangements we obtain

$$\frac{di}{d\varphi_1} = -\frac{Fi\,(i^{\lim} - i)}{RTi^{\lim}}. \tag{35.5}$$

By substituting relation (35.5) in the expression for the capacitance (35.3) we find

$$C_1 = \sqrt{\frac{\varepsilon_1 e}{2\pi}} \; \frac{_i\text{lim} \, N_D - \dfrac{i}{k_1} \, (_i\text{lim} - i)}{_{2i}\text{lim} \, \sqrt{-N_D \cdot \varphi_1 - \dfrac{RT}{F} \cdot \dfrac{i}{k_1}}} \; . \tag{35.6}$$

Let us examine the capacitance as a function of the current passing. It is convenient to pick out two limiting cases, namely, with a current $i \ll {_i}\text{lim}$ and with $i \leqslant {_i}\text{lim}$. In the first case, by neglecting the current i in comparison with the limiting current $_i\text{lim}$ we obtain for the differential capacitance

$$C_1 = \sqrt{\frac{\varepsilon_1 e}{2\pi}} \; \frac{N_D - \dfrac{i}{k_1}}{2 \sqrt{-N_D \cdot \varphi_1 - \dfrac{RT}{F} \cdot \dfrac{i}{k_1}}} \; . \tag{35.7}$$

By using the formula $i = -k_1 p_s$, we obtain the following expression:

$$C_1 = \sqrt{\frac{\varepsilon_1 e}{2\pi}} \; \frac{N_D + P_s}{2 \sqrt{-N_D \varphi_1 + \dfrac{RT}{F} P_s}} \; . \tag{35.8}$$

If $N_D \gg p_s$ and $-\varphi_1 \gg RT/F$, then formula (35.8) changes into the equilibrium expression (12.10) for the capacitance of the depletion layer. If $N_D \approx p_s$, then the capacitance increases with the absolute value of the potential in the following way:

$$C_1 = \sqrt{\frac{\varepsilon_1 e}{2\pi}} \; \frac{N_D + p^0 \cdot e^{\frac{-e\varphi_1}{kT}}}{2 \sqrt{-N_D \varphi_1 + \dfrac{RT}{F} P_s}} \; . \tag{35.9}$$

Let us now turn to an examination of the behavior of the capacitance with a current tending toward the limit. In this case, $_i\text{lim} - i \to 0$, and we obtain equation (35.10). At potentials which satisfy the condition $N_D \, \varphi_1 \gg (RT/F) i^{\text{lim}}(1/k_1)$, we again obtain the formula for the capacitance of the depletion layer (12.10).

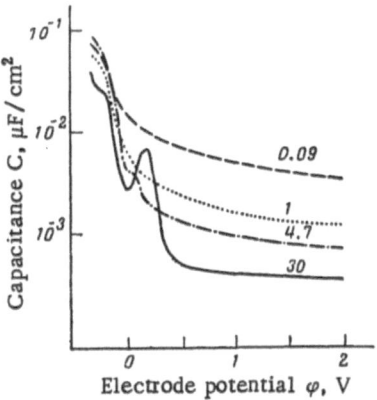

Fig. 75. Relation of the capacitance C of
an n-type germanium electrode to the
electrode potential φ [90]. The parameter
is the specific resistance of the germanium
(given on the curves in $\Omega \cdot$ cm); 0.1 N
H_2SO_4 solution.

$$C_1 = \sqrt{\frac{\varepsilon_1 e}{2\pi}} \; \frac{N_D}{2\sqrt{-N_D \, \varphi_1 - \frac{RT}{F} i \lim . \frac{1}{k_1}}} \; . \tag{35.10}$$

In the calculations carried out, no account was taken of
the potential drop in the Helmholtz layer. To obtain general
formulas it is necessary to use the expression for the concentration of holes (29.2). The expression for $dp_s/d\varphi_1$, which appears in formula (35.2), has the form

$$\frac{dp_s}{d\varphi_1} = -\frac{e}{kT} p^0 \left(1 - \frac{i}{i\lim}\right) e^{-\frac{e\varphi_1}{kT}} + \frac{p^0 e^{-\frac{e\varphi_1}{kT}}}{i\lim} \frac{di}{d\varphi_1} . \tag{35.11}$$

The derivative $di/d\varphi_1$ may be found by using equation (29.8).
However, we then obtain cumbersome formulas, which we will
not give here.

Figure 75 gives $C - \varphi$ curves plotted on an n-type germanium electrode in the region of anode currents [90]. Let us first examine the curve for a relatively high resistance sample (30 $\Omega \cdot$ cm). Close to the steady potential where the anode current is considerably less than the limiting current, on the $C - \varphi$ curve there is a minimum, which is characteristic of equilibrium conditions. Evidently here the concentration of holes over the space charge region p_1 hardly changes under the effect of the current and equals the equilibrium concentration in the bulk of the sample p^0.

With sufficiently high anode polarization (about 0.1 V relative to a normal calomel electrode) the rise in capacitance with potential ceases and with a further shift in the potential, the capacitance falls (cf. Fig. 74) and here there is a linear relation between the reciprocal square of the capacitance and the potential, which is characteristic of the depletion layer [equation (12.11)].

With low-resistance samples both close to the steady potential and with anode polarization the relation of the capacitance to the potential is described qualitatively by equation (12.11). Consequently, here the "equilibrium" region of the depletion layer changes into the "nonequilibrium" region, omitting the inversion layer. The reason for this is that for these samples the surface potential at which the "equilibrium" depletion layer changes into the inversion layer (i.e., Γ_p becomes greater than $N_D \cdot L_l$) would have been reached with an extremely positive electrode potential, i.e., under conditions where the anode current is already close to the limiting current. *

It is possible to increase the capacitance by increasing by some means p_1 in the range of currents close to the limiting current. In [90] it was shown that on illumination of an electrode (and also on injection of holes by means of a p-n junction on the opposite side of the electrode, see Fig. 64b), on the $C - \varphi$

---

*Another reason for the fall in the differential capacitance on anode polarization of an n-type semiconductor, which is not connected with the current through the interphase, will be examined in § 42.

curve of low-resistance germanium there appears a rise in capacitance, corresponding to the inversion layer.

A fall in capacitance in the range of currents close to the limiting current of free electrons is also observed on cathode polarization of p-type germanium.

§36. Simultaneous Participation of Free and Valence Electrons in Oxidation — Reduction Reactions on a Germanium Electrode

From §§24 and 25 it follows that in the general case both energy bands of a semiconductor may participate simultaneously in oxidation—reduction reactions (as we already saw in the anode solution of germanium). Thereupon the fractions of carriers of each type in the total current are determined by their concentrations at the electrode surface and the distribution of the overvoltage at the interphase. It should be emphasized that there has been no quantitative check on the theory up to now. This is connected particularly with the fact that a germanium electrode, which is most convenient for such investigations, corrodes in solutions of oxidants. Therefore the oxidation—reduction reactions are complicated by the simultaneous dissolution of the germanium.

The participation of valence electrons in reduction reactions was first investigated by Gerischer and Beck [91]. It was found that in the presence of potassium ferricyanide the limiting current of anode solution of n-type germanium increases (Fig. 76). Since the limiting current is determined by the flow of holes to the interphase, this suggests that a new source of holes appears in the presence of ferricyanide. In [91] the hypothesis was put forward that this source is the reduction of $Fe(CN)_6^{3-}$ ions at the electrode, which proceeds simultaneously with anode solution and also through the valence band, i.e., it is accompanied by the injection of holes into germanium. These holes are used by the anode reaction and its limiting current increases.

Analogous results were obtained in the case of the other oxidants Ce^{4+}, Fe^{3+}, and MnO_4^- in an acid medium, and O_2 in an

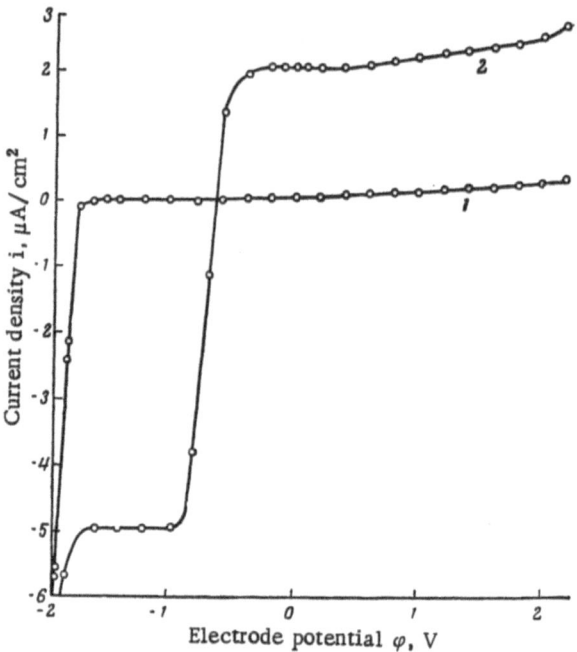

Fig. 76. Effect of the addition of $K_3Fe(CN)_6$ on the limiting current of anode solution of n-type germanium in NaOH solution [91]: 1) 0.05 N NaOH; 2) 0.05 N NaOH + 5 g/liter $K_3Fe(CN)_6$.

alkaline medium [73, 92].* This combination of the reduction of an oxidizing agent with the anode solution of germanium is discussed in detail in Chapter IV, so now we will examine it simply as a method of demonstrating the participation of valence electrons in reduction reactions.

---

*In the presence of ions of metals more noble than germanium, namely gold and copper, the limiting anode current on n-type germanium is also increased with the simultaneous deposition of metal [93, 94]. It is possible that in this case there is also injection of holes. An alternative explanation is an increase in the rate of surface recombination under the influence of the deposition of copper and gold on the electrode (see § 21).

However, this method has a substantial drawback. In actual fact, anode and cathode processes occur on the same sections of the surface and this makes it impossible to determine unequivocally the mechanism of their interaction. In particular, it is impossible to distinguish reliably electrical and chemical phenomena.

The thin double-sided electrode described in §33 was used in [76] to investigate the participation of the valence band and the conductivity band in the reduction of oxidizing agents. On one side (indicator) of the electrode (see Fig. 68), anode solution occurs and on the other side (investigated side), cathode reduction of the oxidizing agent. * Since these reactions are separated in space here, they can interact only through the diffusion of holes across the germanium plate. An increase in the rate of anode solution on the indicator side as a result of the reduction of $K_3Fe(CN)_6$ and $KMnO_4$ on the side of the electrode investigated gave a reliable demonstration of the injection of holes and, consequently, the participation of valence electrons in the reduction of these substances [79,95]. † By the same method it was shown that the liberation of hydrogen simultaneously with the reduction of an oxidizing agent with high cathode polarization increases the surface recombination rate at the germanium—solution interphase and reduces the effect of the injection of holes [47,79,100]. This phenomenon is illustrated in Fig. 77, which shows the relation of the increase in the indicator current to the potential on the side investigated.

Direct measurements of the fraction γ of valence electrons in the reduction current for a series of oxidants were made in [76,100-102] by means of a thin germanium electrode

---

*The advantage of the latter method is shown particularly clearly here: it is possible to investigate the interaction of anode and cathode processes which occur at different electrode potentials. Such processes cannot be realized simultaneously on a normal electrode.

†The injection of holes into anthracene from an electrolytic contact was observed with a similar device [96-99].

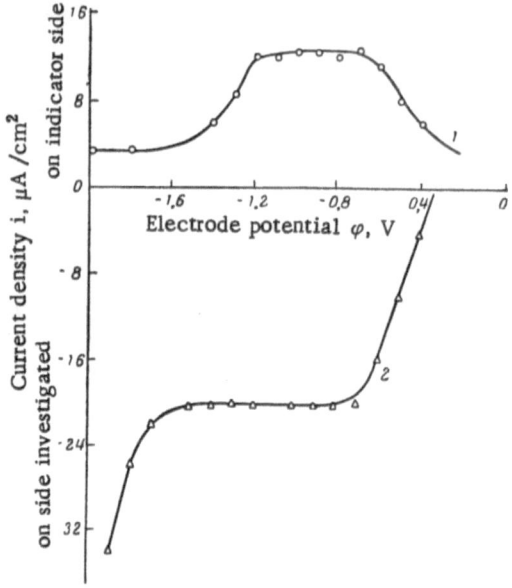

Fig. 77. Relation of the density of the indicator current of a thin double-sided germanium electrode (1) and the cathode current on the side investigated (2) to the electrode potential φ of the side investigated in the solution 0.1 M NaOH + 0.05 M $K_3Fe(CN)_6$ [47].

Table 5.   Fraction of Valence Electrons in Reductions of Some Substances

Substance reduced	Solution of indifferent electrolyte	Fraction of valence electrons	Potential φ, V	$\varphi - \varphi_{rb}$, V
HNO_3	—	1	—	—
$KMnO_4$	H_2SO_4	0.8—0.9	0.2	0.1
$K_3Fe(CN)_6$	KOH	0.6—0.8	—0.3	0.3
KI_3	KI	0.4	0.2	—
Quinone	H_2SO_4	0.4	0	—0.1
$K_2Cr_2O_7$	H_2SO_4	0.03—0.08	—0.3	—0.4
H_2O_2	K_2SO_4	0	—0.7	—0.7 *
H^+	—	$\leqslant 0.1$	—	—

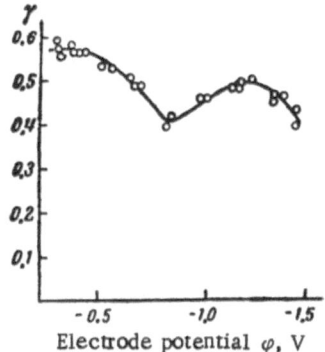

Fig. 78. Relation of the "apparent" value of the fraction γ of valence electron in the reduction current for $K_3Fe(CN)_6$ on a germanium cathode to the electrode potential φ [101].

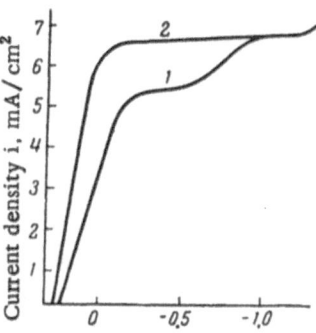

Electrode potential φ, V

Fig. 79. Polarization curves of the reduction of KI_3 on a germanium cathode [104]. 1) p-Type; 2) n-type. Solution $2 \cdot 10^{-2}$ M KI_3 + 1 M KI.

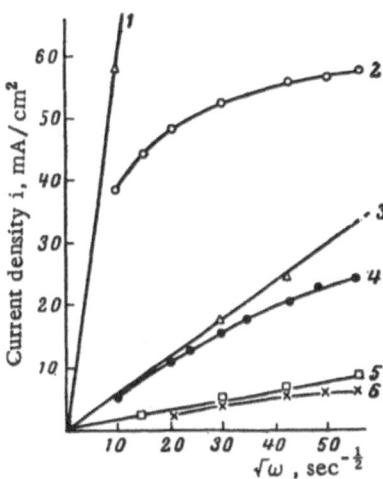

Fig. 80. Relation of the current density i for KI_3 reduction on a rotating disk germanium electrode at a potential of 0.4 V to the square root of the angular velocity of rotation of the electrode $\sqrt{\omega}$ [104]. 1) 0.095 N KI_3, n-type; 2) the same, p-type; 3) $8 \cdot 10^{-3}$ N KI_3, n-type; 4) the same, p-type; 5) $2.2 \cdot 10^{-3}$ N KI_3, n-type; 6) the same, p-type.

with a p-n junction, described in §33b (Fig. 64b). Table 5
gives the values of γ and also the potential at which they were
measured, relative to the flat-band potential φ_{fb}. The latter
was determined for alkaline and acid solutions containing no
oxidant in [103]. As a rough approximation it was assumed that
the introduction of oxidants into the solution does not change the
flat-band potential. The same table gives the values of γ for
the reduction of HNO_3 and the liberation of H_2, which were
found by the method of measurement of the floating potential of
the p-n junction described in §33a [25].

As the table shows, γ has values from 1 to 0. Between
γ and $\varphi - \varphi_{fb}$ there is a definite relation, namely, the more
positive $\varphi - \varphi_{fb}$, the greater is the fraction of valence electrons in the reduction. This agrees with equation (24.10): with
an increase in the surface concentration of free electrons, the
electron component of the current increases, while the hole
component decreases, and vice versa.

The relation of the "apparent" value of γ to the potential
for the reduction of $K_3Fe(CN)_6$ is given in Fig. 78 [101]. With
an increase in the cathode polarization γ falls, though much
more slowly than is required by theory. The minimum in the
"apparent" value of γ at -0.7 V is apparently connected with
the increase in the surface recombination rate in an alkaline solution at this potential [63].

It is interesting to compare the value of γ with the kinetic
peculiarities of the reaction. The kinetics of the reduction of
the oxidants listed in Table 5 on germanium has been investigated with a rotating disk electrode [104]. On the polarization
curves of the reduction of $KMnO_4$, $K_3Fe(CN)_6$, quinone, $K_2Cr_2O_7$,
and H_2O_2 on n- and p-type germanium there was a quite well-
expressed limiting current. The magnitude of the limiting current did not change on illumination and was independent of the
type of conductivity of the germanium. It was proportional to
the square root of the angular velocity of the electrode and,
therefore, was determined by the rate of diffusion of the ions
or molecules reduced from the bulk of the solution to the electrode surface [105]. The lack of a relation between the limit-

ing reduction rate and the semiconductor properties of the electrode may be explained in the following way. The reduction of
$K_3Fe(CN)_6$ and $KMnO_4$ proceeds mainly with the participation of
valence electrons and therefore is not retarded on p-type germanium. $K_2Cr_2O_7$ and H_2O_2 are reduced predominantly with the
participation of free electrons and on p-type germanium at a
high overvoltage the reduction rate is less than on n-type germanium. However, the limiting current of these reactions is
reached at very negative potentials at which the rate of surface
recombination is increased because of the adsorption of hydrogen on germanium. In the limiting current region the free
electrons required for the reduction apparently arise by generation on the electrode surface.

Most interesting results were obtained in an investigation
of the kinetics of iodine reduction in potassium iodide solution.
While on n-type germanium the polarization curve has the form
characteristic of a process whose kinetics are determined by
the diffusion of ions in solution, on the polarization curve
plotted on p-type germanium there are two regions of limiting
current (Fig. 79). In the region of potentials from -0.3 to -0.5
V the current is less than on n-type germanium and changes
more slowly with an increase in the electrode rotation rate (Fig.
80). On illumination, and also with preliminary hydrogen treatment of a p-type electrode, the polarization curve has the same
form as on n-type germanium. In combination with the measured value $\gamma = 0.4$ this makes it possible to conclude that at potentials from -0.3 to -0.5 V, the rate of reduction of I_3^- ions is
determined by the rate of generation of free electrons. Since
the reduction current is 1-2 orders higher than the rate of bulk
generation, calculated from equation (27.4), generation on the
surface evidently predominates here. Adsorbed atoms or ions
of iodine are apparently the generation centers. The reduction
current is proportional to the square root of the concentration
of I_3^- ions in the solution at the electrode surface (i.e., the
order of reaction with respect to I_3^- ions is 0.5). If the concentration of recombination centers is proportional to the amount
of iodine adsorbed, this means that the relation of adsorption to
concentration is described by a Freundlich isotherm.

At potentials from ⁻0.9 to −1.2 V, the rate of the process
is independent of the type of conductivity of germanium and is
determined by the diffusion of I_3^- ions in solution. The change
from generation to diffusion kinetics on p-type germanium is
apparently connected with an increase in the rate of surface re-
combination−generation because of the adsorption of hydrogen
on the germanium surface at these potentials.

Another example of a reduction proceeding through the
valence band is the formation of GeH_4 on a germanium cathode:

$$Ge + 4H^+ + me^- \rightarrow GeH_4 + (4 - m)e^+. \tag{36.1}$$

This process is accompanied by the emission of light (produced by the recombination of nonequilibrium holes formed),
whose wavelength corresponds to the width of the forbidden band
of germanium. It is possible to observe the radiation only in
electrolytes prepared with heavy water as ordinary water ab-
sorbs the infrared light. This reaction occurs only with a high
cathode current density (above 0.1 A/cm^2) [106].

Among the few oxidations investigated up to now, the oxidation of divalent vanadium ions on a germanium anode [107] is
of interest. This process occurs simultaneously with the anode
solution of germanium, but in contrast to the latter it does not
require the participation of holes. The rate of oxidation of V^{2+}
ions on n-type germanium exceeds the limiting current of holes
and depends only on the concentration ov V^{2+} in solution. This
reaction was investigated by means of a rotating disk germanium electrode [108]; similar results were obtained with station-
ary electrodes [73].

The systems $TiCl_4/TiCl_3$, $VOCl_2/VCl_3$, $VOSO_4/V_2(SO_4)_3$
in aqueous solutions are inert toward a germanium electrode
[73]. A number of organic and inorganic substances in alcohol
solutions are reduced on a germanium cathode [109].

§37. Kinetics with Slow Generation of Minority
Carriers in the Space Charge Region (Theory)

In the case of semiconductors with a wide forbidden band
it is usually necessary to take into account the generation of

minority carriers in the space charge region. As an example, we will examine the anode solution of n-type silicon. From formula (27.4) it follows that if holes are necessary for the solution of silicon, as for germanium, then the limiting current of silicon dissolution must be considerably less than the limiting current of solution of germanium of the same specific resistance. In actual fact, formula (27.4) includes the value n_i, which is related to the width of the forbidden band by equation (4.5). Hence, it is obvious that with an increase in the width of the forbidden band, the concentration n_i and, consequently, the limiting current of the diffusion of minority carriers, must fall sharply. However, in an experiment on the solution of silicon, a current of the same order as in the solution of germanium was found. Flynn [60] put forward the hypothesis that the holes necessary for the solution of silicon are generated in the space charge region. A quantitative calculation on the anode solution of a semiconductor taking into account generation in the Debye region was carried out in [110] on the assumption that recombination may be described by the Shockley−Read theory [61].

At low currents when the supply of holes to the contact is not the slow stage of the process, the volt-ampere characteristics, as in the case of germanium, are described by the Tafel relation. However, at high currents, the volt-ampere characteristics will be determined gy generation in the space charge region. Before giving an account of the exact theory, we present semiquantitative estimates and show primarily that the generation current in the Debye region in a number of cases may considerably exceed the limiting current of minority carriers arising in the quasineutral region (27.4). For this purpose we use the results in § 6. The number of carriers generated in unit volume in unit time is determined by relation (6.8). It will be assumed that in an n-type semiconductor there is a depletion layer and that over the whole range of potentials examined it may be assumed that $C_n(n + n') \gg C_p(p + p')$. Then equation (6.8) is written in the form

$$R = \frac{C_p(n \cdot p - n_i^2)}{(n + n')} \, . \qquad (37.1)$$

As a result of the smallness of the concentration of electrons, in the depletion layer practically everywhere the relation $n < n'$ holds, so that for R we obtain

$$R = \frac{C_p(n \cdot p - n_i^2)}{n'} = A(n \cdot p - n_i^2).$$

(37.2)

In the case where the limiting current of holes flows from the quasineutral region into the space charge region (simultaneously with the generation of holes in the space charge region), the concentration of holes p_1 at the boundary of these regions tends to zero. Therefore, over the whole of the space charge region it is possible to neglect $n \cdot p$ in comparison with n_i^2:

$$R = -\frac{C_p n_i^2}{n'}.$$

(37.3)

In the steady state all current carriers formed in the Debye region participate in the electrochemical reaction. Therefore, to obtain the density of the hole current in the plane of the electrolyte—semiconductor boundary it is necessary to multiply the value R by the electronic charge and by the thickness of the depletion layer region [equation (26.17)]:

$$\left| i^{gen} \right| \simeq e C_p \frac{n_i^2}{n'} L_1 \sqrt{\left| \frac{e\varphi_1}{kT} \right|}.$$

(37.4)

(Here L_1 is the Debye length, which is close to the thickness of the depletion layer at low values of φ_1.) The concentration n' is related to the value n_i by the equation $n' = n_i e^{(E_t^0 - E_i)/kT}$.

By using formula (6.14) also, we rewrite (37.4) in the form

$$\left| i^{gen} \right| \simeq \frac{e}{\tau_{po}} n_i e^{\frac{E_i - E_t^*}{kT}} L_1 \sqrt{\left| \frac{e\varphi_1}{kT} \right|}.$$

(37.5)

In order to compare the generation current in the Debye region (37.5) with the limiting current of generation in the quasineutral region (27.4), it is convenient to rewrite these expressions in the form

$$i^{\text{gen}} \sim \frac{n_i}{\tau_{p0}}, \quad i^{\lim} \sim \frac{n_i^2}{\sqrt{\tau_p}}.$$

(37.6)

It is readily seen that with a decrease in n_i (i.e., with an increase in the width of the forbidden band) the current $i^{\lim}$ falls much more rapidly than i^{gen}. Therefore, in semiconductors with a relatively narrow forbidden band (for example, in germanium) generation in the quasineutral region normally predominates, while in semiconductors with a wide forbidden band, generation in the space charge region predominates.

Let us now compare $i^{\lim}$ and i^{gen} for the same semiconductor (i.e., with given values of n_i and τ_p) and find conditions under which the generation current in the Debye region is the main one, i.e., $i^{\lim}/i^{\text{gen}} \ll 1$. By using formulas (27.4) and (37.4) we obtain

$$\frac{i^{\lim}}{i^{\text{gen}}} = \frac{en_i^2 D_p}{N_D L_p} \div \frac{en_i^2 L_1 \cdot \sqrt{\left|\frac{e\varphi_1}{kT}\right|}}{\tau_{p0} n'} = \frac{n' D_p \tau_{p0}}{N_D L_p L_1 \sqrt{\left|\frac{e\varphi_1}{kT}\right|}} \ll 1. \quad (37.7)$$

When relation (37.7) holds, formula (37.5) determines the form of the volt-ampere characteristics of the contact at high currents. It is important that in equation (37.5) the Debye length $L_1 \sim 1/\sqrt{n^0}$ appears. If n' is independent of n^0, then

$$i^{\text{gen}} \sim \frac{1}{\sqrt{n^0}} \sim \sqrt{\rho}.$$

(37.8)

Thus, the current i^{gen} is proportional to the square root of the specific resistance of the semiconductor ρ.

By an approximate method we have obtained the volt-ampere characteristics at high currents. Let us now turn to the exact theory, which makes it possible to find the volt-ampere characteristics at any currents. Firstly it is assumed that holes are necessary for anode solution of the semiconductor material. We will assume that the relation between the total solution current and the concentration of holes, as in the solution of germanium, is given by expression (27.1). Neglecting the potential drop in the Helmholtz layer we write this relation in the form

$$i = - i^0 \frac{p_s}{p_s^0} .$$

(37.9)

By taking logarithms of formula (37.9), substituting the value of p_s from relation (29.2), and using the expression

$$p_s^0 = p^0 e^{-\frac{e\varphi_1^0}{kT}} ,$$

we obtain

$$\varphi_1 - \varphi_1^0 = - \frac{kT}{e} \ln \left(- \frac{i}{i^0} \right) + \frac{kT}{e} \ln \left(1 - \frac{i_{p1}}{i_p^{\lim}} \right) ,$$

(37.10)

where i_{p1} is the hole current at the boundary of the quasineutral and space charge regions.

To find the volt-ampere characteristics of the system it is necessary to relate the current i_{p1} to the potential and the total current i. For this we examine the change in the hole current in the Debye region with the coordinate

$$\frac{di_p}{dx} = - R.$$

(37.11)

In accordance with formula (37.2) we write the latter expression in the form

$$\frac{di_p}{dx} = - A (n \cdot p - n_i^2).$$

(37.12)

By substituting the concentrations of electrons and holes at the surface [expressions (29.2) and (26.11)] in formula (37.12) we obtain

$$\frac{di_p}{dx} = - A n_i^2 \left[\left(1 - \frac{i_{p1}}{i_p^{\lim}} \right) \left(1 - \frac{\sqrt{2}}{i_n^{\lim}} \int\limits_0^{\sqrt{\frac{-e\varphi_1}{kT}}} i_n e^{z^2} \, dz \right) - 1 \right].$$

(37.13)

The latter relation may also be rewritten in the form

$$\frac{di_p}{dx} = A\, n_i^2 \left[\frac{i_{p1}}{i_p^{\lim}} + \left(1 - \frac{i_{p1}}{i_p^{\lim}} \right) \frac{\sqrt{2}}{i_n^{\lim}} \int_0^{\sqrt{\frac{-e\varphi_1}{kT}}} i_n e^{z^2}\, dz \right].$$

(37.14)

The second term in formula (37.14) is connected with the deviation in the electron concentration from equilibrium. We will first eliminate this term in the calculations and then show that this may be done over a wide range of potentials and currents.

The derivative di_p/dx now assumes the form

$$\frac{di_p}{dx} = An_i^2 \frac{i_{p1}}{i_p^{\lim}}.$$

(37.15)

Integrating with respect to x from x = 0 to x = L_{eff}, we obtain

$$i_{ps} - i_{p1} = -\, An_i^2 \frac{i_{p1}}{i_p^{\lim}} L_{eff} \quad,$$

(37.16)

where i_{ps} is the value of the hole current at the electrode surface.

Assuming that in the region adjacent to the contact there arises a depletion layer, and using expression (26.17) for L_{eff}, we obtain

$$i_{ps} - i_{p1} = -\, An_i^2 L_1 \sqrt{-\frac{2e\varphi_1}{kT}} \frac{i_{p1}}{i_p^{\lim}}.$$

(37.17)

It will be assumed that the following inequality holds:

$$\frac{An_i^2 L_1}{i_p^{\lim}} \sqrt{-\frac{2e\varphi_1}{kT}} \gg 1,$$

which is equivalent to the inequality (37.7). Then (37.17) is written in the form

$$i_{ps} = -\, An_i^2 L_1 \sqrt{-\frac{2e\varphi_1}{kT}} \frac{i_{p1}}{i_p^{\lim}}.$$

(37.18)

By substituting the value of i_{p1} from (37.18) in (37.10) we find

$$\varphi_1 - \varphi_1^0 = -\frac{kT}{e}\ln\left(-\frac{i}{i^0}\right) + \frac{kT}{e}\ln\left(1 + \frac{i_{ps}}{An_i^2 L_1 \sqrt{-\dfrac{2e\varphi_1}{kT}}}\right). \quad (37.19)$$

Assuming that relation (27.2) between the hole and electron currents at the surface holds, we can express i_{ps} in terms of the total solution current

$$\varphi_1 - \varphi_1^0 = -\frac{kT}{e}\ln\left(-\frac{i}{i^0}\right) + \frac{kT}{e}\ln\left(1 + \frac{i}{\left(1+\dfrac{m}{r}\right)An_i^2 L_1 \sqrt{-\dfrac{2e\varphi_1}{kT}}}\right).$$

$$(37.20)$$

Formula (37.20) gives the relation between the potential applied to the contact φ_1 and the current i, i.e., the volt-ampere characteristics of the system. With small currents, when the following relation holds:

$$i \bigg/ \left(1 + \frac{m}{r}\right) An_i^2 L_1 \sqrt{-\frac{2e\varphi_1}{kT}} \ll 1, \quad (37.21)$$

the second term in formula (37.20) may be neglected and the relation of the current to the potential is given by the expression

$$\varphi_1 - \varphi_1^0 = -\frac{kT}{e}\ln\left(-i/i^0\right), \quad (37.22)$$

i.e., the potential depends logarithmically on the current. With a further increase in the current, the value of $i[1 + (m/r)] \cdot An_i^2 L_1 \sqrt{-2e\varphi_1/kT}$ approaches −1 and the current i is given approximately by [cf. equation (37.4)]

$$i = -An_i^2 L_1 \left(1 + \frac{m}{r}\right)\sqrt{-\frac{2e\varphi_1}{kT}}. \quad (37.23)$$

The latter relation may also be written in the form

$$-\varphi_1 = \frac{kT}{2e}\frac{i^2}{A^2 n_i^4 L_1^2 \left(1+\dfrac{m}{r}\right)^2}. \quad (37.24)$$

The volt-ampere curve described by equation (37.23) will be a good approximation for equation (37.20) if ξ, which is determined by the relation

$$i = - An_i^2 L_1 \left(1 + \frac{m}{r}\right) \sqrt{-\frac{2e\varphi_1}{kT}} + \xi \qquad (37.25)$$

is considerably less in absolute value than

$$An_i^2 L_1 \left(1 + \frac{m}{r}\right) \sqrt{-\frac{2e\varphi_1}{kT}}.$$

By substituting relation (37.25) in formula (37.20) and neglecting $\ln(-i/i^0)$ in comparison with

$$\ln \left[1 + \frac{i}{\left(1 + \frac{m}{r}\right) An_i^2 L_1 \sqrt{-\frac{2e\varphi_1}{kT}}}\right],$$

we obtain for ξ the expression

$$\xi = \left(1 + \frac{m}{r}\right) An_i^2 L_1 \sqrt{-\frac{2e\varphi_1}{kT}} \cdot e^{\frac{e\left(\varphi_1 - \varphi_1^0\right)}{kT}}.$$

Thus, provided that the condition

$$\exp\left[\frac{e(\varphi_1 - \varphi_1^0)}{kT}\right] \ll 1$$

holds, formula (37.23) will be a good approximation for expression (37.20).

From a comparison of equations (37.22) and (37.24) it is obvious that the volt-ampere characteristics consist of two regions. In the first region (at low currents) the potential changes slowly with the current by a logarithmic law, while in the second region it changes rapidly by a parabolic law.

Finally, it remained to check the validity of the approximation made previously. In particular, we assumed that in formula (37.14) it is possible to omit the term

$$\frac{\sqrt{2}}{i_n^{\lim}}\left(1 - \frac{i_{p1}}{i_p^{\lim}}\right)\int_0^{\sqrt{-\frac{e\varphi_1}{kT}}} i_n e^{z^2}\, dz,$$

(37.26)

which is small in comparison with $i_{p1}/i_p^{\lim}$. We should note the following: from equation (37.10) it is evident that when $i_{p1}/i_p^{\lim} \ll 1$ the equation of the volt-ampere curve has the form of (37.22), independent of the accuracy with which the relation between i_{ps} and i_{p1} was found. Therefore, expression (37.26) should be estimated in this case when $i_{p1}/i_p^{\lim}$ approaches 1, i.e., in the parabolic region [equation (37.24)]. In this case, from equation (37.10) it is evident that

$$1 - \frac{i_{p1}}{i_p^{\lim}} = e^{\frac{e\left(\varphi_1 - \varphi_1^0\right)}{kT}}.$$

(37.27)

We will now estimate the integral which appears in expression (37.26). We should note that the electron current i_n forms only part of the total current and, therefore, we only increase the integral if we replace the electron current i_n by the total current.

We then take out the exponent e^{z^2} from the integral sign and take the exponent with the maximum value $z = \sqrt{-e\varphi_1/kT}$; then we obtain

$$\frac{1}{i_n^{\lim}}\left(1 - \frac{i_{p1}}{i_p^{\lim}}\right)\int_0^{\sqrt{-\frac{e\varphi_1}{kT}}} i e^{z^2}\, dz \approx \frac{e^{\frac{e\left(\varphi_1 - \varphi_1^0\right)}{kT}}}{i_n^{\lim}} i\, e^{-\frac{e\varphi_1}{kT}} \int_0^{\sqrt{-\frac{e\varphi_1}{kT}}} dz =$$

$$= e^{-\frac{e\varphi_1^0}{kT}} \cdot \sqrt{-\frac{e\varphi_1}{kT}} \cdot \frac{i}{i_n^{\lim}}.$$

(37.28)

Thus, the condition of applicability of the theory developed above has the form

$$e^{-\frac{e\varphi_1^0}{kT}} \sqrt{-\frac{e\varphi_1}{kT}} \frac{i}{i_n^{\lim}} \leqslant 1.$$

(37.29)

Since the currents in electrochemical systems are many orders less than the limiting electron currents in an n-type semiconductor, relation (37.29) will no longer hold only with very strong initial impoverishment of the surface in electrons.

§38. Kinetics with Slow Generation of Minority Carriers in the Space Charge Region (Experiment)

a. Anode Polarization Curves. On a germanium electrode the effect of generation in the space charge region is not normally noticeable against the background of large diffusion currents of minority carriers. On the other hand, on silicon and other semiconductors with a wide forbidden band, this mechanism becomes predominant.

Generation in the space charge region was first observed by Flynn in an investigation of anode solution of n-type silicon in hydrofluoric acid [60] and was then investigated in a series of studies [49, 111, 112]. As Fig. 81 shows, with a current density less than 10^{-5} A/cm^2, the rate of anode solution is independent of the type of conductivity and increases with potential by an exponential law. With a further increase in the current density on p-type silicon, Tafel's law still applies, while n-type samples are strongly polarized.

The overvoltage on an n-type silicon anode reaches some tens of volts and is apparently localized in the space charge region and not in the Helmholtz layer. When the electrode is illuminated the current density of anode solution of n-type silicon increases. All this made it possible to conclude that the anode solution of silicon, like germanium, requires holes and on n-type samples where holes are the minority carriers, the solution rate is limited by the supply of holes to the electrode surface. The volt-ampere characteristics of n-type silicon are described qualitatively by formula (37.20). A change from a logarithmic to a quadratic relation of the current to the potential occurs at 10^{-5}-10^{-4} A/cm^2. As was mentioned in §37, the experimental currents of the anode solution of silicon are several orders higher than the limiting diffusion current calculated

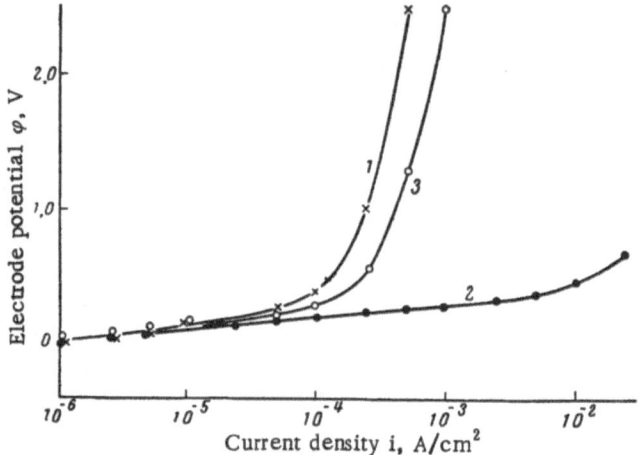

Fig. 81. Relation of the anode current to the potential of a silicon anode [49]. 1) n-Type, 10 $\Omega \cdot$ cm; 2) p-type, 10 $\Omega \cdot$ cm, 2.5 N HF solution; 3) n-type, 10 $\Omega \cdot$ cm, 2.5 N HF + 0.05 N $K_3Fe(CN)_6$.

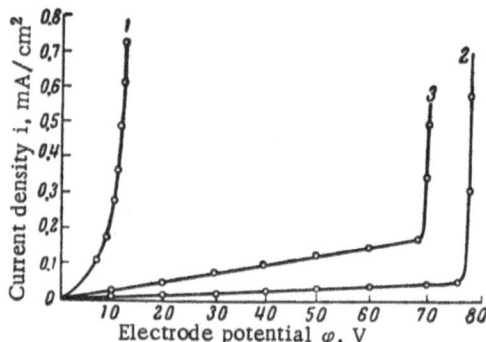

Fig. 82. Anode polarization curves on a silicon electrode [114]. 1) p-Type, 1 $\Omega \cdot$ cm, 10 N KOH solution; 2) n-type, 1 $\Omega \cdot$ cm, 10 N KOH solution; 3) the same, 1 N H_2SO_4 solution.

from equation (27.4). Such a high rate of solution is difficult to explain by surface generation of minority carriers since it would be necessary to assume an improbably high value for the surface recombination rate (more than 10^6 cm/sec). It is obvious that in the anode solution of silicon the holes are generated in the space charge region at the silicon—electrolyte interphase. This is supported by the fact that the solution rate is independent of the sample thickness over a range from a few microns to tenths of a millimeter [111].

When potassium ferricyanide and other oxidants are introduced into the solution the anode current density on n-type silicon increases (Fig. 81, curve 3). The authors of [49, 111] suggested (in analogy with a germanium electrode) that the reduction of the $Fe(CN)_6^{3-}$ ion on silicon involves valence electrons and is accompanied by the injection of holes into the semiconductor. The reduction proceeds simultaneously with anodic dissolution and increases its rate. Analogous results were obtained in [113].

Until now there has been no quantitative check of the theory given in §37. For this it is necessary to plot the volt-ampere characteristics and to measure the parameters of the generation centers (energy position and concentration) on the same sample. Flynn [60] estimated the current with formula (37.5) and obtained the right order of magnitude. (He used literature data on the characteristics of the generation centers in silicon samples.)

In solutions which do not contain hydrofluoric acid, even with a slight anode polarization (0.2-0.3 V) silicon is passivated toward anode solution. In the passive region the volt-ampere characteristics are independent of the type of conductivity (see §39).* Figure 82 gives polarization curves plotted in NaOH and H_2SO_4 solutions with overvoltages up to 80 V. With p-type silicon the current increases rapidly with potential, while with n-type samples in NaOH solution the current density is only $2 \cdot 10^{-4}$ A/cm^2 at 70 V. As was shown in [114], in the region of high overvoltages there is a linear relation between the square of

---

* At 15-20 V, breakdown of the oxide film occurs and with a further increase in polarization the rate of the anode process, i.e., the liberation of oxygen and further growth of the oxide film, again differs markedly on n- and p-type samples [52, 114, 115].

the current and the potential and apparently the reaction rate is also determined by the generation of holes in the space charge region. In H_2SO_4 solutions the current is higher than in alkali (Fig. 82) and shows a more complex relation to the potential. It is possible that in acid solutions the reaction rate is determined to a great extent by the state of the electrode surface.

Polarization curves of the anode solution of gallium arsenide are similar to curves plotted on active silicon. While the current increases exponentially with the potential on p-type samples, n-type samples are strongly polarized. Thus, with an overvoltage of 1 V, the current density is about 10^{-4} A/cm^2. On sufficiently pure samples the polarization reaches 5-6 V and more. With a constant potential the current increases when the electrode is illuminated. Estimation from formula (37.5) gives for the beginning of the "quadratic" region a current density of 10^{-6} A/cm^2 (with $L_1 = 10^{-5}$ cm, $\tau_p = 10^{-10}$ sec, $n_i = 10^8$ cm^{-3}), which is somewhat lower than the values observed experimentally. The current density depends on the state of the electrode surface [53]. There is apparently multiplication of current on an anode of n-type gallium arsenide (α' is 1.5-2) [116].

The anode decomposition of n-type cadmium sulfide also involves holes. The current density is 10^{-8}-10^{-5} A/cm^2 for different samples. It is determined by the rate of generation of holes in the space charge region and probably on the electrode surface. By means of equation (37.5), from the magnitude of the current it is possible to estimate the position of the generation centers in the forbidden band of cadmium sulfide, namely, 0.9-1.1 eV above the threshold of the valence band [117].

Let us examine the relation between the generation rate and specific resistance of the semiconductor ρ. According to equation (37.8), the generation current must be proportional to the square root of ρ. In actual fact, the rate of anode solution of n-type silicon in hydrofluoric acid increases with an increase in the specific resistance [111]. However, in the case of cadmium sulfide the rate of the anode process is practically independent of ρ [117]. As will be shown below for the cathode liberation of hydrogen on p-type germanium, an increase in the specific resistance may even be accompanied by a fall in the generation rate. The reason for this apparent contradiction is evidently as follows. Formula (37.8) was derived on the assumption that a change in the specific resistance of a semiconductor affects the generation current only through a change in the width of the space charge region (i.e., the region in which minority carriers are generated) and the concentration of generation centers remains constant. In actual fact, as a rule, the doping of a semiconductor material with donors or acceptors is accompanied by its contamination by foreign substances. In contrast to the main doping impurities, they may not affect the electrical conductivity (if their energy levels are far from edges of the forbidden band), but they affect the lifetime of minority carriers in the bulk and their generation rate at contacts. The concentration of crystallographic defects in heavily doped crystals usually is also increased. The generation current

$$i_{gen} \sim N_{gen} \cdot L_1 \sim N_{gen} \cdot \sqrt{\rho},$$

where N_{gen} is the concentration of generation centers. If N_{gen} changes with ρ more rapidly than $\sqrt{\rho}$, then the effect of the change in the concentration of generation centers predominates over the effect of the change in the width of the space charge region and with an increase in the specific resistance, the generation current falls. If $N_{gen} \sim 1/\sqrt{\rho}$, then the two effects compensate each other.

To summarize the above material, it may be stated that the anode processes examined up to now on semiconductors with a wide forbidden band proceed with the participation of holes and on samples with n-type conductivity the rate of the process is determined by the rate of generation of holes in the space charge region at the contact with the electrolyte.

b. Structure of Space Charge Region. By measuring the capacitance it was shown that in the cases examined above, on polarization of the electrode, the space charge region was a depletion layer in which practically the whole of the overvoltage was concentrated.* On cadmium sulfide and gallium arsenide electrodes, a linear relation between the square of the reciprocal capacitance and the potential, which is characteristic of the depletion layer, is observed with anode polarization right up to the breakdown potential and not merely close to the steady potential, where the current through the interphase is very small (§ 16). The slope of $(1/C^2) - \varphi$ lines is close to that calculated from the bulk concentration of donors [equation (12.11)] [118, 119]. On a silicon electrode with n-type conductivity the capacitance also falls on anode polarization, as is characteristic of a depletion layer [120, 121].

Other methods have been used for investigating the space charge in addition to measurement of the capacitance. Thus, Williams [122, 123] measured experimentally the width of the space charge region on a cadmium sulfide electrode and compared it with the value calculated from equation (26.17). The measurements were made with the device illustrated in Fig. 83. A thin plate of cadmium sulfide with ohmic contacts at the ends was placed in a plastic reservoir filled with KCl solution. The contacts remained outside the reservoir and were used to measure the conductivity along the sample (a small constant potential was applied to them during the experiment). An auxiliary electrode was placed in the solution for polarization of the cadmium sulfide. With an increase in the anode polarization the depletion layer at the cadmium sulfide − solution interphase became thicker. When a certain critical potential was reached it occupied the whole of the sample and the current along the crystal

---

*In this case, the name "depletion layer" is not quite accurate. With the limiting current of minority carriers (for example, in the anode solution of an n-type semiconductor), the conditions for the depletion layer $n \ll N_D$ and $p \ll N_D$ do not hold in a certain part of the space charge region close to the contact. In actual fact, the concentration of minority carriers at the contact p_s, which determines the magnitude of the current, may be comparable with N_D or even exceed it. However, we will use this term because of its brevity, bearing in mind the fact that the true depletion layer does not occupy the whole of the space charge region.

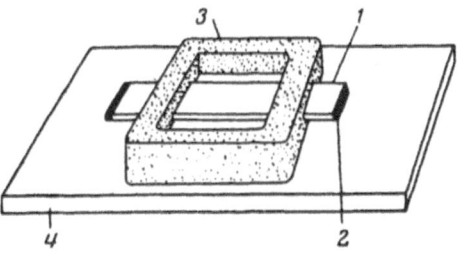

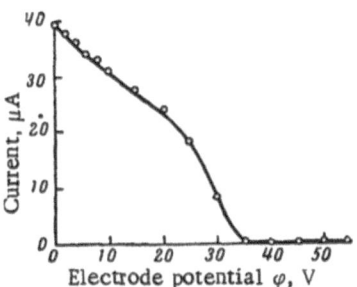

Fig. 83. Device for measuring the thickness of the depletion layer in cadmium sulfide crystals at the boundary with the electrolyte. 1) Cadmium sulfide crystal; 2) ohmic contact; 3) reservoir; 4) glass plate.

Fig. 84. Relation of the current along a flat crystal (see Fig. 83) to the electrode potential φ [122].

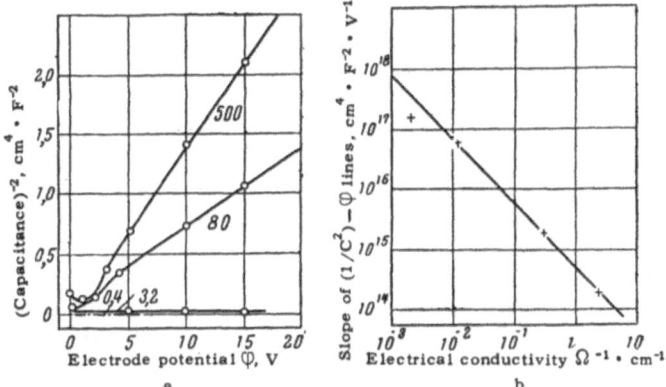

Fig. 85. Relation of the reciprocal square of the capacitance to the electrode potential φ of a silicon cathode (p-type) (a); the specific resistance (in $\Omega \cdot$ cm) is given on the curves. Relation of the slope of $1/C^2 - \varphi$ lines (from Fig. 85a) to the specific electrical resistance of the silicon (b) [90].

fell to an insignificantly low value. At this moment the thickness of the space charge region equaled half the thickness of the sample, which could be measured readily. This method uses the phenomenon of modulation of the conductivity along the sample by a transverse field. (The field triode is based on the same principle.)

The results of measurements for one sample are given in Fig. 84. Complete shutoff of the measuring current along the sample occurred at a cadmium sulfide potential of about 35 V relative to the auxiliary electrode. For samples with a specific resistance of the order of 1 $\Omega \cdot$ cm the thickness of the space charge region measured in this way coincides with the calculated value with an accuracy of $\pm 10\%$.

An analogous method was used in the case of a silicon electrode with the only difference that the indicator was not the longitudinal conductivity of the sample, but the electrode capacitance. At the moment that the space charge regions on opposite sides of the sample merged, the whole of the electrode immersed in the solution was insulated from the current lead because of the high resistance of the depletion layer. Therefore, the capacitance of the electrode changed abruptly to some low value and with a further increase in polarization it was independent of both the potential and the area of the silicon—solution contact. In this case also, good agreement was obtained between the direct measurement of the thickness of the space charge region and calculation [124].

A depletion layer is also formed on cathode polarization of p-type silicon [90]. The square of the reciprocal capacitance changes with polarization by a linear law (Fig. 85a) with the slope of $(1/C^2) - \varphi$ lines inversely proportional to the electrical conductivity and, consequently, the concentration of acceptors in the samples (Fig. 85b).

c. Cathodic Liberation of Hydrogen on p-Type Silicon. Flynn [60] put forward the hypothesis that the current density on a cathode of p-type silicon is determined by the generation of free electrons in the space charge region. In actual fact, a barrier effect was observed in [60, 125, 126] in the cathode liberation of hydrogen on p-type silicon. While on n-type silicon the current and potential followed a Tafel relation with a slope from 0.09 [125] to 0 18 V [50], on p-type cathodes it is possible to reach an overvoltage of 40–50 V and above with a current of 10^{-3} A/cm^2 [52]. The cathodic current increases on illumination of the electrode.

It was observed in [112] that the current falls with an increase in the specific resistance of silicon and this does not agree with equation (37.8). A possible reason for this was considered on p. 247.

It was found that the current increases with potential more rapidly than in proportion to $\sqrt{\varphi}$. At a constant potential the current normally increases with time. On the other hand, a peak is observed on galvanostatic charging curves, i.e., at the moment of switching on the cathodic current the electrode acquires a very negative potential, which then gradually becomes less negative with a direct current. On cathodes with n-type conductivity this peak is absent and the potential shifts monotonically toward negative values [51, 127]. All these phenomena may be associated with the adsorption of atomic hydrogen on the electrode or its penetration into the crystal lattice of the silicon. * Thereupon there apparently arise new recombination-generation centers and there is an increase in the rate of generation of free electrons, while there is also a change in the distribution of the potential between the space charge region and the Helmholtz region.

On the whole, cathode processes on silicon have not been investigated adequately to draw definite conclusions on their mechanism.

To conclude this section, we should examine the theory of retardation of the cathodic process on semiconductor electrodes proposed in [50] and then used in some other papers. It was observed that cathodes of both p-type and n-type silicon are strongly polarized at current densities of 10^{-3} and 10^{-2} A/cm^2, respectively, in KOH, H$_2$SO$_4$, and HF solutions. At lower current densities (10^{-6}-10^{-4} A/cm^2) the liberation of hydrogen obeys the Tafel law. The difference in the polarization curves on n-type and p-type silicon and also the phenomenon of strong

---

*The presence of hydrogen or silicon hydrides on the surface of a silicon cathode was detected by means of curves of the potential drop after the cathode current had been switched off [52].

polarization of the electrode itself were associated by the authors of [50] not with the participation of free electrons in the cathode reaction, but with the ohmic potential drop in the impoverished layer of the semiconductor, which is included in the measured value of the overvoltage. It was assumed that the thickness and, consequently, the resistance of this layer were maximal in the case of low-resistance p-type silicon and minimal with low-resistance n-type silicon.

However, the space charge region cannot be represented as an ohmic resistance. In actual fact, the potential drop in this region φ_1 does not have an ohmic character. Its increase with an increase in the polarization of the electrode is accompanied by an exponential increase in the current [equation (24.14)] if minority carriers participate in the reaction. Therefore, we cannot expect strong polarization of a cathode of an n-type semiconductor if free electrons participate in the liberation of hydrogen. On cathodes of both n- and p-type, the high polarization may be connected with the participation of valence electrons in the reaction if the potential drop in the Helmholtz region changes little on polarization. However, the latter may occur over a limited range of potentials (somewhat less than the width of the forbidden band of the semiconductor). With sufficiently high cathodic polarization, the potential drop in the Helmholtz layer begins to increase (§ 10), while the current increases with the overvoltage approximately exponentially. Finally, the electrode is strongly polarized if minority carriers participate in the reaction. In this case there is formed at the contact a depletion layer, in which practically the whole of the overvoltage is concentrated, and the current is related to the potential by equation (27.3) or (37.20).

The reason for the retardation of the cathodic liberation of hydrogen on n-type silicon described in [50, 51] remains unclear.

d. Breakdown of the Surface Barrier. On all the electrodes examined above, with a sufficiently high anodic polarization there is breakdown of the surface barrier and the current increases sharply. This phenomenon was investigated in detail with cadmium sulfide electrodes by Williams [122, 123,

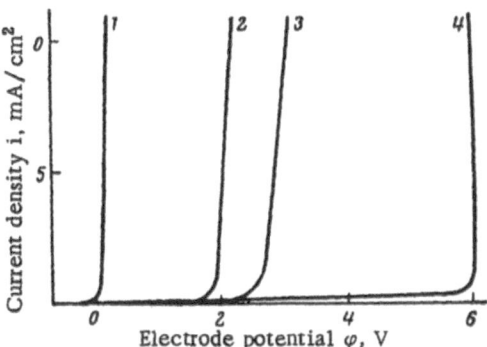

Fig. 86. Anode polarization curves for p-type gallium arsenide [53]. Donor concentration: 1) 10^{18}
cm^{-3}; 2) $5 \cdot 10^{17}$ cm^{-3}; 3) $9 \cdot 10^{16}$ cm^{-3}; 4) $5 \cdot 10^{15}$
cm^{-3}.

128]. The breakdown potential was 80–100 V. The field strength
at the surface on breakdown depends on the concentration of
donors N and varies from 10^6 to 2.5 · 10^6 V/cm with a change
in N_D from 10^{16} to 2.4 · 10^{17} cm^{-3}. By injecting holes with light
and comparing the current of injected holes with the photocur-
rent through the interphase, Williams found that they were equal
to each other at all potentials. This means that impact ionization does not occur in the case of cadmium sulfide. Breakdown
is apparently connected with the transition of electrons into the
conductivity band from the valence band from deep traps or
from the solution under the action of the strong field.

In the case of very heavily doped crystals of cadmium sul-
fide ($\rho \approx 0.1 \ \Omega \cdot$ cm), breakdown occurs at 20–25 V and is accompanied by luminescence [129]. The intensity of the lumin-
escence is proportional to the current, while its spectral distribution depends on the composition of the solution. For example,
in NaH_2PO_4 solution a maximum in the luminescence spectrum
is observed at 580 mμ, while in KBr solution it is at 640 mμ.

Breakdown occurs at much lower voltages in gallium arsenide (Fig. 86), but this may be connected with inadequate
purity of the samples used. With an increase in the donor con-

centration, the breakdown voltage falls, but more slowly than the thickness of the space charge region. Consequently, the critical value of the field strength at which the surface barrier breaks down increases with an increase in N_D [53].

In n-type silicon the breakdown voltage depends on the composition of the solution and in an alkaline solution it is close to 80 V (Fig. 82).

§ 39. Formation of Oxide Layers on Semiconductor Electrodes

As in the case of metal electrodes, if the products of anode oxidation of a semiconductor are insoluble or only slightly soluble in the electrolyte, they are deposited on the electrode surface. Oxide films are formed particularly readily on silicon as silicon oxides are insoluble in most electrolytes. In acid and neutral aqueous solutions silicon is always covered with an oxide film, which passivates it toward anode solution and dissolution. Therefore, silicon does not dissolve in water with the liberation of hydrogen, though its equilibrium potential is much more negative than the hydrogen potential (see, for example, [130]). Silicon oxides are soluble in hot concentrated alkalis and the amount of oxygen on the surface at the steady potential and with cathode polarization is apparently low. The silicon is active, as is shown, in particular, by rapid dissolution with the liberation of hydrogen (this process will be analyzed in § 54). However, even under these conditions a shift in the potential of 0.2-0.3 V in an anode direction produces passivation of the silicon [113, 131].

The potentiostatic anode curve for n-type silicon in 10 N KOH solution is given in Fig. 87. It is similar to the passivation curves of metal electrodes. At the potential of the maximum the electrode is already passivated toward dissolution. In the region of the maximum the magnitude of the current depends on the type of conductivity and the specific resistance of the silicon, while at more positive potentials the current is independent of the semiconductor properties of the electrode, but is determined by the rate of transfer of electric charges through

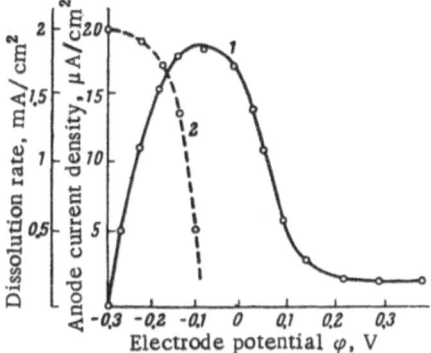

Fig. 87. Anode passivation of silicon in 10 N KOH
solution [132]. 1) Anode polarization curve; 2) relation of silicon dissolution rate (in current density
units) to potential.

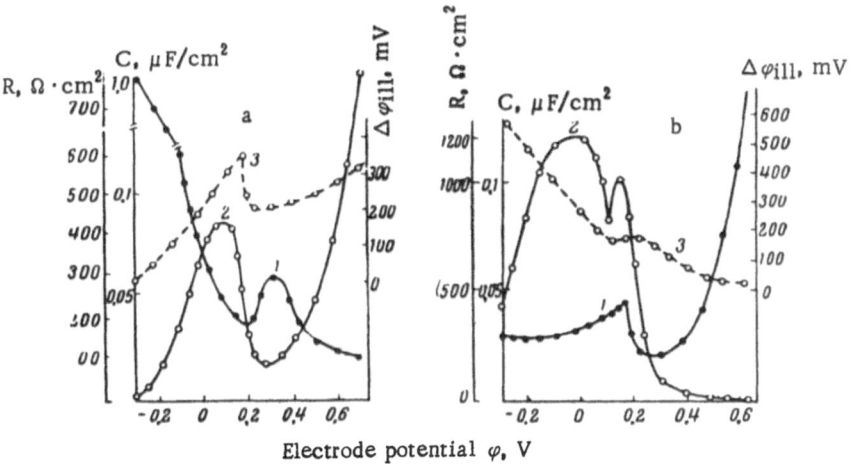

Electrode potential φ, V

Fig. 88. Anode passivation of silicon (a — n-type, b — p-type) in 10 N KOH solution. Relation of capacitance C (1), resistance R (2), and photopotential $\Delta \varphi_{ill}$ (3)
to electrode potential φ [132].

the film [115]. At 15-20 V the liberation of oxygen begins on
the passive electrode.

In investigating the structure of the double layer on active and passive silicon, Izidinov [115, 132] observed that in the
passivation process there is a change in the distribution of the

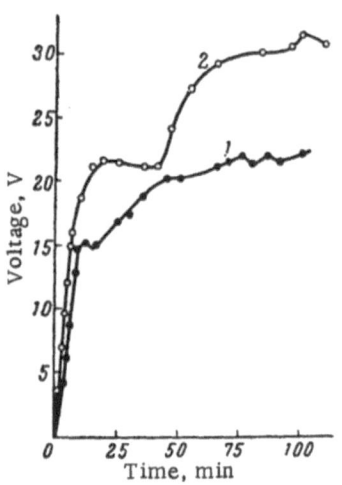

Fig. 89. Galvanostatic "voltage—time" curve in the formation of an oxide film on a silicon anode in water [133]. 1) n-Type (0.08 Ω · cm); 2) p-type (30 Ω · cm). Current density $0.35 \cdot 10^{-3}$ A/cm².

potential at the silicon—solution interface. At a potential of about 0.2 V (i.e., somewhat more positive than the potential at the maximum on the polarization curve) the regular increase in the photopotential and fall in the capacitance, which are characteristic of the depletion layer on an n-type semiconductor, are disturbed (Fig. 88a). The anomalies observed may be explained by assuming that at this potential there is an abrupt change in the potential drop in the Helmholtz layer or in the oxide film (as occurs on germanium, see § 14). This shift is about 0.2 V. In actual fact, as Fig. 88 shows, at an electrode potential of 0.2 V the capacitance and photopotential again assume values corresponding to a potential of 0 V. The change in the interphase potential drop φ_0 is connected with a change in the coverage of the surface by oxygen or with a change in the character of the silicon—oxygen bond [115]. These observations agree with the results in [113], where a shift in the $C - \varphi$ curve toward negative potentials was observed on activation of a silicon electrode by addition of sodium fluoride to the H_2SO_4 solution in which the silicon was passivated. The surface oxides of silicon dissolve in solutions of fluorides and the silicon changes into an active state.

In solutions which do not contain alkalis or fluoride ions, on anode polarization of a silicon electrode a thick oxide layer (up to several thousand angstroms) is formed on its surface. The kinetics of its growth are approximately the same as on the barrier-layer metals aluminum, tantalum, etc. With a constant current density the voltage on the cell changes linearly

with time (Fig. 89). At some critical voltage, which is called
the "sparking voltage," the linear relation is disrupted and
breakdown of the oxide film soon occurs.

The difference between silicon and the barrier-layer metals lies in the fact that at the beginning of the growth of the
oxide layer the rate of the process depends on the semiconductor properties of the backing and is apparently determined to a
considerable extent by the properties of the energy barrier at
the silicon—oxide interface [115, 133].

On silicon samples coated with a thin layer of amorphous
oxide before anodizing, light emission is observed at a potential of about 1 V. The addition of oxidants (permanganate, bichromate, etc.) to the solution reduces its intensity. With a
change to the linear section of the φ—t curve, the emission
ceases. It is not observed in solutions of alkalis and fluorides,
which dissolve the oxide film [134].

With thickening of the oxide layer the effect of the silicon—
oxide interphase on the growth kinetics is reduced and with a
thickness above 400 Å, the limiting stage becomes the transport
of material through the oxides [135].

Aqueous solutions of sulfuric and boric acids, in which
the breakdown voltage does not exceed 200 V, may be used as
electrolytes for anodizing. The best results are obtained with
solutions of nitrates in methylacetamide [136, 137]. The n-type
samples are anodized with light emission. At first the anodization conditions are galvanostatic (current density of 10 mA/cm^2)
and then at 260 V there is a change to potentiostatic conditions
and the current falls with time. Then anodization is carried
out at a constant current density up to 560 V. The field strength
inside the film is independent of its thickness and is about 2.6
· 10^7 V/cm.

The current yield for SiO$_2$ formation does not exceed 1%
and the rest of the current is consumed in the liberation of hydrogen. However, the introduction of halogen ions into the solution raises the current yield to 50%.

An electron microscope investigation showed [138] that nonporous oxide layers are formed in methylacetamide. Films grown in aqueous solutions of H_2SO_4 and H_3BO_3 are porous with the diameter of the pores increasing with an increase in the formation voltage. However, under these conditions on the actual surface of the silicon there is a very thin (< 100 Å) solid oxide layer.

In its electrical properties the oxide film on silicon is similar to oxide layers on barrier-layer metals [139-141]. In particular, the silicon−oxide−solution system has rectifying properties. The back direction of the current is anodic: as long as the voltage remains less than the formation voltage of the film, the current through the interphase is very low. In the case of n-type silicon, on cathode polarization the system has a very low resistance.* On p-type silicon the cathode direction is the back direction, but here the barrier at the silicon−oxide interphase plays a part: the transmission of a cathode current requires electrons, which are the minority carriers in p-type silicon [140].

A layer of SiO_2 on silicon has good dielectric properties and may be used in electrolytic capacitors, for example, for producing systems with a nonlinear capacitance [143]. In addition, anodic oxide films are used for masking the surface of silicon in the production of semiconductor instruments and also for stabilization and protection of instruments.

Water dissolves germanium oxides and therefore aqueous solutions are unsuitable for producing oxide layers on this ma-

---

*There is a rectifying action only in protonic solvents (water and liquid ammonia). In an aprotic solvent (liquid sulfur dioxide) it is very markedly reduced or is not observed at all. According to one of the rectification theories [142], the penetration of protons into the oxide layer under the action of the field during cathodic polarization increases the electrical conductivity of the oxide and is responsible for the low resistance of the system in the forward direction. Under anodic polarization the protons are removed from the oxide and the resistance of the system again increases (back direction).

terial. Nonaqueous solvents such as nitrates in methylacetamide [136] and sodium acetate in acetic acid [144] are used for this purpose. The best results are obtained in a solution with the following composition: $0.6 \cdot 10^{-3}$ M $LiNO_3$, 0.1 M CH_3COOH, 0.28 M H_2O, 10^{-5} M GeO_2 in acetic anhydride. The current density is from 25 to 250 μA cm^2. Layers of crystalline germanium dioxide are formed with a thickness up to 7000 Å with a current yield up to 80%. The breakdown voltage is 150 V [145, 146]. The kinetics of growth of an oxide layer on germanium are the same as on silicon. The field strength in the film is about 10^6 V/cm and the specific resistance 10^{10} $\Omega \cdot$ cm.

An oxide layer on indium antimonide may be obtained by anodization in 0.1 N KOH solution. With a low current density the oxidation rate depends on the crystallographic orientation of the indium antimonide and at a high current density it is determined by the rate of transfer of material through the oxide [147, 148]. The film formed shows photosensitivity [149, 150].

The semiconductor properties of oxides of nickel, manganese, and other metals are of great importance in the electrochemistry of oxide electrodes, which are used as chemical current sources (see, for example, [151]).

§ 40. Photoeffect under Conditions of the Limiting Current of Minority Carriers

The theory of the photopotential given in § 17 is based on the assumption that the diffusion length of minority carriers L_p is much greater than the thickness of the space charge region L_1. This condition is used in the conversion from the calculated photocurrent of minority carriers to the photopoential. If the space charge region in a semiconductor is a depletion layer, as occurs under conditions of the limiting current of minority carriers, * there is another possible way of calculating the photopotential in which the condition that $L_p \gg L_1$ is not necessary.

---

*See footnote on p. 248.

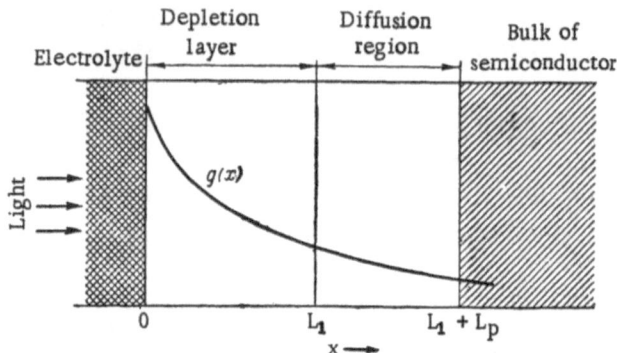

Fig. 90. Diagram of light absorption in a semiconductor
and origin of the photoeffect in the depletion layer.

Let us consider an electrolyte – semiconductor interphase
(as a concrete example, an n–type semiconductor) illuminated
from the electrolyte side, * and calculate the photocurrent [152].
The density of the energy dissipated on absorption of the light,
as a function of the coordinate x, is described by the expression

$$g(x) = \Phi a e^{-ax}, \tag{40.1}$$

where a is the coefficient of light absorption by the semiconductor and $\Phi = \int_0^\infty g(x)\,dx$ is the intensity of the light flux falling on

the semiconductor surface.

The function g(x) is illustrated by the solid curve in Fig.
90. On absorption of photons in the semiconductor, electron-
hole pairs are generated and in the simplest case the generation
function coincides with g(x). If excitons are formed as an intermediate stage in the absorption of light, then the generation
function of nonequilibrium carriers changes with the coordinate

---

*This examination is also valid for a metal–semiconductor contact and for a p-n
junction if the layer of the metal or the p-type semiconductor is thin enough and
transparent to light.

more slowly than g(x) (and the more slowly, the longer the life-time of the exciton).

Let a depletion layer of thickness L_1 exist in the surface region of the semiconductor. When $1/a \ll L_1$, the light is practically completely absorbed in the depletion layer. When $1/a \approx L_1$, a considerable part of the incident light is absorbed beyond the space charge region in the neutral bulk of the semiconductor.

Let us examine the behavior of nonequilibrium carriers generated by the light. In the space charge region the holes move under the action of the electric field toward the surface; then they recombine through surface recombination centers and pass into the electrolyte as a result of an electrochemical process or are held in surface traps (in the latter case the process is nonsteady). It is assumed that nonequilibrium holes do not recombine in the space charge region. (This assumption is justified as the concentration of free electrons in the depletion layer is very low.) With sufficiently high fields the space charge region is a potential well for holes formed in the neutral bulk of the semiconductor and diffusing to the edge of the space charge region ($x = L_1$). The width of the diffusion region from which the space charge region "collects" nonequilibrium holes is equal in order of magnitude to the diffusion length of holes L_p. Thus, the photocurrent consists of two components:

$$i_\Phi = i_1 + i_B, \tag{40.2}$$

where i_1 is the generation current of holes in the space charge region and i_B is the diffusion current of holes from the neutral bulk.

Photoelectrons move under the action of the electric field from the surface to the depth of the semiconductor.

Let us assume for simplicity that the generation function of holes is described by equation (40.1) and we obtain for the current i the following expression:

$$i_1 = -e \int_0^{L_1} g(x)\,dx = e\Phi\,(e^{-aL_1} - 1).$$

$$(40.3)$$

For holes in the neutral bulk we may write the equation of continuity

$$D_p \frac{d^2 p}{dx^2} - \frac{p - p^0}{\tau_p} + g(x) = 0,$$

$$(40.4)$$

where D_p is the diffusion coefficient of holes, p^0 is their dark concentration, and τ_p is the lifetime.

This equation differs from (17.11) in the last term, which takes into account the generation of holes by light. The following boundary conditions were selected in [152]*:

$$p = p^0 \quad \text{when } x = \infty, \tag{40.5a}$$

$$p = 0 \quad \text{when } x = L_1. \tag{40.5b}$$

A solution for equation (40.4) with the limiting conditions (40.5a) and (40.5b) is given by the following expression:

$$p = p^0 - (p^0 + Ae^{-aL_1})\,e^{(L_1 - x)/L_p} + Ae^{-ax}, \tag{40.6}$$

where

$$L_p = \sqrt{D_p \tau_p} \text{ and } A = \frac{\Phi}{D_p}\,\frac{aL_p^2}{(1 - a^2 L_p^2)}.$$

---

*The condition (40.5b) is too rigid. The correct treatment is given in [153], which is devoted to calculation of the kinetics of the photoeffect in the depletion layer. We should adopt as the limiting condition

$$D\frac{dp}{dx}\bigg|_{x=L_1} = p(L_1) \cdot v,$$

$$(40.5c)$$

where v is a parameter determined by the properties of the depletion layer and the mechanism of the disappearance of holes at the contact. In particular, for high fields, v equals half the thermal velocity of holes. When $v \to \infty$, condition (40.5c) changes into (40.5b).

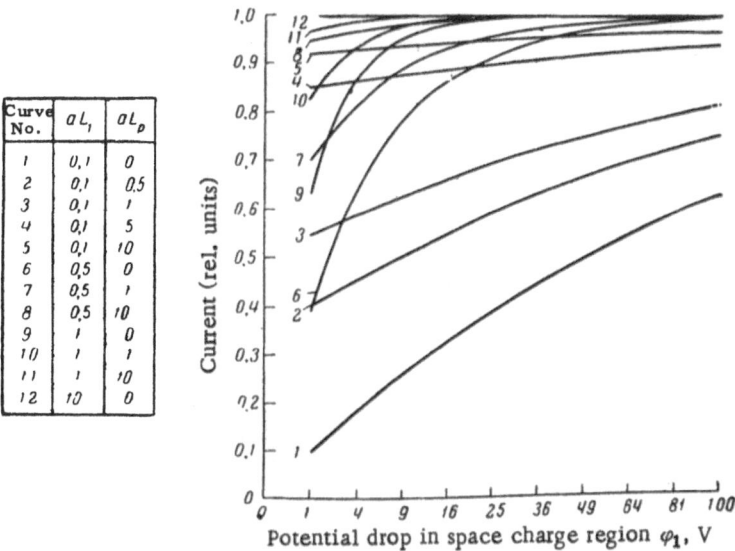

Curve No.	aL_1	aL_p
1	0,1	0
2	0,1	0,5
3	0,1	1
4	0,1	5
5	0,1	10
6	0,5	0
7	0,5	1
8	0,5	10
9	1	0
10	1	1
11	1	10
12	10	0

Fig. 91. Theoretical curves of the relation of the photocurrent to the potential drop in the space charge region φ_1 at various values of aL_1 and aL_p [152].

Hence, for the current i_B we obtain (when $x = L_1$)

$$i_B = -e\Phi \frac{aL_p}{1 + aL_p} e^{-aL_1}. \tag{40.7}$$

From equations (40.2), (40.3), and (40.7) it follows that

$$i_\Phi = -e\Phi [1 - e^{-aL_1}/(1 + aL_p)]. \tag{40.8}$$

Thus, the photocurrent is a function of the light-absorption coefficient, the diffusion length of holes, and the thickness of the depletion layer, which depends on the potential [see equation (26.17)]. Figure 91 shows the photocurrent, normalized relative to its maximum value, in relation to the potential drop in the depletion layer φ_1 (a and L_p are the parameters). Here it is assumed that $L_1 = L_1^0 \sqrt{\varphi_1}$, where L_1^0 is the width of the space charge region when $\varphi_1 = 1$ V. The figure shows that with values of L_p and $1/a$ which are not too small and not too great in comparison with L_1, the photocurrent depends strongly on the potential.

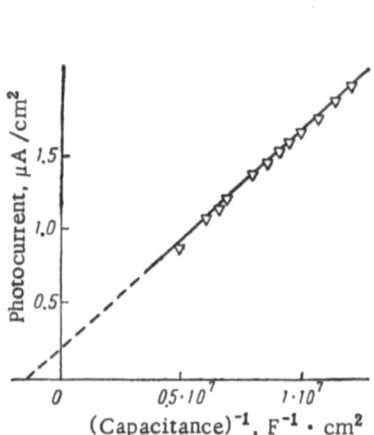

Fig. 92. Relation of photocurrent to the reciprocal capacitance of a cadmium sulfide electrode in K_2SO_4 solution [154].

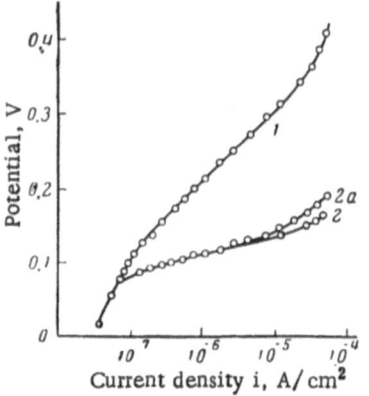

Fig. 93. Distribution of potential on a germanium electrode (n-type, 1.1 $\Omega \cdot$ cm) under anode polarization in 1 N H_2SO_4 solution [162]. 1) Change in electrode potential $\Delta \varphi$; 2,2a) change in potential in Helmholtz layer $\Delta \varphi_0$, measured at frequencies of 5 and 20 kHz, respectively.

As an example, let us examine the photoeffect at a cadmium sulfide – solution contact [154]. The wavelength of the light was chosen so that the condition $1/a > L_1$ held. The results of measuring the photocurrent i in relation to the reciprocal of the capacity, which is proportional to the width of the depletion layer L_1, are illustrated in Fig. 92. Extrapolation of the photocurrent to $1/C \to 0$ (i.e., to zero width of the space charge region) gives the value of the diffusion component of the photocurrent. The diffusion length of holes L_p was determined by extrapolation of the relation of i_Φ to $1/C$ to $i_\Phi = 0$. The lifetime of the holes calculated from L_p was 10^{-9}–10^{-10} sec. This method of measuring short lifetimes was used previously for p-n junctions in crystals of gallium arsenide and phosphide [155, 156].

For cadmium sulfide, $L_1 \lesssim L_p$ and, therefore, the photocurrent depends substantially on the potential. On the other hand, for germanium, the photocurrent hardly changes with polarization [157], since $L_1 \ll L_p$ and the photocurrent is determined wholly by the diffusion current of holes i_B, which is independent of the potential.

Thus, electrons and holes formed by absorption of light are separated by the field of the space charge. Under galvanostatic conditions this leads to charging of the double layer. Assuming that nonequilibrium holes are trapped close to the surface, we may write for the photopotential

$$\Delta\varphi = \frac{Q}{\overline{C}},$$

(40.9)

where Q is the magnitude of the separated charge and $\overline{C}$ is the integral capacitance of the depletion layer.

Thus, the photo emf is related to the electrode potential through the capacitance. These concepts were confirmed by investigation of the photopotential of cadmium sulfide under a direct current [154, 158]. From the values of $\Delta\varphi$ and Q measured experimentally, by means of formula (40.9) the capacitance was determined and this was found to be very close to the value measured by a direct method (charging the electrode with square current pulses).*

The specific resistance of the semiconductor affects the photopotential (with a given electrode potential) through the change in L_1 and, consequently, the capacitance.

There has been no quantitative comparison of the photopotential or photocurrent with the capacitance on other semiconductor materials. It has only been observed that over a wide range the photocurrent is proportional to the intensity of illumination. Analysis of work on silicon (for example, [115]) shows that with an increase in the capacitance the photopotential falls so that qualitatively Q ≈ const [see equation (40.9)], as should be observed for semiconductors with a high diffusion length

---

*The idea that the photopotential could be treated as the result of charging an electrochemical capacitance by a photocurrent was put forward by Veselovskii [159].

$(L_p \gg L_1)$. The spectral distribution of the photopotential on an n-type germanium anode and a p-type germanium cathode has been investigated [160]. The kinetics of the photopotential during the anodic solution of n-type germanium were examined in [121,161].

To conclude this section we will examine the method of investigating the distribution of the potential at a germanium—electrolyte interface [162], which is based on the change in the potential drop in the space charge layer on illumination. The method is applicable under conditions where a depletion layer arises in the semiconductor. Its principle is as follows. In the dark the capacitance of the depletion layer falls with an increase in the polarization of the electrode because of the increase in φ_1 [equation (12.10)]. On the other hand, illumination reduces φ_1 and increases the capacitance with a constant electrode potential. It is assumed that with weak illumination, as in the dark, the relation of the capacitance to the potential is described by formula (12.10) (i.e., the concentration of free carriers in the region of the depletion layer remains less than the concentration of ionized impurities), so that illumination affects the capacitance only through the change in φ_1. Then C_1 and φ_1 are unequivocally related and we may use measurement of the capacitance for following φ_1. It is essential that polarization and illumination have opposite effects on the capacitance under the conditions of the depletion layer.

The measurements are made in the following way. The electrode potential is shifted from some initial value so that the concentration of free carriers at the surface decreases. The capacitance thereupon decreases and simultaneously the potential drops in the space charge region and the Helmholtz layer changes. Then the illumination of the electrode is adjusted so that the capacitance returns to the original value. Then the potential drop in the space charge region φ_1 also assumes the previous (dark) value and the difference between the electrode potential before polarization in the dark and the potential after polarization with illumination equals the change in the Helmholtz potential drop. Figure 93 shows an experimental polarization curve (current—electrode potential) and the "corrected" polarization curve (current—overvoltage in Helmholtz layer) found by the method described for n-type germanium with weak anode polarization. In this work it was observed that the change in the potential difference in the Helmholtz layer is comparable with the change in the potential drop in the space charge layer (see § 13).

This method is particularly effective in the case of low-resistance (heavily doped) samples as in this case the region of the depletion layer is retained under equilibrium conditions over a sufficiently wide range of potentials (on the equilibrium $C_1 - \varphi$ curve it lies between the flat-band potential and the minimum capacitance potential). The method is not applicable to samples whose conductivity approaches the intrinsic level. Its advantage over the method described in § 13 of comparing experimental and theoretical $C - \varphi$ curves lies in the fact that it requires no knowledge of either the absolute value of the capacitance (and consequently, the true surface of the electrode) or the theoretical $C - \varphi$ relation for the given sample. On the other hand, this is also a drawback of the method as it is difficult to be certain that the measured capacitance actually is the capacitance of the space charge (which is unequivocally determined by the value φ_1) and does not include, for example, the capacitances of fast surface states. However, as the latter also depend on the value φ_1, the limitations on the concentration of fast states on the electrode surface in this method are apparently less rigid than in the method of comparing $C - \varphi$ curves. In actual fact, the results obtained in [162] were practically independent of frequency over the range of 0.4-20 kHz.

Literature Cited

1. A.N. Frumkin, V.S. Bagotskii, Z.A. Ioffa, and B.N. Kabanov, Kinetics of Electrode Processes, Izd. Mosk. Gos. Univ. (1952).
2. Yu.A. Vdovin, V.G. Levich, and V.A. Myamlin, Collection: Some Problems in Theoretical Physics, Atomizdat, Moscow (1958), p.3.
3. Yu.A. Vdovin, V.G. Levich, and V.A. Myamlin, Dokl. Akad. Nauk SSSR, 124 : 350 (1949).

4. M. Green, Solid-State Physics in Electronics and Tele-
    communications, London (1959), p. 619.
5. J.F. Dewald, Ann. N.Y. Acad. Sci., 101:872 (1963).
6. M. Green, New Problems in Modern Electrochemistry,
    Vol. 2 [Russian translation], IL, Moscow (1962), p.377.
7. R.M. Lazorenko-Manevich and S.O. Izidinov, Dokl.
    Akad. Nauk SSSR, 140:172 (1961).
8. R.A. Marcus, Trans. Symposium on Electrode Processes
    (ed. E. Yeager), Wiley, New York (1961), p.239.
9. J. Dewald, Semiconductors (ed. N. Hannay), Reinhold,
    New York (1959).
10. H. Gerischer, Z. Phys. Chem., Frankfort, 26:223,325
    (1960); 27:48 (1961).
11. H. Gerischer, Surface Chemistry of Metals and Semi-
    conductors (ed. H.C. Gatos), New York, Wiley (1960),
    p.177.
12. H. Gerischer, Advances in Electrochemistry and Electro-
    chemical Engineering, Vol. 1 (ed. P. Delahay),
    Intersciences Publishers, New York–London (1961), p. 139
13. R.R. Dogonadze and Yu.A. Chizmadzhev, Dokl.Akad.
    Nauk SSSR, 144:1077 (1962); 145:849 (1962); 150:333
    (1963).
14. R.R. Dogonadze, A.M. Kuznetsov, and Yu.A.
    Chizmadzhev, Zh.Fiz.Khim., 38:1195 (1964).
15. A.M. Kuznetsov and R.R. Dogonadze, Izv. Akad. Nauk
    SSSR, Ser. Khim., (1964), p. 2140.
16. W. Shockley, Theory of Electronic Semiconductors [Rus-
    sian translation], IL, Moscow (1953).
17. Yu.A. Vdovin, B.M. Grafov, and V.A. Myamlin, Dokl.
    Akad. Nauk SSSR, 129:827 (1959).
18. Yu.A. Vdovin, V.G. Levich, and V.A. Myamlin, Some
    Problems in Theoretical Physics, Atomizdat, Moscow
    (1958), p. 10.
19. Yu.A. Vdovin, V.G. Levich, and V.A. Myamlin, Dokl.
    Akad. Nauk SSSR, 126:1296 (1959).
20. V.A. Myamlin, Dokl. Akad. Nauk SSSR, 140:870 (1961).
21. E A. Efimov and I.G. Erusalimchik, Zh.Fiz.Khim.,
    32:413 (1958).

22. E.A. Efimov and I.G. Erusalimchik, Zh. Fiz. Khim.,
    32 : 1103 (1958).

23. V.A. Tyagai, Elektrokhimiya, 1: 387 (1965).

24. A.M. Goodman, Surface Sci., 1: 54 (1964).

25. J.F. Dewald, Surface Chemistry of Metals and Semiconductors (ed. H.C. Gatos), Wiley, New York, (1960), p. 205.

26. D.R. Turner, J. Electrochem. Soc., 103 : 252 (1956).

27. M.N. Platonova, Zh. Prikl. Khim., 35 : 334 (1962); V.M.
    Kochegarov, E.A. Zyablova, and F.I. Zaburdaeva, Zh.
    Prikl. Khim., 37 : 1494 (1964).

28. F. Jirsa, Z. Anorg. Chem., 268 : 84 (1952).

29. M. Green and P.H. Robinson, J. Electrochem. Soc.,
    106 : 253 (1959).

30. Lj. Duić, Z. Kovač, and B. Lovreček, Croat. Chem.
    Acta, 32 : 213 (1960).

31. F. Beck and H. Gerischer, Z. Electrochem., 63 : 500 (1959).

32. K. Bohnenkamp and H.-J. Engell, Z. Elektrochem.,
    61 : 1184 (1957).

33. R.M. Lazorenko-Manevich, Dokl. Akad. Nauk SSSR, 144 :
    1094 (1962).

34. A.A. Yakovleva, T.I. Borisova, and V.I. Veselovskii,
    Dokl. Akad. Nauk SSSR, 133 : 889 (1960).

35. A.A. Yakovleva, T.I. Borisova, and V.I. Veselovskii,
    Dokl. Akad. Nauk SSSR, 145 : 373 (1962).

36. T.P. Hore, New Problems in Modern Electrochemistry
    [Russian translation], Vol. 2, IL, Moscow (1962), p. 284.

37. H. Gerischer, Record Chem. Progr (Kresge-Hooker
    Sci. Libr.), 23 : 134 (1962); J.A. Harrison and H.
    Gerischer, Z. Elektrochem., 66 : 762 (1962); P.J. Boddy,
    J. Electrochem. Soc., 111 : 1136 (1964).

38. H. Gobrecht, O. Meinhardt, and W. Mindt., Ber.
    Bunsenges., 68 : 190 (1964).

39. V.R. Erdélyi and M. Green, Nature, 182 : 1592 (1958).

40. S. Sheff, H.C. Gatos, and S. Zwerdling, Rev. Scient.
    Instrum., 29 : 531 (1958).

41. I.V. Borovkov, Zh. Fiz. Khim., 34 : 2682 (1960).

42. A.G. Pecherskaya and V.V. Stender, Zh. Fiz. Khim.,
    24 : 856 (1950).

43.   B.N. Zuev, Collection: Metallurgical and Chemical In-
       dustry of Kazakhstan, Central Institute for Scientific and
       Technical Information, Alma-Ata, 6(10) : 82 (1960).
44.   R.M. Lazorenko-Manevich, Dissertation, L. Ya. Kar-
       pova Physicochemical Institute, Moscow (1963).
45.   D.P. Zosimovich and N.E. Nechaeva, Proceedings of
       the Fourth Conference on Electrochemistry, Moscow
       (1959), p. 541.
46.   J. Mieluch, Bull. Acad. Polon. Sci., 7: 151 (1959).
47.   H. Gerischer, Anal. Real. Soc. Espan. Fis. Quim.,
       B-56 : 535 (1960).
48.   E.A. Efimov and I.G. Erusalimchik, Zh. Fiz. Khim.,
       32 : 1967 (1958).
49.   E.A. Efimov and I.G. Erusalimchik, Zh. Fiz. Khim.,
       35 : 384 (1961).
50.   E.A. Efimov and I.G. Erusalimchik, Dokl. Akad. Nauk
       SSSR, 124 : 609 (1959).
51.   E.A. Efimov, I.G. Erusalimchik, and G.P. Sokolova,
       Zh. Fiz. Khim., 36 : 1005 (1962).
52.   M. Seipt and H. Fischer, Anal. Real. Soc. Espan. Fis.
       Quim., B-56 : 443 (1960).
53.   Yu. V. Pleskov, Dokl. Akad. Nauk SSSR, 143 : 1399 (1962).
54.   A.N. Frumkin, L.I. Bogulavskii, and V.S. Serebrennikov,
       Dokl. Akad. Nauk SSSR, 142 : 878 (1962).
55.   W.H. Brattain and C.G.B. Garrett, Bell System Techn.
       J., 34 : 129 (1955).
56.   W.H. Brattain and C.G.B. Garrett, Physica, 20 : 885
       (1954).
57.   W.H. Brattain and C.G.B. Garrett, Phys. Rev., 94 : 750
       (1954).
58.   S.G. Ellis, Phys. Rev., 100 : 1140 (1955).
59.   A. Uhlir, Bell System Techn. J., 35 : 333 (1956).
60.   J.B. Flynn, J. Electrochem. Soc., 105 : 715 (1958).
61.   W. Shockley and W.T. Read, Phys. Rev., 87 : 835 (1952).
62.   P.P. Konorov and M.N. Kolbin, Fiz. Tverd. Tela, 3 :
       1553 (1961).
63.   V.A. Tyagai and Yu. V. Pleskov, Fiz. Tverd. Tela, 4 :
       343 (1962).

64. W.W. Harvey, Quart. Progr. Rept. Solid-State Res. Lincoln Lab. MIT, Cambridge, Mass., Jan., 1961, p. 11.

65. Yu. V. Pleskov, Dokl. Akad. Nauk SSSR, 132 : 1360 (1960).

66. H. Gerischer and F. Beck, Z. Phys. Chem., Frankfort, 24 : 378 (1960).

67. E. A. Efimov and I. G. Erusalimchik, Zh. Fiz. Khim., 38 : 589 (1964).

68. E. A. Efimov and I. G. Erusalimchik, Dokl. Akad. Nauk SSSR, 128 : 124 (1959).

69. E. A. Efimov and I. G. Erusalimchik, Zh. Fiz. Khim., 35 : 543 (1961).

70. E. A. Efimov and I. G. Erusalimchik, Zh. Fiz. Khim., 36 : 1791 (1962).

71. T. Gabor, J. Appl. Phys., 32 : 1361 (1961); A. Wolkenberg, Roczinki Chem., 39 : 291 (1965).

72. G. Dejardin, G. Mesnard, and A. Dolce, Compt. Rend., 246 : 1016 (1958).

73. F. Beck and H. Gerischer, Z. Elektrochem., 63 : 943 (1959).

74. R. J. Flannery, J. E. Thomas, and D. Trivich, J. Electrochem. Soc., 110 : 1054 (1963).

75. Z. Trouzil, Czech. J. Fiz., 4 : 238 (1954).

76. Yu. V. Pleskov, Dokl. Akad. Nauk SSSR, 130 : 362 (1960).

77. E. A. Efimov and I. G. Erusalimchik, Dokl. Akad. Nauk SSSR, 122 : 632 (1958).

78. F. Berz and D. C. Emmony, Phys. Letters, 2 : 197 (1962).

79. Yu. V. Pleskov, Dokl. Akad. Nauk SSSR, 126 : 111 (1959).

80. W. W. Harvey, J. Phys. Chem. Solids, 14 : 82 (1960).

81. O. V. Romanov and P. P. Konorov, Fiz. Tverd. Tela, 4 : 2276 (1962).

82. H. Gobrecht, O. Meinhardt, and L. Schulz, Ber. Bunsenges., 67 : 156 (1963).

83. E. A. Efimov, I. G. Erusalimchik, and E. I. Gorgoraki, Zh. Fiz. Khim., 38 : 720 (1964).

84. E. A. Efimov and I. G. Erusalimchik, Zh. Fiz. Khim., 34 : 2804 (1960).

85. R. M. Lazorenko-Manevich, N. A. Aladzhalova, and V. I. Veselovskii, Dokl. Akad. Nauk SSSR, 133 : 620 (1960).

86. E.N. Paleolog, A. Z. Fedotova, and N.D. Tomashov, Dokl. Akad. Nauk SSSR, 137:900 (1961).

87. N.D. Tomashov, E.N. Paleolog, and A. Z. Fedotova, Zh. Fiz. Khim., 34:832 (1960).

88. H. Gobrecht, R. Kuhnkies, and A. Tausend, Z. Elektrochem., 63:541 (1959).

89. R.M. Lazorenko-Manevich, Zh. Fiz. Khim., 38:1235 (1964); Electrochim. Acta, 10:141 (1965).

90. H. Gobrecht and O. Meinhardt, Ber. Bunsenges., 67:142 (1963); R. Memming, Philips Res. Repts., 19:323 (1964); V.I. Veselovskii, T.I. Borisova, A.A. Yakovleva, and S. O. Izidinov, Electrochim. Acta, 10:325 (1965).

91. H. Gerischer and F. Beck. Z. Phys. Chem., Frankfort, 13:389 (1957).

92. E.N. Paleolog, A. Z. Fedotova, and N.D. Tomashov, Dokl. Akad. Nauk SSSR, 129:623 (1959).

93. W. Mehl, H.F. Gossenberger, and E. Helpert, J. Electrochem. Soc., 110:239 (1963).

94. V.V. Eletskii and Yu.V. Pleskov, Elektrokhimiya, 1:194 (1965).

95. W.W. Harvey, J. Phys. Chem., 65:1641 (1961).

96. H. Kallmann and M. Pope, Nature, 185:753 (1960).

97. M. Silver, M. Swicord, R.C. Jarnagin, A. Many, S.C. Weisz, and M. Simhony, J. Phys. Chem. Solids, 23:419 (1962).

98. W. Helfrich and P. Mark, Z. Phys., 168:495 (1962).

99. R.C. Jarnagin, J. Gilliland, and J.S. Kim, J. Chem. Phys., 39:573 (1963).

100. Yu.V. Pleskov, Proceedings of International Conference on Semiconductor Physics, Prague (1961), p. 573.

101. Yu.V. Pleskov, Zh. Fiz. Khim., 35:2576 (1961).

102. Yu.V. Pleskov, Dissertation, Inst. Fiz.-Khim., Akad. Nauk SSSR, Moscow (1960).

103. Yu.V. Pleskov and V.A. Tyagai, Dokl. Akad. Nauk SSSR, 141:1135 (1961).

104. Yu.V. Pleskov, Zh. Fiz. Khim., 35:2540 (1961).

105. V.G. Levich, Physicochemical Hydrodynamics, Fizmatgiz, Moscow (1959).

106.   P. F. Schmidt and C. H. Church, J. Electrochem. Soc., 108 : 296 (1961).

107.   Yu. V. Pleskov and B. N. Kabanov, Dokl. Akad. Nauk SSSR, 123 : 884 (1958).

108.   Yu. V. Pleskov, Zh. Fiz. Khim., 34 : 623 (1960).

109.   M. Balkanski, J. Bardeleben, and A. F. Bogenschutz, J. Chim. Phys., 57 : 507 (1960); C. Krischer, Dissertation, Techn. Hochschule, Stuttgart (1963); E. A. Efimov and I. G. Erusalimchik, Zh. Fiz. Khim., 38: 2868 (1964).

110.   V. A. Myamlin, Dokl. Akad. Nauk SSSR, 139 : 1153 (1961).

111.   E. A. Efimov and I. G. Erusalimchik, Dokl. Akad. Nauk SSSR, 130 : 353 (1960); E. A. Efimov, I. G. Erusalimchik, and G. P. Sokolova, Zh. Fiz. Khim., 38 : 2172 (1964).

112.   E. N. Paleolog, K. S. Korotkova, and N. D. Tomashov, Dokl. Akad. Nauk SSSR, 133 : 170 (1960).

113.   R. M. Hurd and P. T. Wrotenbery, Ann. N. Y. Acad. Sci., 101 : 876 (1963).

114.   S. O. Izidinov, T. I. Borisova, and V. I. Veselovskii, Zh. Fiz. Khim., 36 : 1246 (1962).

115.   S. O. Izidinov, Dissertation, L. Ya. Karpova Physico-chemical Institute, Moscow (1963).

116.   S. A. Greenberg and R. N. Sanders, J. Electrochem. Soc., 110 : 188 C (1963).

117.   V. A. Tyagai, Izv. Akad. Nauk SSSR, Ser. Khim. (1963), 1956.

118.   T. P. Birintseva and Yu. V. Pleskov, Izv. Akad. Nauk SSSR, Ser. Khim. (1965), p. 251.

119.   V. A. Tyagai, Izv. Akad. Nauk SSSR, Ser. Khim. (1964), 34.

120.   V. I. Zvyagin and A. S. Lyutovich, Izv. Akad. Nauk UzSSR, Ser. Fiz.-Matem. Nauk (1959), p. 25.

121.   H. Gobrecht, O. Meinhardt, and B. Reinicke, Ber. Bunsenges, 67 : 493 (1963).

122.   R. Williams, Phys. Rev., 123 : 1645 (1961).

123.   R. Williams, J. Phys. Chem. Solids, 22 : 129 (1961).

124.   K. Böke, Z. Naturforsch., 15a : 550 (1960).

125.   G. Greger, Z. Naturforsch., 16a : 284 (1961).

126.   M. Seipt, Z. Naturforsch., 14a : 926 (1959).

127. E.A. Efimov, I.G. Erusalimchik, and G.P. Sokolova, Zh.Fiz.Khim., 36:1219 (1962).

128. R. Williams, Phys.Rev., 125:850 (1961).

129. E. Maruyama, J.Phys.Soc.Japan, 16:2341 (1961).

130. J.I. Carasso and M.M. Faktor, Electrochemistry of Semiconductors (ed. P.J.Holmes), Academic Press, London–New York (1962), p.205.

131. S.O. Izidinov, T.I. Borisova, and V.I. Veselovskii, Dokl.Akad.Nauk SSSR, 133:392 (1960).

132. S.O. Izidinov, T.I. Borisova, and V I. Veselovskii, Dokl.Akad.Nauk SSSR, 145:598 (1962).

133. L.A. Dubrovskii, V.G. Mel'nik, and L.O. Odynets, Zh.Fiz.Khim., 36:2199 (1962).

134. A. Gee, J. Electrochem. Soc., 107:787 (1960); S.P. Maminova and L.L. Odynets, Electrokhimiya, 1:365 (1965); W. Waring and E.A. Benjamini, J. Electrochem. Soc., 111:1256 (1964).

135. P.F. Schmidt and W. Michel, J. Electrochem. Soc., 104:230 (1957).

136. P.F. Schmidt, US Patent 2,909,470 (1959); Russ.Zh. Khim., 1:175 (1961); P.F. Schmidt and A.E. Owen, J. Electrochem. Soc., 111:682 (1964); P.F. Schmidt, Halbleiter und Phosphore, Braunschweig (1958), p.570.

137. E.F. Dufték, C.Mylroie, and E.A. Benjamini, J. Electrochem. Soc., 111:1042 (1964); S.P. Maminova and L.L. Odynets, Zh.Fiz.Khim., 39:531 (1965); P.F. Schmidt, R. Stickler, G.D. Rose, and A.N. Knopp, Electrochem. Technol., 3:49 (1965).

138. A. Politycky and E. Fuchs, Z.Naturforsch., 14a:271 (1959); A. Revesz and K. Zaininger, J.Phys., 25:66 (1964).

139. L. Young, Anodic Oxide Films, Academic Press, London–New York (1961).

140. P.F. Schmidt, F. Huber, and R.F. Schwarz, J.Phys. Chem.Solids, 15:270 (1960).

141. D.A. Vermilyea, Advances in Electrochemistry and Electrochemical Engineering, Vol. 3 (ed. P.Delahay), Interscience Publishers, New York–London (1963), p. 211; N.G. Bardina, Usp.Khim., 33:602 (1964).

142. P. F. Schmidt, J. Appl. Phys., 28:278 (1957).

143. L. S. Berman, Nonlinear Semiconductor Capacitance, Fizmatgiz, Moscow (1963).

144. S. Zwerdling and S. Sheff, J. Electrochem. Soc., 107: 338 (1960).

145. R. D. Wales, J. Electrochem. Soc., 110:914 (1963).

146. R. D. Wales, J. Electrochem. Soc., 111:478 (1964).

147. J. F. Dewald, J. Electrochem. Soc., 104:244 (1957).

148. M. C. Lavine, A. J. Rosenberg, and H. C. Gatos, J. Appl. Phys., 29:1131 (1958).

149. J. D. Venables and R. M Broudy, J. Appl. Phys., 30: 1110 (1959).

150. J. D. Venables and R. M. Broudy, J. Electrochem. Soc., 107:296 (1960).

151. P. D. Lukovtsev, Dissertation, Inst. Fiz.-Khim., Akad. Nauk SSSR (1952).

152. W. W. Gärtner, Phys. Rev., 116:84 (1959).

153. V. A. Tyagai, Zh. Fiz. Khim., 38:2472 (1964).

154. V. A. Tyagai, Fiz. Tverd. Tela, 6:1602 (1964).

155. R. A. Logan and A. G. Chynoweth, J. Appl. Phys., 33: 1649 (1962).

156. R. A. Logan, A. G. Chynoweth, and R. G. Cohen, Phys. Rev., 128:2518 (1962).

157. Yu. V. Pleskov, Elektrokhimiya, 1:4 (1965).

158. R. Williams, Phys. Rev., 117:1487 (1960).

159. V. I. Veselovskii, Dissertation, Moscow (1949).

160. S. Hemilä, J. Chem. Phys., 40:37, 3739 (1964).

161. P. J. Boddy and W. H. Brattain, Ann. N. Y. Acad. Sci., 101:683 (1963).

162. R. M. Lazorenko-Manevich, Dokl. Akad. Nauk SSSR, 144:1094 (1962).

Additional Literature

Alpatova, N. M., Yu. M. Kessler, and A. I. Gorbanev, Anode behavior of silicon in some nonaqueous solutions, Elektrokhimiya, 1:844, 1344 (1965).

Collard, J. R., and F. Sterzer, Dependence of the space-charge capacity of the reverse biased p-n junction on the current value, Appl. Phys. Letters, 5:165 (1964).

Dogonadze, R. R., A. M. Kuznetsov, and A. A. Chernenko, Theory of homogeneous and heterogeneous electron processes in liquids, Usp. khim., 34:1779 (1965).

Dogonadze, R. R., and A. M. Kuznetsov, Kinetics of oxidation-reduction reactions in the system impurity semiconductor-electrolyte solution, Elektrokhimiya, 1:742 (1965).

Efimov, E. A., and I. G. Erusalimchik, Anode solution of gallium arsenide, Elektrokhimiya, 1:818 (1965).

Eletskii, V. V., and Yu. V. Pleskov, Photoelectrochemical method of approximate measurement of diffusion length in semiconductors, Elektrokhimiya, 2 (in press).

Gerischer, H., Über den Mechanismus der anodischen Auflösung der Galliumarsenid, Ber. Bunsenges., 69: 578 (1965).

Kashcheeva, T. P., and Yu. V. Pleskov, Differential capacitance of a heavily doped germanium-electrolyte interphase, Elektrokhimiya, 2 (in press).

Konorov, P. P., and O. V. Romanov, Volt-ampere characteristics of surface barriers in germanium at the interface with electrolytes, Vestn. Leningr. Gos. Univ., Ser. Fiz. Khim., 22:65 (1965).

Lovrecek, B., and K. Moslavac, The anodic dissolution of germanium, Elektrochim. Acta, 10:627 (1965).

Maninova, S. P., and L. L. Odynets, Electrochemical oxidation of silicon in ethylene glycol, Elektrokhimiya, 2:346 (1966).

Many, A., Tunneling processes across the CdS-electrolyte interface, J. Phys. Chem. Solids, 26:587 (1965).

Mehl, W., Redoxreaktionen an Antracenelektroden, Ber. Bunsenges., 69:583 (1965).

Memming, P., and G. Schwandt, Anodic dissolution of silicon in hydrofluoric acid solutions, Surface Sci., 4:109 (1966).

Reid, W. E., Some electrochemical aspects of germanium dissolution, J. Phys. Chem., 69:2269, 3168 (1965).

Story, J. B., Mechanism of anodic germanium oxide film formation, J. Electrochem. Soc., 112:1107 (1965).

Wolkenberg, A., Wptyw naswietlania na rozpuszczanie sie elektrody germanowej w 0.1 M NaNO$_3$, Arch. Hutnictwa, 10:287 (1965).

Chapter III

Transmission of a Sinusoidal Current Through a Semiconductor—Electrolyte System

§ 41. Basic Concepts and Definitions

In previous sections we used the concept of the differential capacitance $C = -dQ/d\varphi$ to describe the semiconductor—electrolyte contact (see § 12). Subsequently we will use another, more general definition of capacitance.

Let us examine the properties of a semiconductor—electrolyte system when a sinusoidal current of low amplitude is passed through it. To characterize such a system it is normal to introduce the concepts of capacitance and resistance in analogy with the capacitance and resistance of linear electrical circuits. However, the electrolyte—semiconductor system is not linear and therefore its description by means of the concepts of resistance and capacitance is approximate. Let us examine a circuit consisting of two elements AB and BC in series, in which there is a potential drop. These elements may be, for example, the space charge region in a semiconductor, the Helmholtz layer, a homogeneous semiconductor, etc. The total electric current $\vec{I}$ consists of the current of free carriers $\vec{i}$ and the displacement current $(1/4\pi)(\partial\vec{D}/\partial t)$, where $\vec{D}$ is the electrostatic induction

$$\vec{I} = \vec{i} + \frac{1}{4\pi}\frac{\partial\vec{D}}{\partial t}.\qquad(41.1)$$

279

The total current $\vec{I}$ is independent of the coordinates. This means that with a change in the coordinates, for example, from point A to point B, each of the values $\vec{i}$ and $(1/4\pi)(\partial\vec{D}/\partial t)$ changes in such a way that the total current remains constant at all points. *

Let there flow through the system a direct current $\overline{I}$, on which is superimposed an alternating current of small amplitude $\tilde{I}$ for measuring the capacitance. (Here and below all values that are independent of time have a straight line above them, while all time-dependent values are marked by the sign ~.) The total current is given by

$$I = \overline{I} + \tilde{I}, \qquad (41.2)$$

where

$$\overline{I} = \overline{i},$$
$$\tilde{I} = \tilde{i} + \frac{1}{4\pi}\frac{\partial\tilde{D}}{\partial t}. \qquad (41.3)$$

---

*In actual fact, in accordance with Maxwell's equation (see, for example, [1]),

$$\operatorname{curl}\vec{H} = \frac{4\pi}{c}\vec{i} + \frac{1}{c}\frac{\partial\vec{D}}{\partial t}$$

(c is the velocity of light).

By taking the divergence from the right to the left part of the equation, and taking into account the fact that div curl $\vec{H} = 0$, we obtain

$$\operatorname{div}\left(\vec{i} + \frac{1}{4\pi}\frac{\partial\vec{D}}{\partial t}\right) = 0,$$

whence, in the one-dimensional case it follows that:

$$\frac{d}{dx}\left(i + \frac{1}{4\pi}\frac{\partial D}{\partial t}\right) = 0.$$

Integration with respect to the coordinate x leads to the expression

$$i + \frac{1}{4\pi}\frac{\partial D}{\partial t} = \text{const.}$$

In all the following discussion we will consider the one-dimensional case.

Let us assume that the circuit AB is a linear electrical circuit. This means that the potential drop φ_{AB} is related to the current $\tilde{I}$ by the expression

$$\tilde{\varphi}_{AB} = \tilde{I}Z_{AB}, \tag{41.4}$$

where Z_{AB} is the impedance of the circuit AB.

If the circuit AB is not a linear circuit, then the measurements are made by applying an alternating current of low amplitude so that the values $\tilde{i}$ and $\tilde{D}$ may be expanded into series with respect to the small parameter $e\tilde{\varphi}_{AB}/kT$. In this case, by taking the first term of the series, we have

$$\tilde{i} = a\tilde{\varphi}_{AB},$$
$$\tilde{D} = b\tilde{\varphi}_{AB}. \tag{41.5}$$

Below we will examine the transmission of a sinusoidal current, i.e., it will be assumed that $\tilde{\varphi}_{AB}$ is proportional to $e^{j\omega t}$, where ω is the frequency of the periodic signal and j is an imaginary number.

By substituting relation (41.5) in equation (41.3), we obtain

$$\tilde{I} = \left(a + \frac{j\omega b}{4\pi}\right)\tilde{\varphi}_{AB}. \tag{41.6}$$

Relation (41.6) may be written in the form (41.4) by introducing the concept of the impedance of the circuit AB.

If the circuit BC is also linear (or has been linearized by the method given above), in analogy with formula (41.4) it is possible to write the following:

$$\tilde{\varphi}_{BC} = \tilde{I}Z_{BC}. \tag{41.7}$$

We introduce the total impedance of the circuit Z through the relation

$$\tilde{\varphi} = \tilde{I}Z. \tag{41.8}$$

Since the total potential $\tilde{\varphi} = \tilde{\varphi}_{AB} + \tilde{\varphi}_{BC}$, then, by using formulas (41.4) and (41.7) it is possible to obtain for the total impedance of the circuit

$$Z = Z_{AB} + Z_{BC}. \qquad (41.9)$$

This relation defines the combination rule for impedances of circuits in series.

The impedance of the circuit is a complex value and, therefore, relation (41.8) may also be rewritten in the form

$$\tilde{I} = (l + jd)\,\tilde{\varphi}, \qquad (41.10)$$

where l and d are actual values.

The concepts of the resistance and capacitance of the circuit (in contrast to the impedance) are not unequivocal: they are determined by the selection of the equivalent circuit.

If the system is described by an equivalent circuit with a capacitor and a resistor in parallel, then

$$\tilde{I} = \left(\frac{1}{R} + j\omega C\right)\tilde{\varphi}. \qquad (41.11)$$

By comparing equations (41.11) and (41.7) we obtain a relation of the impedance of the system to the capacitance C and the resistance R:

$$Z^{-1} = \frac{1}{R} + j\omega C. \qquad (41.12)$$

The resistance and capacitance are determined by the coefficients l and d:

$$l = \frac{1}{R}\; ; d = \omega C. \qquad (41.13)$$

If we represent the system by an equivalent circuit with a capacitor C and resistor R in series, then in this case

$$\tilde{I} = \frac{\tilde{\varphi}}{R - \dfrac{j}{\omega C}}. \qquad (41.14)$$

The relation (41.14) may also be written in the following way:

$$\tilde{I} = \frac{R + \dfrac{i}{\omega C}}{R^2 + \dfrac{1}{\omega^2 C^2}} \, \tilde{\varphi} . \qquad (41.15)$$

The capacitance and resistance are now found from given values of l and d by the relations:

$$l = \frac{R\omega^2 C^2}{1 + R^2\omega^2 C^2}; \qquad d = \frac{\omega C}{1 + R^2\omega^2 C^2} . \qquad (41.16)$$

The definition of the differential capacitance (12.1) is a particular case of the definitions derived here. In the vicinity of point A let the current flow through a small ohmic resistance, such that practically the whole of the current $\tilde{I}$ is a current of free carriers. The relation between the current and the potential is given by Ohm's law:

$$\tilde{I} = \sigma \tilde{E}_A, \qquad (41.17)$$

where σ is the conductivity of the system in region A.

Let us assume that in the region of point B there is no electron current, so that the total current is wholly a displacement current. In accordance with equation (41.3), the current $\tilde{I}$ may be written in the form

$$\tilde{I} = \frac{j\omega \tilde{D}_B}{4\pi}, \qquad (41.18)$$

where D_B is the induction in the region of point B.

The alternating charge $\tilde{Q}$, which is in the region AB, is related to the fields $\tilde{E}_A$ and $\tilde{E}_B$ by the expression

$$\varepsilon_A \tilde{E}_A - \varepsilon_B \tilde{E}_B = 4\pi \tilde{Q}. \qquad (41.19)$$

From a comparison of equations (41.17) and (41.18) it is obvious that with a sufficiently high conductivity σ the value $\tilde{E}_A$ may be neglected in comparison with $\tilde{D}_B = \varepsilon_B \tilde{E}_B$. Hence it follows that:

$$\tilde{E}_B = -\frac{4\pi}{\varepsilon_B}\tilde{Q}.$$

By substituting the expression for $\tilde{D}_B$ in equation (41.18), we obtain

$$\tilde{I} = -j\omega\tilde{Q}. \tag{41.20}$$

To the region AB is applied a constant potential differential $\overline{\varphi}_{AB}$ and an alternating potential $\tilde{\varphi}_{AB}$, which for brevity we will subsequently denote by $\overline{\varphi}$ and $\tilde{\varphi}$, respectively. The total charge Q is given by

$$Q = \overline{Q}(\overline{\varphi}) + \tilde{Q}(\overline{\varphi},\ \tilde{\varphi}).$$

We will then assume that this charge is unequivocally defined by the total potential $\overline{\varphi} + \tilde{\varphi}$. The total charge Q may be expanded into a series with respect to the small parameter $e\tilde{\varphi}/kT$:

$$Q(\overline{\varphi} + \tilde{\varphi}) = \overline{Q} + \tilde{Q} = \overline{Q}(\overline{\varphi}) + \frac{\partial Q}{\partial\tilde{\varphi}}\bigg|_{\tilde{\varphi}=0}\tilde{\varphi}. \tag{41.21}$$

By using the equation

$$\frac{\partial Q\,(\overline{\varphi} + \tilde{\varphi})}{\partial\tilde{\varphi}}\bigg|_{\tilde{\varphi}=0} = \frac{\partial Q\,(\overline{\varphi} + \tilde{\varphi})}{\partial\overline{\varphi}}\bigg|_{\tilde{\varphi}=0} = \frac{d\overline{Q}}{d\overline{\varphi}},$$

we obtain the following expression for $\tilde{Q}$:

$$\tilde{Q} = \frac{d\overline{Q}}{d\overline{\varphi}}\tilde{\varphi}. \tag{41.22}$$

By substituting expression (41.22) in equation (41.20) and using relation (41.11), it is possible to obtain an expression for the differential capacitance:

$$C = -\frac{d\overline{Q}}{d\overline{\varphi}}, \tag{41.23}$$

which coincides with expression (12.1).

It should be noted that the total charge is not always de-
fined unequivocally by the sum $(\overline{\varphi} + \widetilde{\varphi})$. For example, when diffusion is present there arises delayed movement of free carri-
ers and the expression for the capacitance is more complex.

Moreover, in the derivation of formula (41.23) we assumed that there was no current of free carriers at the contact
(point B). When there is a reaction the current of free carriers
(41.3) may make an additional contribution to the capacitance.
These phenomena will be examined in § 45.

§ 42. Capacitance of the Space Charge Region in a Semiconductor [2-5]

Expression (41.23) for the static capacitance reflects the
relation between the equilibrium charge and the potential drop
in the Debye region and is valid for infinitely slow charging of
the contact (i.e., when the frequency $\omega = 0$). Let us examine
the characteristics of the charging of the Debye region by an alternating current of finite frequency.

In Fig. 94 the broken curves show the theoretical relation
of the capacitance of the space charge region to the potential
drop in the Debye region of an n-type semiconductor for various
frequencies. Curve 1 corresponds to the frequency $\omega = 0$ and
is described by formula (41.23). Curves 2 and 3 correspond to
finite values of the frequency and $\omega_3 > \omega_2$. The figure shows
that in the potential range where the contact is enriched in majority carriers (accumulative layer), the capacitance is independent of frequency. In the potential range where the contact
is enriched in minority carriers (inversion layer), there is
sharply expressed dispersion of the capacitance and, in particular, the capacitance falls substantially with an increase in frequency. These effects may be understood by considering the
following qualitative arguments. With a change in the potential
drop in the Debye region (because of the application of an alternating signal), the space charge also changes. If we neglect the
transfer of charges through the semiconductor—electrolyte interphase and also recombination-generation processes in the
Debye region and on the surface, the change in the space charge

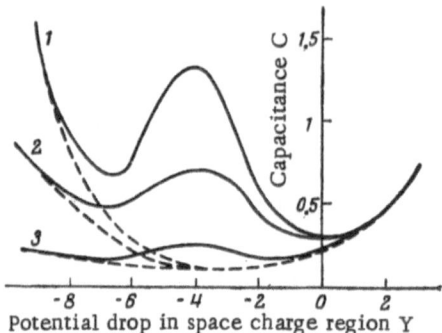

Fig. 94. Relation of the capacitance of a semiconductor electrode C to the potential drop in the space charge region Y and
the frequency (inversion and depletion layer) [4]. Broken lines)
in the absence of surface levels; solid lines) in the presence of
surface levels ($N_t = 9.6 \cdot 10^{10}$ cm^{-2}, $E_t^0/kT = -4$, $C_n = 0$). The
capacitance is expressed in units of $10C^*$ (see p. 298). The
frequency $\omega \gg 1/\tau_p$. The values of $(n^0/p^0)^3 A \sqrt{\omega/\omega_r}$ are:
1) 0; 2) 1; 3) 5.

may occur only as a result of the exchange of charges between
the Debye region and the neutral volume of the semiconductor.
Thereupon, in the depth of the semiconductor, the majority current carriers move under the influence of the electric field and
the minority carriers, as a result of diffusion. Therefore, in
considering the movement of minority carriers, as in studying
the principles of transmission of a direct current (see § 17), it
is advantageous to divide the semiconductor into two regions,
the space charge region and the quasineutral (or diffusion) region.

In the case of the transmission of an alternating current,
the thickness of the diffusion region $l_p(\omega)$ is dependent on the
frequency. As will be shown below, this dependence is described by the following relation:

$$l_p(\omega) = \operatorname{Re} \sqrt{\frac{D_p \tau_p}{1 + j\omega\tau_p}}$$

(42.1)

Thus, at low frequencies ($\omega \tau_p \ll 1$) the diffusion length equals $\sqrt{D_p \tau_p}$, while at high frequencies ($\omega \tau_p \gg 1$) the diffusion length decreases with an increase in frequency according to the law:

$$l_p(\omega) = \sqrt{\frac{D_p}{\omega}} \, .$$

(42.2)

The minority carrier current density at the boundary of the quasineutral and space charge regions has the following order of magnitude:

$$\tilde{i}_p \sim \frac{eD_p \tilde{p}_1}{l_p(\omega)} \, ,$$

(42.3)

where $\tilde{p}_1$ is the change in the concentration of holes at the boundary of the regions considered produced by the alternating current.

The change in the space charge of minority carriers at high frequencies during a period $T = 2\pi / \omega$ is given by

$$\tilde{Q}_p \sim T\tilde{i}_p \sim T \frac{eD_p \tilde{p}_1}{l_p(\omega)} = \frac{2\pi e \sqrt{D_p} \tilde{p}_1}{\sqrt{\omega}} \, .$$

(42.4)

Since, in the total concentration,

$$p_1 = p^0 + \tilde{p}_1$$

(42.5)

the alternating part may not exceed in absolute value the constant part, the numerator in equation (42.4) consequently is limited, while the denominator in this expression increases with frequency. Therefore, the alternating space charge and capacitance decrease with frequency at high frequencies.

Thus, in the region of the inversion layer diffusion processes are appreciable and this leads to the appearance of the frequency dependence of the capacitance. When the space charge region is enriched in majority carriers, diffusion processes are absent and, therefore, the change in the space charge of majority carriers does not lag behind the phase of the change in potential and the capacitance is independent of frequency.

Let us turn to quantitative calculations. Let us examine phenomena occurring in the quasineutral region. To determine the relation of the concentration of holes to the coordinate and the frequency it is necessary to solve equation (17.8).

As has already been mentioned, the alternating concentration of holes is proportional to $e^{j\omega t}$. By substituting expression (42.5) in equation (17.8) we obtain an equation for p:

$$j\omega\tilde{p} = -\frac{\tilde{p}}{\tau_p} + D_p \frac{d^2\tilde{p}}{dx^2} . \tag{42.6}$$

In the depth of the semiconductor $\tilde{p}$ obviously equals zero. The solution satisfying the limiting condition $\tilde{p} \to 0$ when $x \to \infty$ has the form

$$\tilde{p} = Ae^{-x\sqrt{\frac{i\omega}{D_p} + \frac{1}{\tau_p D_p}}} \cdot e^{j\omega t}, \tag{42.7}$$

where A is the integration constant.

From formula (42.7) it is evident that quenching of the oscillations in the concentration of holes occurs at a length $l(\omega)$, which was introduced in equation (42.1) [cf. (17.13)]. By substituting formula (42.7) in the expression for the current

$$\tilde{i}_p = -eD_p \frac{d\tilde{p}}{dx} ,$$

we obtain

$$\tilde{i}_p = eD_p \sqrt{\frac{i\omega}{D_p} + \frac{1}{\tau_p D_p}} A \exp\left\{ -x\sqrt{\frac{i\omega}{D_p} + \frac{1}{\tau_p D_p}} \right\} e^{j\omega t}. \tag{42.8}$$

All subsequent calculations will be carried out on the assumption that the diffusion length remains greater than the Debye length L_1:

$$l_p(\omega) \gg L_1. \tag{42.9}$$

In combination with formula (42.2) this means that the frequency of the alternating current satisfies the condition

$$\omega \ll \frac{D}{L_1^2}, \quad \text{i.e.,} \quad T \gg \frac{L_1^2}{D} = t_d, \tag{42.10}$$

where T is the period of the oscillations.

For aperiodic signals the fulfillment of condition (42.10) (with the time of change of the signal t substituted for the period T) means that during the time t the diffusion region is able to form and, consequently, the space charge is in a state of quasi-equilibrium with the bulk of the semiconductor.

Expression (42.8) for the current $\tilde{i}_p$ is rewritten in the form

$$\tilde{i}_p = \frac{eD_p A}{L_p(\omega)} e^{-\frac{x}{L_p(\omega)}} e^{j\omega t}, \tag{42.11}$$

where

$$L_p(\omega) = \sqrt{\frac{D_p \tau_p}{1 + j\omega \tau_p}} .$$

Let us find the values of the concentration $\tilde{p}_1$ and the current $\tilde{i}_{p_1}$ at the boundary of the quasineutral and Debye regions. For this it is necessary to substitute $x = L_1$ in formulas (42.7) and (42.8). Taking into account relation (42.9) we may write

$$\tilde{p}_1 = A e^{j\omega t},$$

$$\tilde{i}_{p_1} = \frac{eD_p A}{L_p(\omega)} e^{j\omega t}. \tag{42.12}$$

By eliminating the constant A from expressions (42.12) we obtain

$$\tilde{i}_{p_1} = \frac{eD_p}{L_p(\omega)} \tilde{p}_1. \tag{42.13}$$

Let us now examine the space charge region. When the frequency of the alternating signal is less than the reciprocal of the time to establish equilibrium in the space charge region t_d, it is possible to use Boltzmann's distribution for determining the concentrations of free carriers. In germanium the relaxation time of the space charge t_d is of the order of 10^{-10} sec. Therefore, condition (42.10) is not rigid. When this condition is fulfilled, the concentrations of electrons and holes are determined by the relations

$$n(x) = n_1 e^{\frac{e\varphi(x)}{kT}}, \; p(x) = p_1 e^{-\frac{e\varphi(x)}{kT}},$$

(42.14)

where n, p, and φ are the total concentrations of electrons and holes and the total potential at the point x, respectively.

These values may be represented as the sums of constant and alternating components

$$n = \bar{n} + \tilde{n}; \quad p = \bar{p} + \tilde{p}; \quad \varphi = \bar{\varphi} + \tilde{\varphi}.$$

(42.15)

Let us write Poisson's equation in the form (29.5). By integrating with respect to the space charge region, we obtain

$$\frac{1}{2} \varepsilon_1 (E_s^2 - E_B^2) = 4\pi kT \left(n_1 e^{\frac{e\varphi_1}{kT}} - n_1 + p_1 e^{-\frac{e\varphi_1}{kT}} - p_1 \right) - 4\pi e N_D \varphi_1.$$

(42.16)

Let us then represent the values $n_1 e^{e\varphi/kT}$ and $p_1 e^{-e\varphi/kT}$ which appear in relation (42.16) as sums of variable and constant components in accordance with expressions (42.15). We will assume that $e\tilde{\varphi}/kT \ll 1$ and restrict ourselves to the first order of smallness with respect to the alternating values. Then

$$n = (n^0 + \tilde{n}_1) e^{\frac{e(\bar{\varphi} + \tilde{\varphi})}{kT}} = \bar{n} + \bar{n}\frac{e\tilde{\varphi}}{kT} + \tilde{n}_1 e^{\frac{e\bar{\varphi}}{kT}} = \bar{n} + \tilde{n},$$

(42.17)

whence

$$\tilde{n} = \bar{n}\frac{e\tilde{\varphi}}{kT} + \tilde{n}_1 e^{\frac{e\bar{\varphi}}{kT}}.$$

(42.18)

Analogously, for the alternating concentration of holes, we obtain

$$\widetilde{p} = -\overline{p}\,\frac{e\widetilde{\varphi}}{kT} + \widetilde{p}_1 e^{-\frac{e\overline{\varphi}}{kT}}.$$

(42.19)

By substituting the electric field in the form $\overline{E} + \widetilde{E}$ in relation (42.16), the constant and alternating components in this equation are taken separately. For the alternating values we obtain the following expression:

$$\varepsilon_1 \overline{E}_s \widetilde{E}_s = 4\pi e(\overline{n}_s - \overline{p}_s - N_D)\widetilde{\varphi}_1 + 4\pi kT \widetilde{p}_1 \left(e^{\frac{e\overline{\varphi}_1}{kT}} + e^{-\frac{e\overline{\varphi}_1}{kT}} - 2\right). \quad (42.20)$$

Here we used the condition of neutrality in the diffusion region and assumed that $\widetilde{n}_1 = \widetilde{p}_1$. By solving equation (42.20) with respect to $\widetilde{E}_s$ we obtain

$$\widetilde{E}_s = \frac{4\pi e}{\varepsilon_1 \overline{E}_s} \left[(\overline{n}_s - \overline{p}_s - N_D)\widetilde{\varphi}_1 + \frac{kT}{e}\,\widetilde{p}_1 \left(e^{\frac{e\overline{\varphi}_1}{kT}} + e^{-\frac{e\overline{\varphi}_1}{kT}} - 2\right)\right]. \quad (42.21)$$

To determine the impedance of the system it is necessary to find the relation between $\widetilde{p}_1$ and $\widetilde{\varphi}_1$. For this we use the condition of the conservation of holes. Since, according to equation (42.9), the recombination of holes in the space charge region is insignificantly small while there is no current through the contact, holes flowing into the Debye region are consumed solely in increasing the space charge. In accordance with the equation of continuity,

$$\frac{\partial p}{\partial t} + \frac{d}{dx}\,i_p = 0$$

it is possible to write the following:

$$\widetilde{i}_{p1} - \widetilde{i}_{ps} = -e\frac{\partial}{\partial t}\Gamma_{\text{п}},$$

(42.22)

where $\widetilde{i}_{ps}$ is the hole current at the point x = 0.

The value Γ_p is the excess amount of holes in the Debye region and is defined by the relation

$$\Gamma_p = \int [p(\bar{\varphi} + \tilde{\varphi}, \ p^0 + \tilde{p}_1) - (p^0 + \tilde{p}_1)] \, dx, \qquad (42.23)$$

where integration is carried out for the space charge region. Since holes by stipulation do not cross the semiconductor–electrolyte interphase, the hole current $\tilde{i}_{ps} = 0$, and hence,

$$\tilde{i}_{p1} = -e \frac{\partial}{\partial t} \Gamma_p. \qquad (42.24)$$

By carrying out the differentiation in equation (42.24) we obtain

$$\tilde{i}_{p1} = -j\omega e \tilde{\Gamma}_p, \qquad (42.25)$$

where $\tilde{\Gamma}_p$ is the alternating component of Γ_p.

By using relations (42.13) and (42.25) it is possible to write

$$\frac{eD_p}{L_p(\omega)} \tilde{p}_1 = -j\omega e \tilde{\Gamma}_p. \qquad (42.26)$$

Let us turn to the calculation of $\tilde{\Gamma}_p$. By expanding Γ_p into a series in small variables, we obtain

$$\tilde{\Gamma}_p = \left(\frac{\partial}{\partial \tilde{\varphi}_1} \Gamma_p \right)_{\tilde{\varphi}_1 = 0} \cdot \tilde{\varphi}_1 + \tilde{p}_1 \left(\frac{\partial}{\partial \tilde{p}_1} \Gamma_p \right)_{\tilde{p}_1 = 0}$$

As is evident from formula (42.23), Γ_p depends only on the sums $\bar{\varphi}_1 + \tilde{\varphi}_1$ and $p^0 + \tilde{p}_1$ and, therefore, we may write

$$\tilde{\Gamma}_p = \left(\frac{\partial \Gamma_p}{\partial \bar{\varphi}_1} \right)_{\tilde{\varphi}_1 = 0} \cdot \tilde{\varphi}_1 + \tilde{p}_1 \left(\frac{\partial \Gamma_p}{\partial p^0} \right)_{\tilde{p}_1 = 0}.$$

Hence we find

$$\tilde{\Gamma}_p = \left(\frac{\partial \bar{\Gamma}_p}{\partial \bar{\varphi}_1} \right) \tilde{\varphi}_1 + \left(\frac{\partial \bar{\Gamma}_p}{\partial p^0} \right) \tilde{p}_1 \equiv I_1 \tilde{\varphi}_1 + I_2 \tilde{p}_1. \qquad (42.27)$$

By substituting equation (42.27) in (42.26) we find the alternating concentration of holes:

$$\tilde{p}_1 = - \frac{j\omega e l_1}{\dfrac{eD_p}{L_{p(\omega)}} + j\omega e l_2} \tilde{\varphi}_1.$$

$$(42.28)$$

By introducing $\tilde{p}_1$ into expression (42.21), we obtain the value of the electric field at the contact:

$$\tilde{E}_s = \frac{4\pi e}{e_1 \bar{E}_s} \left[(\bar{n}s - \bar{p}_s - N_D)\tilde{\varphi}_1 - \frac{kT j\omega I_1 \left(e^{\frac{e\bar{\varphi}_1}{kT}} + e^{-\frac{e\bar{\varphi}_1}{kT}} - 2 \right)}{\dfrac{eD_p}{L_{p(\omega)}} + j\omega e l_2} \tilde{\varphi}_1 \right]. \quad (42.29)$$

Relation (42.29) may also be written in the form

$$\tilde{E}_s = \frac{4\pi e}{e_1 \bar{E}_s} \left[(\bar{n}_s - \bar{p}_s - N_D) - \frac{4kT j\omega I_1 \, \mathrm{sh}^2 \dfrac{e\bar{\varphi}_1}{2kT}}{\dfrac{eD_p}{L_{p(\omega)}} + j\omega e l_2} \right] \tilde{\varphi}_1. \quad (42.30)$$

In accordance with equation (41.3), the total current $\tilde{I}$ is given by

$$\tilde{I} = \tilde{i}_s + \frac{e_1}{4\pi} \frac{\partial \tilde{E}_s}{\partial t} = \frac{j\omega e}{\bar{E}_s} \left[(\bar{n}_s - \bar{p}_s - N_D) - \frac{4kT j\omega I_1 \, \mathrm{sh}^2 \dfrac{e\bar{\varphi}_1}{2kT}}{\dfrac{eD_p}{L_{p(\omega)}} + j\omega e l_2} \right] \tilde{\varphi}_1. \quad (42.31)$$

According to the definition of impedance (41.4),

$$\frac{1}{Z} = \frac{j\omega e}{\bar{E}_s} \left[(\bar{n}_s - \bar{p}_s - N_D) - \frac{4kT j\omega I_1 \mathrm{sh}^2 \dfrac{e\bar{\varphi}_1}{2kT}}{\dfrac{eD_p}{L_{p(\omega)}} + j\omega e l_2} \right]. \quad (42.32)$$

By describing the semiconductor—electrolyte system in terms of the equivalent circuit with a capacitance and a resistance in parallel, we find the capacitance of the system in accordance with equations (41.12) and (42.32)

$$C_1 = \frac{e}{\overline{E}_s}(\overline{n}_s - \overline{p}_s - N_D) -$$

$$\frac{4kT}{\overline{E}_s} \text{sh}^2\left(\frac{e\overline{\varphi}_1}{2kT}\right)\frac{I_1}{I_2} \; \frac{I_2\dfrac{\omega^2}{\omega_{\text{эф}}^2}\sqrt{\dfrac{2\omega_{\text{eff}}}{D_p}}\left(1 + I_2\sqrt{\dfrac{2\omega_{\text{eff}}}{D_p}}\right)}{1 + \dfrac{\omega^2}{\omega_{\text{eff}}^2}\left(1 + I_2\sqrt{\dfrac{2\omega_{\text{eff}}}{D_p}}\right)^2}, \qquad (42.33)$$

where $\omega_{\text{eff}} = [1 + \sqrt{1 + (\omega\tau_p)^2}]/\tau_p$.

Let us turn to the discussion of the results obtained. In the limit when the frequency $\omega \to 0$ the capacitance is given by

$$C_1 = \frac{e}{E_s}(\overline{n}_s - \overline{p}_s - N_D)$$

and in accordance with equation (12.13) it is defined by the normal relation

$$C_1 = -\frac{dQ_1}{d\varphi_1}.$$

The second term in formula (42.33) is connected with the diffusion retardation of the minority carriers. In actual fact, if, in formula (42.33) the diffusion coefficient D_p formally tends to infinity, i.e., diffusion retardation is neglected, then the second term in formula (42.33) becomes zero and the capacitance is again reduced to the form

$$C_1 = -\frac{dQ_1}{d\varphi_1}.$$

It may also be shown that the second term in formula (42.33) is small in the range of potentials where the contact is enriched in majority carriers and also in the region of the depletion layer.

In the case of the inversion layer when $p_s > N_D$, formula (42.33) may be simplified. For this purpose we calculate approximately the values I_1 and I_2, defined by equation (42.27).

In the integral

$$I_1 = \frac{\partial}{\partial \overline{\varphi}} \int_0^{L_1} p \, dx$$

we convert from the integration variable x to the variable φ and then we obtain

$$I_1 = \frac{\partial}{\partial \overline{\varphi}_1} \int_{\varphi_1}^0 \frac{p d\varphi}{\frac{d\varphi}{dx}} = \frac{\partial}{\partial \overline{\varphi}_1} \int_0^{\overline{\varphi}_1} \frac{p d\varphi}{E} = \frac{\overline{p}_s}{\overline{E}_s} \, .$$

$$(42.34)$$

If in the expression for I_2 we carry out analogous rearrangements and take into account the formula $p = p^0 e^{-e\varphi/kT}$, then it assumes the form

$$I_2 = \frac{\partial}{\partial p^0} \int_0^{L_1} p \, dx = \frac{\partial}{\partial p^0} \int_0^{\overline{\varphi}_1} \frac{p^0 \left(e^{-\frac{e\varphi}{kT}} - 1 \right) d\varphi}{E} \, .$$

$$(42.35)$$

By using formula (10.3) and bearing in mind the fact that $E < 0$, we write I_2 in the form

$$I_2 = \frac{\partial}{\partial p^0} \int_0^{\overline{\varphi}_1} \frac{p^0 \left(e^{-\frac{e\varphi}{kT}} - 1 \right)}{E(\varphi)} \, d\varphi =$$

$$-\frac{1}{\sqrt{\frac{8\pi kT}{\varepsilon_1}}} \frac{\partial}{\partial p^0} \int_0^{\overline{\varphi}_1} \frac{p^0 \left(e^{-\frac{e\varphi}{kT}} - 1 \right) d\varphi}{\sqrt{n^0 \left(e^{\frac{e\varphi}{kT}} - 1 \right) + p^0 \left(e^{-\frac{e\varphi}{kT}} - 1 \right) - N_D \frac{e\varphi}{kT}}} \, .$$

$$(42.36)$$

The integral under the derivative sign $\partial / \partial p^0$ may be represented in the form of two components:

$$-\frac{e}{kT}\int_0^{\overline{\varphi_1}} \frac{p^0\left(e^{-\frac{e\varphi}{kT}}-1\right)d\varphi}{\sqrt{n^0\left(e^{\frac{e\varphi}{kT}}-1\right)+p^0\left(e^{-\frac{e\varphi}{kT}}-1\right)-N_D\frac{e\varphi}{kT}}}=$$

$$2\sqrt{n^0\left(e^{\frac{e\varphi_1}{kT}}-1\right)+p^0\left(e^{-\frac{e\varphi_1}{kT}}-1\right)-N_D\frac{e\varphi_1}{kT}}-$$

$$\int_0^{\varphi_1}\frac{n^0\left(e^{\frac{e\varphi}{kT}}-1\right)e\,d\varphi}{kT\sqrt{n^0\left(e^{\frac{e\varphi}{kT}}-1\right)+p^0\left(e^{-\frac{e\varphi}{kT}}-1\right)-N_D\frac{e\varphi}{kT}}}. \tag{42.37}$$

The first component in the right-hand part of equation (42.37)
approximately equals $2\sqrt{p^0e^{-e\varphi_1/kT}}$. For the calculation of
the second component in the right-hand part of equation (42.37)
we introduce the potential φ_i, defined by the equation

$$p^0e^{-\frac{e\varphi_i}{kT}}=-N_D\frac{e\varphi_i}{kT}+n^0\left(e^{\frac{e\varphi_i}{kT}}-1\right). \tag{42.38}$$

Taking into account the fact that $p^0 \ll n^0$ and $|\varphi_1| > |\varphi_i| >$
kT/e, we obtain approximately

$$-\frac{e}{kT}\int_0^{\varphi_1}\frac{n_0\,(e^{e\varphi/kT}-1)\,d\varphi}{\sqrt{n^0\,(e^{e\varphi/kT}-1)+p^0\,(e^{-e\varphi/kT}-1)-N_D\frac{e\varphi}{kT}}}=$$

$$-\frac{e}{kT}\int_0^{\varphi_i}\frac{n^0\left(e^{\frac{e\varphi}{kT}}-1\right)d\varphi}{\sqrt{n^0\left(e^{\frac{e\varphi}{kT}}-1\right)-N_D\frac{e\varphi}{kT}}}+\frac{e}{kT}\int_{\varphi_i}^{\varphi_1}\frac{n^0\,d\varphi}{\sqrt{p^0e^{-\frac{e\varphi}{kT}}}}=$$

$$-2\sqrt{n^0\left(e^{-\frac{e\varphi_i}{kT}}-1-\frac{e\varphi_i}{kT}\right)}-\frac{2n^0}{\sqrt{p^0}}e^{\frac{e\varphi_i}{2kT}}. \tag{42.39}$$

For the value I_2 we find

$$I_2=\frac{kT}{e\sqrt{\frac{8\pi kT}{\varepsilon_1}}}\frac{\partial}{\partial p^0}\left[2\sqrt{p^0}e^{-\frac{e\overline{\varphi_1}}{2kT}}-2\left(\frac{n^0}{\sqrt{p^0}}e^{-\frac{e\varphi_i}{2kT}}+\sqrt{p^0}e^{\frac{e\varphi_i}{2kT}}\right)\right].$$

According to equation (42.38) we have

$$-\frac{ep^0}{kT}\frac{d\varphi_i}{dp^0}=\frac{-\frac{e\varphi_i}{kT}-1}{2+e\varphi_i/kT}.$$

Hence we find the value I_2:

$$I_2=-\frac{kT}{e}\frac{e^{-\frac{e\bar{\varphi}_1}{kT}}}{\bar{E}_s}+\frac{\frac{kT}{e}\sqrt{n^0\varepsilon_1}}{\sqrt{8\pi kTp^0}}\frac{1}{\sqrt{\left(-\frac{e\varphi_i}{kT}-1\right)}}. \tag{42.40}$$

In the range of potentials examined, the second component in expression (42.40) is much less than the first and is independent of $\bar{\varphi}_1$. By using the calculated values for I_1 and I_2, and taking into account the fact that when $|e\bar{\varphi}_1/kT|\gg1$ the value $4\,\mathrm{sh}^2(e\bar{\varphi}/2kT)\approx e^{-e\bar{\varphi}_1/kT}$, we can represent the coefficient $(4kT/E_s)\,\mathrm{sh}^2\,(e\bar{\varphi}_1/2kT)(I_1/I_2)$ in equation (42.33) in the form

$$\frac{4kT}{E_s}\,\mathrm{sh}^2\left(\frac{e\bar{\varphi}_1}{2kT}\right)\frac{I_1}{I_2}=-\frac{ep_s}{E_s}-\frac{e\sqrt{n^0\varepsilon_1}}{\sqrt{8\pi kT}}\frac{1}{\sqrt{-\frac{e\varphi_i}{kT}-1}}=C^+-C_\infty. \tag{42.41}$$

Here $C^+=-ep_s/E_s$ is a value which may be called the hole part of the capacitance, while C_∞ denotes the value

$$C_\infty=e\sqrt{\frac{n^0\varepsilon_1}{8\pi kT}}\frac{1}{\sqrt{-e\varphi_i/kT-1}}.$$

Taking into account the fact that in the range of potentials examined it is possible to neglect the value $(n_s-n_D)/E_s$, and also the fact that $C^+>C_\infty$, we may write the expression for the capacitance (42.33) in the form

$$C_1=\frac{C^+\left[1+\left(\frac{\omega}{\omega_{\mathrm{eff}}}\right)^2\left(1+\frac{C^+}{C^*}\left(\frac{n^0}{p^0}\right)^{3/4}\sqrt{\frac{\omega_{\mathrm{eff}}}{\omega_r}}\right)\right]+C_\infty\left(\frac{\omega}{\omega_{\mathrm{eff}}}\right)^2\left(\frac{C^+}{C^*}\left(\frac{n^0}{p^0}\right)^{3/4}\sqrt{\frac{\omega_{\mathrm{eff}}}{\omega_r}}\right)^2}{1+\left(\frac{\omega}{\omega_{\mathrm{eff}}}\right)^2\left(1+\frac{C^+}{C^*}\left(\frac{n^0}{p^0}\right)^{3/4}\sqrt{\frac{\omega_{\mathrm{eff}}}{\omega_r}}\right)^2}. \tag{42.42}$$

Here, $C^* = \varepsilon_1/4\pi L_D$; $\omega_r = D_p/(2L_1)^2$.

Curves illustrating the relation of the capacitance to the potential at various frequencies are given in Fig. 94 (broken lines).

From equation (42.42) it is readily seen that at sufficiently low frequencies which satisfy the condition $\omega \ll \omega_\varphi$, where ω_φ is defined by the equation

$$\frac{C^+}{C^*}\left(\frac{n^0}{p^0}\right)^{3/4}\left(\frac{\omega_\varphi}{\omega_{eff}}\right)^2\sqrt{\frac{\omega_{eff}}{\omega_r}}=1,$$

the capacitance C_1 is independent of the frequency and equals its static value C^+.

When $\omega \gg \omega_\varphi$ dispersion of the capacitance appears and C_1 is substantially less than its static value. By using expression (42.42) we find

$$C_1 = C_\infty\left(1+\sqrt{\frac{\omega_\infty}{\omega_{eff}}}\right),\qquad(42.43)$$

where

$$\omega_\infty = \frac{1}{C_\infty^2}\frac{e^4(p^0)^2}{(kT)^2}\cdot\frac{D_p}{2}.\qquad(42.44)$$

As follows from equation (42.43), the value C_1 is independent of the potential φ_1 and falls with an increase in frequency. In the limit when $\omega \to \infty$, we have $C_1 = C_\infty$.

An analogous result for the case where $\omega \to \infty$ is given in [3].

§43. Frequency Characteristics of the System with Surface Levels Present [2-4]

In §12 we found the static capacitance of the surface levels and this was determined by the relation

$$C_t = -\frac{dQ_t}{d\varphi_1},$$

where Q_t is the charge in the levels and φ_1 is the potential drop in the space charge region.

In this section we examine the effect of the levels on the impedance of the contact when an alternating current of finite frequency is passed and find the relation of the capacitance to the frequency. For this we calculate the density of the alternating charge arising in the surface levels:

$$\tilde{Q}_t = - eN_t\tilde{f}_t, \tag{43.1}$$

where $\tilde{f}_t$ is the alternating component of the filling function.

The law of preservation of charge has the form

$$\frac{\partial \tilde{Q}_t}{\partial t} = - e \left(\tilde{R}_n - \tilde{R}_p\right),$$

where $\tilde{R}_n$ and $\tilde{R}_p$ are the numbers of capture acts of electrons from the conductivity band and holes from the valence band, respectively, by the levels in unit time.

By substituting the expression for $\tilde{Q}_t$ from equation (43.1) in the latter we obtain

$$j\omega N_t\tilde{f}_t = \tilde{R}_n - \tilde{R}_p. \tag{43.2}$$

To determine the values of $\tilde{R}_n$ and $\tilde{R}_p$ we use the Shockley-Read recombination theory, which is presented in §6. By using relations (6.5) and (6.6), and assuming that the variables are small so that it is possible to take only expressions that are linear with respect to the variables, we obtain

$$\tilde{R}_n = C_n [(1 - \overline{f}_t) \tilde{n}_s - (\overline{n}_s + n')\tilde{f}_t], \tag{43.3}$$

$$\tilde{R}_p = C_p [\overline{f}_t\tilde{p}_s + (\overline{p}_s + p')\tilde{f}_t]. \tag{43.4}$$

By substituting these values for $\tilde{R}_n$ and $\tilde{R}_p$ in equation (43.2) and solving the equation obtained for $\tilde{f}_t$, we find

$$\tilde{f}_t = \frac{C_n (1 - \overline{f}_t) \tilde{n}_s - C_p \overline{f}_t\tilde{p}_s}{(\Omega + j\omega) N_t}, \tag{43.5}$$

where Ω is defined by the expression

$$\Omega = \frac{C_n\,(\bar{n}_s + n') + C_p\,\bar{p}_s + p')}{N_t}.$$

(43.6)

By substituting the value of $\tilde{f}_t$ from equation (43.5) in relation (43.4) we find the alternating current of holes to the levels

$$\tilde{R}_p = C_p\bar{f}_t\tilde{p}_s + \frac{C_p\,(p' + \bar{p}_s)\,C_n\,(1 - \bar{f}_t)\,\tilde{n}_s - C_p^2\,(\bar{p}_s + p')\,\bar{f}_t\tilde{p}_s}{N_t\,(\Omega + j\omega)}.$$

(43.7)

By eliminating the alternating potential $\tilde{\varphi}$ from relations (42.18) and (42.19) we find the relation between $\tilde{n}_s$, $\tilde{p}_1$, and $\tilde{p}_s$.

By using the condition of quasineutrality $\tilde{n}_1 = \tilde{p}_1$, and also the inequality $p^0 \ll n^0$, we obtain

$$\tilde{n}_s = \frac{\bar{n}_s}{\bar{p}_s}\left(\tilde{p}_1 e^{-\frac{e\overline{\varphi}_1}{kT}} - \tilde{p}_s\right).$$

(43.8)

By substituting the value $\tilde{n}_s$ from relation (43.8) in formula (43.7) we write the expression for $\tilde{R}_p$ in the form

$$\tilde{R}_p = \mu\tilde{p}_s + \nu\tilde{p}_1,$$

(43.9)

where

$$\mu = C_p\bar{f}_t - \frac{C_n C_p\,(p' + \bar{p}_s)}{N_t\,(\Omega + j\omega)}\,(1 - \bar{f}_t)\,\frac{\bar{n}_s}{\bar{p}_s} - \frac{C_p^2\,(p' + \bar{p}_s)}{N_t\,(\Omega + j\omega)}\bar{f}_t,$$

$$\nu = \frac{C_n \cdot C_p\,(p' + \bar{p}_s)}{N_t\,(\Omega + j\omega)}\,\frac{(1 - \bar{f}_t)\,\bar{n}_s}{\bar{p}_s}\,e^{-\frac{e\overline{\varphi}_1}{kT}}.$$

(43.10)

By using relations (5.4) and (5.5) we rearrange (43.10) to the form

$$\mu = C_p\bar{f}_t - \frac{C_p^2}{N_t\,(\Omega + j\omega)} - \frac{C_n C_p\bar{n}_s}{N_t\,(\Omega + j\omega)},$$

$$\nu = \frac{C_n C_p n^0}{N_t\,(\Omega + j\omega)}.$$

(43.11)

For the future we need to establish the relation between the concentrations $\tilde{p}_1$ and the potential $\tilde{\varphi}_1$. By using the condition of preservation of holes (42.22) and relation (42.27) we obtain

$$\tilde{i}_{p1} - \tilde{i}_{ps} = - j\omega e \,(I_1 \tilde{\varphi}_1 + I_2 \tilde{p}_1). \qquad (43.12)$$

By expressing $\tilde{i}_{p1}$ in terms of $\tilde{p}_1$ by means of relation (42.12) and assuming that $i_{ps} = -eR_p$, we write equation (43.12) in the form

$$\frac{D_p}{L_p(\omega)}\,\tilde{p}_1 + \mu \left(\tilde{p}_1 - p^0 \frac{e\tilde{\varphi}_1}{kT} \right) e^{-\frac{e\overline{\varphi}_1}{kT}} + v\tilde{p}_1 = - j\omega \,(I_1 \tilde{\varphi}_1 + I_2 \tilde{p}_1). \quad (43.13)$$

Hence we find

$$\tilde{p}_1 = - \beta p^0 \frac{e\tilde{\varphi}_1}{kT}, \qquad (43.14)$$

where the value β is defined by the relation

$$\beta = \frac{j\omega \dfrac{kT}{ep^0} I_1 - \mu e^{-\frac{e\overline{\varphi}_1}{kT}}}{[D_p/L_p(\omega)] + v + j\omega I_2 + \mu e^{-\frac{e\overline{\varphi}_1}{kT}}}. \qquad (43.15)$$

Then by substituting the value of $\tilde{p}_1$ from equation (43.14) in (42.21), we find the electric field at the contact

$$\tilde{E}_s = \frac{4\pi e}{\varepsilon_1 E_s} \left\{ (\bar{n}_s - \bar{p}_s - N_D) - 4\beta\, p^0 \text{sh}^2\, \frac{e\overline{\varphi}_1}{2\,kT} \right\} \tilde{\varphi}_1. \qquad (43.16)$$

The total current $\tilde{I}$ is independent of the coordinates. Therefore it is convenient to take its value in the Helmholtz region where by stipulation there is no free carrier current, so that the current $\tilde{I}$ equals the displacement current:

$$\tilde{I} = \frac{1}{4\pi} \frac{\partial}{\partial t} \varepsilon_0 \tilde{E}_0. \qquad (43.17)$$

Thereupon the values $\tilde{E}_0$ and $\tilde{E}_s$ are related by the equation

$$\varepsilon_0 \tilde{E}_0 - \varepsilon_1 \tilde{E}_s = - 4\pi Q_t. \qquad (43.18)$$

By eliminating $\varepsilon_0 \tilde{E}_0$ from equation (43.17) by means of relation (43.18) we find for the current $\tilde{I}$:

$$\tilde{I} = \frac{1}{4\pi}\frac{\partial}{\partial t}(\varepsilon_1 \tilde{E}_s + 4\pi \tilde{Q}_t) = Z^{-1}\tilde{\varphi}_1.$$

(43.19)

By substituting $\tilde{E}_S$ from formula (43.16) in equation (43.19) and the value $\tilde{Q}_t = -eN_t \tilde{f}_t$ from formula (43.5), taking into account (42.18) and (42.19) we find the impedance Z:

$$Z^{-1} = \left\{ j\omega\, \frac{e\,(\bar{n}_s - \bar{p}_s - N_D)}{\bar{E}_s} - \frac{4\,j\omega e p^0 \beta}{\bar{E}_s}\,\text{sh}^2\,\frac{e\bar{\varphi}_1}{2\,kT} + \right.$$
$$\left. \frac{j\omega e^2}{kT\,(\Omega + j\omega)}\,[C_n\,(1 - \bar{f}_t)\,\bar{n}_s + C_p \bar{f}_t \bar{p}_s\,(1 + \beta)] \right\}.$$

(43.20)

We will first describe the contact of the equivalent circuit consisting of a resistance and capacitance in parallel. Then, by comparing expressions (41.12) and (43.20) we find expressions for the capacitance and resistance of the circuit:

$$C = \frac{e\,(\bar{n}_s - \bar{p}_s - N_D)}{\bar{E}_s} - 4ep^0 \text{Re}\,\frac{\beta}{\bar{E}_s}\,\text{sh}^2\,\frac{e\bar{\varphi}_1}{2\,kT} +$$
$$\frac{\Omega}{\Omega^2 + \omega^2}\left[C_n(1 - \bar{f}_t)\,\bar{n}_s + C_p\bar{f}_t\bar{p}_s\left(1 + \text{Re}\,\beta + \frac{\omega}{\Omega}\,\text{Im}\,\beta\right)\right]\frac{e^2}{kT};$$

(43.21)

$$\frac{1}{R} = \frac{4ep^0}{\bar{E}_s}\,\text{sh}^2\,\frac{e\bar{\varphi}_1}{2\,kT}\,\omega\,\text{Im}\,\beta + \frac{\omega^2}{\Omega^2 + \omega^2}\,\frac{e^2}{kT}\,[C_n(1 - \bar{f}_t)\,\bar{n}_s +$$
$$C_p\bar{f}_t\bar{p}_s\left(1 + \text{Re}\,\beta - \frac{\Omega}{\omega}\,\text{Im}\,\beta\right)\right].$$

(43.22)

Let us examine the expression for the capacitance. If we substitute $C_n = C = 0$ in expression (43.21), i.e., assume that the levels are absent, then the third term is reduced to zero and relation (43.21) is converted into the expression for the capacitance of the space charge region found in § 42.

If we neglect diffusion, i.e., D_p formally tends to infinity, then, in accordance with relation (43.15), the coefficient β tends to zero and the expression for the capacitance assumes the form

$$C = C_1^0 + \frac{e^2\Omega}{kT\,(\Omega^2 + \omega^2)}\,[C_n\,(1 - \overline{f}_t)\,\overline{n}_s + C_p\overline{f}_t\,\overline{p}_s]. \qquad (43.23)$$

Here the first term determines the capacitance of the space charge region with a frequency equal to zero and the second term, the capacitance of the surface levels. It may be shown that the second term, when $\omega = 0$, is the static capacity of the bulk levels found previously in § 12. In actual fact, if we replace f_t in equation (43.23) by its equilibrium expression from (5.4) and (5.5), then for the capacitance we obtain the expression

$$C = C_1^0 + \frac{e^2 n_i^2}{kT\Omega}\,\frac{[C_n\,(\overline{n}_s + n') + C_p\,(\overline{p}_s + p')]}{(\overline{p}_s + p')\,(\overline{n}_s + n')}.$$

By substituting the value of Ω from formula (43.6) we obtain

$$C = C_1^0 + \frac{e^2 n_i^2 N_t}{kT\,(\overline{p}_s + p')\,(\overline{n}_s + n')} = C_1^0 + \frac{e^2 N_t}{kT\left(\dfrac{1}{\overline{p}_s} + \dfrac{1}{p'}\right)(p_s + p')} =$$

$$C_1^0 + \frac{e^2 N_t}{kT}\,\frac{\overline{p}_s p'}{(\overline{p}_s + p')^2}. \qquad (43.24)$$

The latter expression may also be written in the form:

$$C = C_1^0 + e N_t \frac{\partial}{\partial\varphi_1}\,\frac{p'}{(\overline{p}_s + p')}.$$

Since the charge in the levels equals $\overline{Q}_t = -e N_t \overline{f}_t$, and the expression for $\overline{f}_t$ is given by relation (5.5), we obtain

$$C = C_1^0 - \frac{\partial}{\partial\varphi_1}\overline{Q}_t. \qquad (43.25)$$

It should be emphasized that the surface levels have some characteristic frequency Ω [equation (43.6)], which depends on the applied steady potential $\overline{\varphi}_1$, but it cannot be less than $\Omega_{min} = N_t^{-1}(\sqrt{C_n n'} + \sqrt{C_p p'})^2$. As follows from expression (43.23), at frequencies $\omega \ll \Omega$, the capacitance of the levels is practically independent of frequency, while when $\omega \gg \Omega$, the capacitance of the levels is inversely proportional to ω^2.

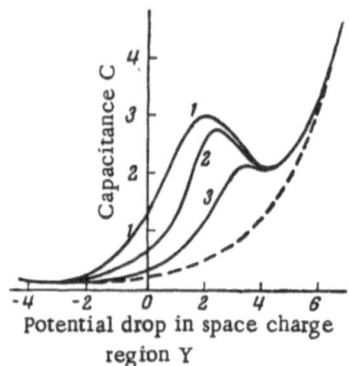

Potential drop in space charge
region Y

Fig. 95. Relation of the capacitance of a semiconductor electrode C to the potential drop in the space charge region Y and
frequency (accumulative layer) [4]. Solid lines) in the presence
of surface levels ($N_t = 2 \cdot 10^{11}$ cm^{-2}, $E_t^0/kT = 2$, $C_p = 0$). The
capacitance is expressed in units of $10C^*$ (see p. 298). The
value $(\omega/\Omega_{min})^2$ equals: 1) 0; 2) 1; 3) 10. The broken line is
in the absence of levels.

The expression (43.23) for the capacitance is illustrated
in Fig. 95. With an increase in the frequency the capacitance
produced by the surface levels falls, while the potential of the
maximum changes. This shift of the maximum is toward positive potentials when $C_p = 0$ and toward negative potentials when
$C_n = 0$.

Finally, let us examine the effect of diffusion of minority
carriers on the capacitance of the surface levels. We will limit
ourselves to the case where $\omega < \Omega$ and $\omega \ll \omega_\infty$, and also, for
simplicity, we assume that $C_n = 0$, i.e., it will be assumed
that the levels trap only holes effectively. (Under these conditions the effect of diffusion of holes appears to the maximum extent.) Moreover, we will assume that $p_s > N_D$. By introducing
the symbol $C_y^+ = (\Omega/\Omega^2 + \omega^2) C_n (1 - \mathcal{F}_t) \bar{n}_s$, in accordance with
expression (43.21) we obtain

$$C = \frac{e(\bar{n}_s - N_D)}{\bar{E}_s} + (C^+ + C_y^+) \frac{1 + \left(\frac{\omega}{\omega_{eff}}\right)^2 \left[1 + \frac{C^+ + C_y^+}{C^*}\left(\frac{n^0}{p^0}\right)^{1/4}\sqrt{\frac{\omega_{eff}}{\omega_r}}\right]}{1 + \left(\frac{\omega}{\omega_{eff}}\right)^2 \left[1 + \frac{C^+ + C_y^+}{C^*}\left(\frac{n^0}{p^0}\right)^{1/4}\sqrt{\frac{\omega_{eff}}{\omega_r}}\right]^2}$$

(43.26)

The relation of the capacitance to the frequency is illustrated by Fig. 94. It is interesting that the height of the maximum produced by surface levels falls with frequency even though the latter remains less than the characteristic frequency of the levels Ω. This situation is connected with the effect of the diffusion of holes from the quasineutral region into the space charge region. With an increase in frequency there is a fall in the hole charge transferred by diffusion during a period and distributed between the Debye region (Q_1) and the surface levels (Q_t). Therefore, an increase in frequency leads to a fall in the capacitance of the levels even when $\omega < \Omega$.

When surface levels are present, the capacitance of the Helmholtz layer may make a substantial contribution to the total impedance of the system. By using the combination rule for impedances (41.9) we note that the total impedance of the system equals the sum of impedances Z_0 of the Helmholtz layer and Z_1 of the space charge layer as the Helmholtz layer is in series with the space charge region. As before, we will assume that there is no electric current of free carriers through the contact and therefore the Helmholtz layer has an infinitely great resistance for an alternating current. By using equation (41.12) we obtain for the impedance of the Helmholtz layer the value

$$\frac{1}{Z_0} = j\omega C_0.$$

The total impedance of the system is given by

$$Z = Z_0 + Z_1 = \frac{1}{j\omega C_0} + \frac{R_1}{1 + j\omega C_1 R_1} = \frac{1 + j\omega C_1 R_1 + j\omega C_0 R_1}{j\omega C_0 (1 + j\omega C_1 R_1)}. \quad (43.27)$$

By introducing the total resistance R and the capacitance of the system C through the relation

$$\frac{1}{Z} = \frac{1}{R} + j\omega C;$$

we obtain the expression

$$C = C_0 \frac{1 + \omega^2 C_1 (C_0 + C_1) R_1^2}{1 + \omega^2 (C_0 + C_1)^2 R_1^2} , \qquad (43.28)$$

which makes it possible to determine the total capacitance of the system.

§ 44. Frequency Characteristics of the System with Levels Present in the Bulk of the Semiconductor [6]

Let us examine the complex resistance of the system n-type semiconductor−electrolyte for the case when in the semiconductor there are two sorts of donor levels (we will denote them arbitrarily by I and II), the bulk concentrations of which equal N_D and N_D', respectively. The shallow levels with a concentration N_D are completely ionized in the bulk of the semiconductor at room temperature. The deep levels with a concentration N_D' are practically completely filled by electrons in the absence of an external field. Therefore, it may be assumed that in the depth of the semiconductor the concentration of electrons n^0 equals approximately N_D. In accordance with the assumptions made, Poisson's equation in the space charge region has the form

$$\varepsilon_1 \frac{dE}{dx} = - 4\pi e \{n^0 e^{\frac{e\varphi}{kT}} - p^0 e^{-\frac{e\varphi}{kT}} - N_D - N_D'(1 - f)\},$$

$$(44.1)$$

where f is the degree of filling of levels of sort II by electrons.

Let us separate the values of the field and the potential into constant and alternating components as was done in § 42:

$$\varepsilon_1 \frac{\overline{dE}}{dx} = - 4\pi e \{n^0 e^{\frac{\overline{e\varphi}}{kT}} - p^0 e^{-\frac{\overline{e\varphi}}{kT}} - N_D - N_D'(1 - \bar{f})\}, \qquad (44.2)$$

$$\varepsilon_1 \frac{\widetilde{dE}}{dx} = - 4\pi e \{n^0 e^{\frac{\overline{e\varphi}}{kT}} \frac{e\widetilde{\varphi}}{kT} + p^0 e^{-\frac{\overline{e\varphi}}{kT}} \frac{e\widetilde{\varphi}}{kT} + N_D' \widetilde{f}\}. \qquad (44.3)$$

In accordance with equation (43.5)

$$N_D'\tilde{f} = \frac{C_n(1-\bar{f})\bar{n} + C_p\bar{f}\bar{p}}{\Omega + j\omega} \frac{\widetilde{e\varphi}}{kT},$$ (44.4)

where C_n and C_p are the trapping cross sections for electrons and holes of type II bulk levels and the value of Ω is defined by relation (43.6).

By means of (5.5) it is readily shown that the value $\tilde{f}$ may be written in the following way:

$$N_D'\tilde{f} = N_D'\frac{\partial\bar{f}}{\partial\bar{\varphi}_1} \frac{\Omega}{\Omega + j\omega} \tilde{\varphi}.$$ (44.5)

Then equation (44.3) is rewritten in the form

$$\varepsilon_1\frac{d\tilde{E}}{dx} = -4\pi e\left\{n^0 e^{\frac{\overline{e\varphi}}{kT}} \frac{\widetilde{e\varphi}}{kT} + p^0 e^{-\frac{\overline{e\varphi}}{kT}} \frac{\widetilde{e\varphi}}{kT} + N_D'\frac{\partial\bar{f}}{\partial\bar{\varphi}_1} \frac{\Omega}{\Omega + j\omega} \tilde{\varphi}\right\}.$$ (44.6)

Let us then examine equation (44.2) for values constant with time. Let the constant potential $\bar{\varphi}_1$ be selected such that a depletion layer is formed at the contact so that $\bar{n}_s \ll N_D$ and $\bar{p}_s \ll N_D$. In this case the first two terms in the right-hand part of equation (44.2) may be neglected in comparison with N_D. The function $\bar{f}$ changes extremely sharply in the vicinity of the characteristic potential φ^*, which satisfies the condition $n(\varphi^*) = \underline{n}'$ (see § 5). Therefore, at potentials $|\bar{\varphi}| < |\varphi^*|$ the function $\bar{f}$ is close to zero, while at potentials $|\bar{\varphi}| > |\varphi^*|$ the value of $\bar{f}$ approaches unity. As a result, equation (44.2) assumes the form

$$\varepsilon_1\frac{d\bar{E}}{dx} = 4\pi e N_D \qquad \text{when } |\bar{\varphi}| \leqslant |\varphi^*|,$$

$$\varepsilon_1\frac{d\bar{E}}{dx} = 4\pi e(N_D + N_D') \quad \text{when } |\bar{\varphi}| > |\varphi^*|.$$ (44.7)

The charge density $\bar{\rho}$ in these two cases equals eN_D and $e(N_D + N_D')$, respectively

By integrating equation (44.7), as was done in § 10, we obtain

$$\bar{E} = -\sqrt{-\frac{8\pi e}{\varepsilon_1} N_D \bar{\varphi}} \qquad \text{when } |\bar{\varphi}| \leqslant |\varphi^*|$$

$$\bar{E} = -\sqrt{-\frac{8\pi e}{\varepsilon_1} [N'_D (\bar{\varphi} - \varphi^*) + N_D \bar{\varphi}]} \text{ when} |\bar{\varphi}| > |\varphi^*|.$$

$$(44.8)$$

Let us now turn to an examination of equation (44.5), which relates the alternating values. We use the fact that the function $\partial f / \partial \varphi_1$ has a sharp maximum in the vicinity of the potential φ^*. Therefore we may write that

$$\frac{\partial \bar{f}}{\partial \bar{\varphi}} = \frac{e}{kT} \delta \left[\frac{e}{kT} (\bar{\varphi} - \varphi^*) \right].$$

Here the symbol δ denotes the delta function. Now equation (44.6) is rewritten in the form

$$\varepsilon_1 \frac{d\tilde{E}}{dx} = -4\pi e \left\{ n^0 e^{\frac{e\bar{\varphi}}{kT}} \frac{e\tilde{\varphi}}{kT} + p^0 e^{-\frac{e\bar{\varphi}}{kT}} \frac{e\tilde{\varphi}}{kT} + \frac{N'_D e}{kT} \delta (\bar{\varphi} - \varphi^*) \frac{\Omega}{\Omega + j\omega} \tilde{\varphi} \right\}.$$

$$(44.9)$$

To integrate equation (44.9) we rearrange the derivative $d\tilde{E}/dx = -d^2\tilde{\varphi}/dx^2$. Noting that $d\tilde{\varphi}/dx = -(d\tilde{\varphi}/d\bar{\varphi})\bar{E}$, we rewrite equation (44.9) in the following form:

$$\bar{E}^2 \frac{d^2\tilde{\varphi}}{d\bar{\varphi}^2} - \frac{4\pi\bar{\rho}}{\varepsilon_1} \frac{d\tilde{\varphi}}{d\bar{\varphi}} = \frac{4\pi e}{\varepsilon_1} \left\{ n^0 e^{\frac{e\bar{\varphi}}{kT}} \frac{e}{kT} + p^0 e^{-\frac{e\bar{\varphi}}{kT}} \frac{e}{kT} + \right.$$

$$\left. N'_D \frac{\Omega}{\Omega + j\omega} \frac{e}{kT} \delta \left[\frac{e}{kT} (\bar{\varphi} - \varphi^*) \right] \right\} \tilde{\varphi}. \qquad (44.10)$$

The solution of equation (44.10) is carried out for two regions, namely, where $|\bar{\varphi}| > |\varphi^*|$ and $|\bar{\varphi}| < |\varphi^*|$. In both regions equation (44.10) has the form

$$\bar{E}^2 \frac{d^2\tilde{\varphi}}{d\bar{\varphi}^2} - \frac{4\pi\bar{\rho}}{\varepsilon_1} \frac{d\tilde{\varphi}}{d\bar{\varphi}} = \frac{4\pi e}{\varepsilon_1} N'_D \frac{\Omega}{\Omega + j\omega} \frac{e}{kT} \delta \left[\frac{e}{kT} (\bar{\varphi} - \varphi^*) \right] \tilde{\varphi}. \qquad (44.11)$$

The limiting condition in the region $|\bar{\varphi}| < |\varphi^*|$ is the equation $\tilde{\varphi} = 0$ when $\bar{E} = 0$.

In the region $|\varphi| > |\varphi *|$ equation (44.11) is solved with the limiting conditions: the continuity of $\tilde{\varphi}$ at the point where $\overline{\varphi} = \varphi*$ and the relation

$$(\overline{E}^*)^2 \left(\frac{d\tilde{\varphi}}{d\overline{\varphi}}\bigg|_{\varphi^*+0} - \frac{d\tilde{\varphi}}{d\overline{\varphi}}\bigg|_{\varphi^*-0} \right) = N'_\infty \frac{\Omega}{\Omega + j\omega} \tilde{\varphi}^*,$$

(44.12)

where $\overline{E}^*$ and $\tilde{\varphi}^*$ are, respectively, the constant electric field and the alternating potential at the point where $\overline{\varphi} = \varphi*$. The subscripts $\varphi* + 0$ and $\varphi* - 0$ denote that the derivatives are to the right and to the left of the point $\varphi*$, respectively.

As a result of integration of equation (44.11) with the given limiting conditions we obtain an expression for the electric field $\tilde{E}_s$:

$$\tilde{E}_s = -\frac{4\pi\overline{\rho}}{\varepsilon_1 \overline{E}_s} \left[1 - \frac{\overline{E}^*}{\overline{E}_s} \left(1 - \frac{(N_D + N'_D)(\Omega^* + j\omega)}{N_D(\Omega^* + j\omega) + N'_D \Omega^*} \right) \right]^{-1},$$

(44.13)

where Ω^* is the value of Ω at the point $\overline{\varphi} = \varphi*$.

Describing the system by the equivalent circuit with a capacitance and resistance in parallel and using relation (41.3), we find expressions for the capacitance and resistance:

$$C_1 = -\frac{\overline{\rho}_s}{\overline{E}_s} \frac{(\Omega^*)^2 (N'_D + N_D)^2 + \Omega^2 N_D \left(N_D + \frac{\overline{E}^*}{\overline{E}_s} N'_D \right)}{(\Omega^*)^2 (N'_D + N_D)^2 + \omega^2 \left(N_D + \frac{\overline{E}^*}{\overline{E}_s} N'_D \right)^2},$$

$$\frac{1}{R_1} = -\frac{\overline{\rho}_s}{\overline{E}_s} \frac{\omega^2 (\overline{E}^*/\overline{E}_s) N_D \Omega^* (N'_D + N_D)}{(\Omega^*)^2 (N'_D + N_D)^2 + \omega^2 \left(N_D + \frac{\overline{E}^*}{\overline{E}_s} N'_D \right)^2}.$$

(44.14)

When $N'_D \ll N_D$, expression (44.14) changes into the normal expression for the capacitance of the depletion layer with the capacitance independent of frequency. When $N'_D \gg N_D$, the dependence of the reciprocal capacitance on the potential and the frequency are as shown in Fig. 96. At a frequency $\omega = 0$ the relation $(1/C_1^2) - \varphi_1$ is represented by two straight lines AB and

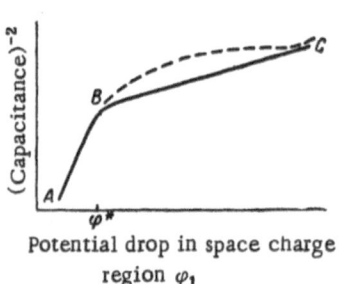

Potential drop in space charge region φ_1

Fig. 96. Relation of the reciprocal square of the capacitance to the potential drop in the space charge region φ_1 for a semiconductor containing bulk levels of two sorts: solid line) at zero frequency; broken line) at high frequency.

BC, which intersect in the vicinity of the point $\varphi_1 = \varphi^*$. Each of these straight lines describes the depletion layer, but the concentration of ionized donors in the region of the depletion layer, which determines the slope of the line of $(1/C_1^2) - \varphi_1$, equals N_D in one case and $N_D + N_D'$ in the other. In the region $|\varphi_1| \lesssim |\varphi^*|$ the capacitance is independent of frequency. At potentials $|\varphi_1| > |\varphi^*|$ the capacitance falls with frequency, whereby at sufficiently high potentials the asymptote of the function $1/C^2$ is the straight line BC.

A similar break on the $(1/C^2) - \varphi$ line is observed in the system gallium arsenide−electrolyte [7] (Fig. 34). The experimental relation of the capacitance to the frequency is qualitatively similar to the calculated relation.

§ 45. Frequency Characteristics of System with Current Close to Limiting Current of Minority Carriers [8]

In §§ 42-44 in the calculation of the impedance we assumed that there was no electrochemical reaction at the contact, i.e., we examined an ideally polarized electrode. Here we examine the effect on the impedance of a current flowing through the electrode−electrolyte interphase and investigate the dependence of the elements of the equivalent circuit on the frequency. The effect of an electric current on the capacitance of a semiconductor electrode has already been examined in §35. However, we used formulas which are valid only for equilibrium conditions. Below we will determined the limits of applicability of the calculations made in §35.

Taking into account the electrode reaction involving minority carriers (as a concrete example, holes) results in the condition of preservation of holes in the space charge region no longer being expressed by equation (42.25) as a current of carriers flows through the contact, i.e., i_{ps} is not zero. Therefore, it is necessary to use the general formula (42.22). The alternating electric current in the plane of the contact may be found by means of the relation (27.1) in which we must substitute $\varphi_0 - \varphi_0^0 = 0$:

$$\tilde{i}_s = - i^0 \frac{\tilde{p}_s}{p_s^0} . \tag{45.1}$$

Here the hole current is only part of the total current:

$$\tilde{i}_{ps} = - \gamma^+ i^0 \frac{\tilde{p}_s}{p_s^0} = - \gamma^+ \tilde{i} \frac{\tilde{p}_s}{\overline{p}_s} , \tag{45.2}$$

where γ^+ is the injection coefficient (i.e., the ratio of the hole current to the total current)

$$\gamma^+ = \frac{|i_p|}{|i|} .$$

The electric current at the boundary of the Debye and diffusion regions is again determined by relation (17.9). The condition of the preservation of holes (42.22) now has the form

$$\frac{eD_p}{L_p(\omega)} \tilde{p}_1 + i^0 \gamma^+ \frac{\tilde{p}_s}{p_s^0} = - e \frac{\partial}{\partial t} \Gamma_p. \tag{45.3}$$

If there passes through the electrolyte — semiconductor boundary a direct current on which is superposed a small alternating current, then Γ_p is determined by relation (42.23) in which the value p^0 should be replaced by p_1. Taking into account also the fact that there is a Boltzmann distribution of holes, we write Γ_p in the form

$$\Gamma_p = \int p_1 (e^{-\frac{e\varphi}{kT}} - 1) dx, \tag{45.4}$$

where $p_1 = \overline{p}_1 + \tilde{p}_1$ is the concentration of holes at the boundary of the Debye and quasineutral regions and $\varphi = \overline{\varphi} + \tilde{\varphi}$ is the total potential.

By expanding the values appearing in equation (45.4) into series in small variables (as in § 42), we obtain

$$\widetilde{\Gamma}_p = I_1\widetilde{\varphi}_1 + I_2\widetilde{p}_1,$$

$$I_1 = \frac{\partial}{\partial\varphi_1}\int_0^{\overline{\varphi}_1}\frac{\overline{p}_1\,(e^{-\frac{e\varphi}{kT}} - 1)\,d\varphi}{\overline{E}\,(\varphi,\,\overline{p}_1)} = \frac{\overline{p}_1\,(e^{-\frac{e\overline{\varphi}_1}{kT}} - 1)}{\overline{E}_s},$$

$$I_2 = \frac{\partial}{\partial\overline{p}_1}\int_0^{\overline{\varphi}_1}\frac{\overline{p}_1\,(e^{-\frac{e\varphi}{kT}} - 1)\,d\varphi}{\overline{E}\,(\varphi,\,\overline{p}_1)}. \tag{45.5}$$

By using equations (45.5) and (45.3) we obtain the condition of preservation of holes in the form

$$\frac{eD_p}{L_p(\omega)}\,\widetilde{p}_1 + i^0\gamma^+\frac{\widetilde{p}_s}{p_s^0} = -\,j\omega e\,(I_1\widetilde{\varphi}_1 + I_2\widetilde{p}_1). \tag{45.6}$$

By using for the determination of the relation between the concentrations p_S and p_1 Boltzmann's relation

$$p_s = p_1 e^{-\frac{e\varphi_1}{kT}}, \tag{45.7}$$

we obtain

$$\widetilde{p}_s = \widetilde{p}_1 e^{-\frac{e\overline{\varphi}_1}{kT}} - \overline{p}_s\frac{e\widetilde{\varphi}_1}{kT}. \tag{45.8}$$

By substituting $\widetilde{p}_1$ from equation (45.8) in formula (45.6), we find the alternating concentration $\widetilde{p}_s$:

$$\widetilde{p}_s = -\,\frac{j\omega kTI_1 + \left(\dfrac{eD_p}{L_p(\omega)} + j\omega eI_2\right)\overline{p}_1}{e^{\frac{e\overline{\varphi}_1}{kT}}\left(\dfrac{eD_p}{L_p(\omega)} + j\omega eI_2\right) + \dfrac{i^0\gamma^+}{p_s^0}}\,\frac{e\widetilde{\varphi}_1}{kT}. \tag{45.9}$$

Now the alternating electric current $\widetilde{i}_{ps}$ may be written in the form

$$\tilde{i}_{ps} = \frac{\gamma^{+}{}^{i0}}{p_s^0} \left\{ \frac{j\omega kTI_1 + \left(\frac{eD_p}{L_p(\omega)} + j\omega el_2\right)\overline{p}_1}{e^{\frac{e\overline{\varphi}_1}{kT}}\left(\frac{eD_p}{L_p(\omega)} + j\omega el_2\right) + \frac{i^0\gamma^+}{p_s^0}} \right\} \frac{e\tilde{\varphi}_1}{kT}. \qquad (45.10)$$

The total current $\tilde{I}$ in the contact equals

$$\tilde{I} = \tilde{i}_s + \frac{j\omega e_1 \tilde{E}_s}{4\pi}.$$

The electric field $\tilde{E}_S$ may be found by means of relation (29.6). The total electric field is given by

$$E_s = -\sqrt{\frac{8\pi e}{\varepsilon_1}\left[-N_D\left(\overline{\varphi}_1 + \tilde{\varphi}\right) + \frac{kT}{e}\left(\overline{p}_1 + \tilde{p}_1\right)e^{-\frac{e\left(\overline{\varphi}_1 + \tilde{\varphi}\right)}{kT}}\right]}. \qquad (45.11)$$

By expanding E_S into a series in the variables $\tilde{\varphi}_1$ and $\tilde{p}_1$ we find

$$\tilde{E}_s = \frac{1}{2\overline{E}_s}\frac{8\pi e}{\varepsilon_1}\left[-N_D\tilde{\varphi}_1 + \frac{kT}{e}\tilde{p}_1 e^{-\frac{e\overline{\varphi}_1}{kT}} - \overline{p}_1 e^{-\frac{e\overline{\varphi}_1}{kT}}\tilde{\varphi}_1\right]. \qquad (45.12)$$

The value $\tilde{p}_1$ may be found from relations (45.6) and (45.8) if $\tilde{p}_S$ is expressed in terms of $\tilde{p}_1$:

$$\tilde{p}_1 = \frac{-j\omega el_1 + \frac{e}{kT}i^0\gamma^+\overline{p}_s/p_s^0}{\frac{eD_p}{L_p(\omega)} + \frac{i^0\gamma^+}{p_s^0}e^{-\frac{e\overline{\varphi}_1}{kT}} + j\omega el_2}\tilde{\varphi}_1. \qquad (45.13)$$

By using equation (45.12), for the value of $\overline{E}_S$ we obtain the following expression:

$$\tilde{E}_s = \frac{\tilde{\varphi}_1}{\overline{E}_s}\frac{4\pi e}{\varepsilon_1}\left[-N_D - \overline{p}_1 e^{-\frac{e\overline{\varphi}_1}{kT}} + \frac{\left(-j\omega kTI_1 + i^0\gamma^+\frac{\overline{p}_s}{p_s^0}\right)e^{-\frac{e\overline{\varphi}_1}{kT}}}{\frac{eD_p}{L_p(\omega)} + \frac{i^0\gamma^+}{p_s^0}e^{-\frac{e\overline{\varphi}_1}{kT}} + j\omega el_2}\right]. \qquad (45.14)$$

The total current is given by

$$\tilde{I} = \frac{\tilde{i}_{ps}}{\gamma^+} + \frac{j\omega e}{\bar{E}_s}\left[-N_D - \bar{p}_1 e^{-\frac{e\overline{\varphi}_1}{kT}} + \frac{\left(-j\omega kTI_1 + i^0\gamma^+ \frac{\overline{p}_s}{p_0^s}\right)e^{-\frac{e\overline{\varphi}_1}{kT}}}{\frac{eD_p}{L_p(\omega)} + \frac{i^0\gamma^+}{p_s^0}e^{-\frac{e\overline{\varphi}_1}{kT}} + j\omega el_2}\right]\tilde{\varphi}_1 \quad . \quad (45.15)$$

By using relation (41.11) we find expressions for the capacitance and resistance of the system

$$C = \frac{i^0}{p_s^0}\,\mathrm{Re}\,\frac{kTI_1 + \left(\frac{eD_p}{L_p j\omega}\sqrt{1+j\omega\tau_p} + el_2\right)\bar{p}_1}{e^{\frac{e\overline{\varphi}_1}{kT}}\left(\frac{eD_p}{L_p}\sqrt{1+j\omega\tau_p} + j\omega el_2\right) + \frac{i^0\gamma^+}{p_s^0}}\,\frac{e}{kT} +$$

$$\mathrm{Re}\,\frac{e}{\bar{E}_s}\left\{-N_D - \bar{p}_1 e^{-\frac{e\overline{\varphi}_1}{kT}} + \frac{\left[-j\omega kTI_1 + i^0\gamma^+ \frac{\overline{p}_s}{p_s^0}\right]e^{-\frac{e\overline{\varphi}_1}{kT}}}{\frac{eD_p}{L_p}\sqrt{1+j\omega\tau_p} + \frac{i^0\gamma^+}{p_s^0}e^{-\frac{e\overline{\varphi}_1}{kT}} + j\omega el_2}\right\}.$$

$$(45.16)$$

$$\frac{1}{R} = \frac{i^0}{p_s^0}\,\mathrm{Re}\,\frac{j\omega kTI_1 + \left(\frac{eD_p}{L_p}\sqrt{1+j\omega\tau_p} + j\omega el_2\right)\bar{p}_1}{e^{\frac{e\overline{\varphi}_1}{kT}}\left(\frac{eD_p}{L_p}\sqrt{1+j\omega\tau_p} + j\omega el_2\right) + \frac{i^0\gamma^+}{p_s^0}}\,\frac{e}{kT} +$$

$$\mathrm{Re}\,\frac{j\omega e}{\bar{E}_s}\left[-N_D - \bar{p}_1 e^{-\frac{e\overline{\varphi}_1}{kT}} + \frac{[-j\omega kTI_1 + i^0\gamma^+\overline{p}_s/p_s^0]\,e^{-\frac{e\overline{\varphi}_1}{kT}}}{\frac{eD_p}{L_p}\sqrt{1+j\omega\tau_p} + \frac{i^0\gamma^+}{p_s^0}e^{-\frac{e\overline{\varphi}_1}{kT}} + j\omega el_2}\right].$$

$$(45.17)$$

For later on, by using formulas (45.1), (45.2), and (17.16) we rewrite the expression for the capacitance

$$C = -\bar{i}\,\mathrm{Re}\,\frac{kTI_1 + \frac{1}{j\omega}(-i_p^{\lim} + \bar{i}_p)\sqrt{1+j\omega\tau_p} + el_2\bar{p}_1}{(-i_p^{\lim} + \bar{i}_p)\sqrt{1+j\omega\tau_p} + j\omega el_2\bar{p}_1 - i_p}\,\frac{e}{kT}$$

$$+ \operatorname{Re} \frac{e}{\overline{E}_s} \left\{ - N_D - \overline{p}_s + \frac{[-j\omega kTl_1 - \overline{i}_p]\overline{p}_s}{(-i_p^{\lim} + \overline{i}_p)\sqrt{1 + j\omega\tau_p} - \overline{i}_p + j\omega el_2\overline{p}_1} \right\}.$$

$$(45.18)$$

We find the value of the capacitance when $\omega = 0$:

$$C = C' + C'',$$

$$C' = \sqrt{\frac{e\varepsilon_1}{2\pi}} \frac{N_D i^{\lim} - \frac{\overline{i}}{k_1} (i^{\lim} - \overline{i})}{2i^{\lim}\sqrt{- N_D\overline{\varphi}_1 - \frac{RT}{F} \frac{\overline{i}}{k_1}}} = \frac{e\left[-N_D i^{\lim} + \frac{\overline{i}}{k_1} (i^{\lim} - \overline{i})\right]}{i^{\lim} \overline{E}_s};$$

$$C'' = \frac{\overline{i}e}{kT} \frac{kTl_1 + el_2\overline{p}_1 + (\overline{i}_p - i_p^{\lim}) \frac{\tau_p}{2}}{i_p^{\lim}}$$

$$(45.19)$$

Here we used the relation $i = -k_1 p_s$ [cf. equation (27.1)].

Thus, the capacitance may be represented in the form of two terms as if the total capacitance of the system consisted of two capacitances C' and C'' connected in parallel. The capacitance C'' is determined by the reaction occurring at the contact. In actual fact, when there is no reaction the capacitance C'' becomes zero. The capacitance C' is the true capacitance of the space charge and is related to the change in the charge in the Debye region. The expression for C' coincides with formula (35.6).

From relation (45.19) it can be shown that formula (35.6) is quite a good approximation for the qualitative description of the behavior of the system at low frequencies with a current flowing. However, C'' becomes of the same order of magnitude as C' when $i \approx i^{\lim}$. The capacitance curve at high frequencies is given in Fig. 74 (curve 3).. The fall in capacitance with frequency occurs as a result of the effect examined in §42. At high frequencies the alternating charge of the holes tends to zero and the hole component of the capacitance of the space charge becomes negligibly small. The capacitance of the reaction also decreases.

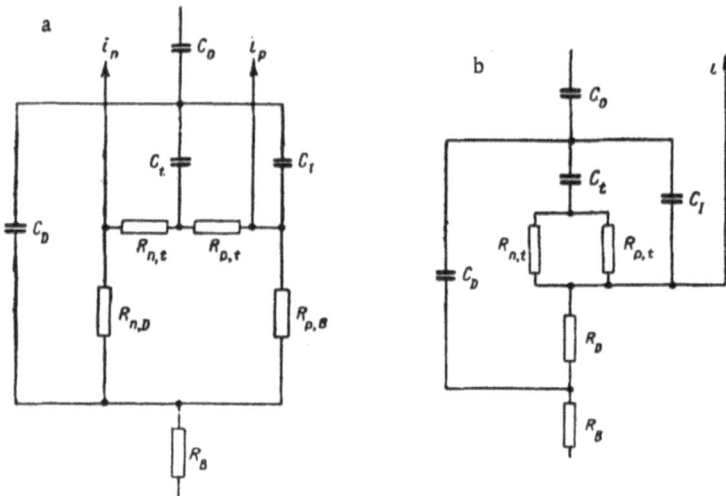

Fig. 97. Equivalent circuit for an electrode with a depletion or inversion [11] a) Surface recombination rate equals zero; b) infinitely high surface recombination rate.

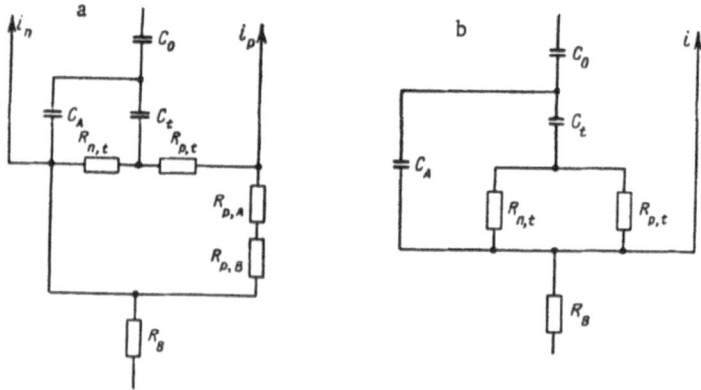

Fig. 98. Equivalent circuit for electrode with accumulative layer [11] a) Surface recombination rate equals zero; b) infinitely high surface recombination rate.

The capacitance of a semiconductor electrode in a solution containing an oxidation−reduction system was calculated in [9].

§ 46. Complete Equivalent Circuit of Semiconductor Electrode

The selection of the equivalent circuit of an electrode, i.e., the representation of the total impedance in the form of a combination of separate elements (capacitances and resistances) depends on both the method of measurement and the method of calculation of the impedance. With the aid of a small alternating (harmonic) current it is possible to measure the true impedance of the system as defined by relation (41.4) (more accurately, the active and reactive components of the complex conductivity of the system); the impedance may then be transformed in accordance with the equivalent circuit selected. When other methods are used, for example, charging of the electrode with square current pulses of low amplitude (see § 23), the concept of impedance cannot be introduced; here the elements of some equivalent circuit are measured directly.

In §§ 42-45 in the calculation of the impedance we used an equivalent circuit consisting of a capacitance and a resistance connected in parallel (in series with which were connected the capacitance of the Helmholtz layer and the ohmic resistance of the semiconductor and the electrolyte) and calculated the relation of the capacitance and the resistance to the frequency. The selection of this equivalent circuit was justified by the fact that in the experimental measurement of the impedance by means of an ac bridge the object measured is often simulated in the arm of the bridge by capacitance and resistance boxes in parallel.

However, another approach to the calculation of the impedance of semiconductor systems has become common in recent years. In [10-12] the semiconductor electrode is represented by a complex equivalent circuit, each element of which (capacitance or resistance) is independent of the frequency and corresponds to a definite process occurring during the charging of the electrode with an alternating current of finite frequency. Thus, the form of the equivalent circuit is unequivocally determined by the physical processes which are considered in the calculation of the impedance.

Figures 97 and 98 give the equivalent circuits of an n-type semiconductor−dielectric−metal contact [11], which may also be used for the system semiconductor−electrolyte. In these systems the capacitance is independent of the frequency; some elements, which are provisionally called resistances, actually have imaginary components. * Figure 97 shows the case of a depletion or inversion layer (the contact is impoverished in majority or enriched in minority carriers) with the rate of recombination at the contact equal to 0 and ∞. Figure 98 shows the case of an accumulative layer (the contact enriched in majority carriers).

In series with the capacitance of the dielectric (or Helmholtz layer) C_0 and the bulk resistance of the semiconductor R_B are connected the capacitances of the surface states (C_t) and the space charge region; the latter is made up of the capacitance of the inversion layer C_1, caused by the change in the hole component of the space charge, and the capacitance of the depletion layer C_D, produced by the change in the thickness of the depletion layer with a change in potential.† The space charge region offers no resistance to the displacement current through the capacitance C_D. At the same time, the transfer of electrons through the depletion layer region to the semiconductor surface is characterized by the resistance $R_{n, D}$. The element of the circuit $R_{p, B}$ describes the generation−recombination of holes in the diffusion region and their diffusion to the boundary of the space charge region and from it ($R_{p, B}$ has active and reactive components). The transfer of electrons from the conductivity band and holes from the valence band to surface levels

---

*These elements, which describe diffusion processes, may be represented in their turn by infinite lines consisting of capacitances and resistances which are independent of frequency.

†With strong enrichment of the contact in minority carriers $C_I \gg C_D$.

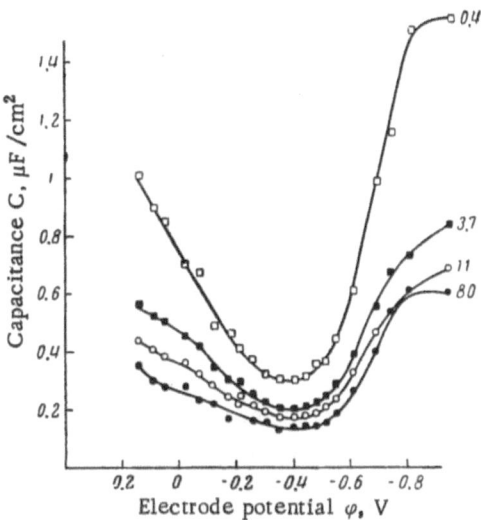

Fig. 99. Relation of the capacitance C of a germanium
electrode (n-type, 30 Ω · cm) to the electrode potential
φ in a solution of KBr in methylformamide. The parameter is the frequency (given on the curves in kilohertz) [14].

is described by the resistances $R_{n,t}$ and $R_{p,t}$, which depend on
the corresponding trapping cross sections. Finally, if there is
loss through the dielectric (or an electrochemical reaction occurs at the electrode) there appear currents through the contact; namely, the electron current i_n and the hole current i_p.
(The capacitance and resistance associated with the flow of the
currents i_n and i_p are not analyzed here.) Thus, for the case
examined where $s \rightarrow 0$, the equivalent circuit consists of two
arms, one of which (left in Fig. 97a) refers to the movement of
majority and the other to the movement of minority carriers
through the quasineutral region, space charge region, and interphase.

In the case of an infinitely high recombination rate at the
contact, the division of the total current into electron and hole
currents has no meaning (except the charging of surface states).
Therefore, the resistances $R_{p,B}$ and $R_{n,D}$ are replaced by one
resistance R_D and the currents i_p and i_n by one current i.

A contact with an accumulative layer is characterized by the capacitance C_A, caused by the change in the electron space charge with a change in the potential. As before, in parallel to it is connected the capacitance of the surface states C_t (with resistances $R_{n,t}$ and $R_{p,t}$). In the absence of surface recombination the transfer of holes through the space charge region for charging the capacitance C_t (and in the case of leakage, for creating a hole current through the contact) is characterized by the resistance $R_{p,A}$, while the element $R_{p,B}$ (as in Fig. 97a) describes the generation—recombination and movement of holes in the quasineutral region. Finally, with an infinitely high recombination rate at the contact the circuit examined (Fig. 98a) changes into the circuit illustrated in Fig. 98b.

The separate components of the equivalent circuits examined above are given in [11] in the form of graphs (for the case with no leakage through the contact). Analogous calculations were made in [12].

§ 47. Frequency Dependence of Impedance of Semiconductor Electrodes (Experiment)

In most experimental investigations, the semiconductor electrode is simulated by the simplest equivalent circuit, namely, a capacitance and resistance connected in parallel. * Let us examine the frequency dependence of the capacitance defined in this way.

Figure 99 shows curves of the capacitance of an n-type germanium electrode in a solution of KBr in methylformamide

---

*Only in [13] was a more complex circuit (Fig. 16) used for a germanium electrode and, as follows from the conditions of the experiment in [13] (inversion layer on anode polarization of n-type germanium and surface recombination rate close to zero) this circuit should be compared with Fig. 97a. The capacitance measured was close to the calculated capacitance of the space charge region C_I and, consequently, no fast surface states were present (i.e., the elements C_t, $R_{n,t}$, $R_{p,t}$, and $R_{n,D}$ should be excluded from the circuit in Fig. 97a). Thus, the resistance R_s' found experimentally (Fig. 16) corresponds to the "resistance" $R_{p,B}$ (Fig. 97a) and apparently describes the diffusion of holes into the quasineutral region. The frequency dependence of the impedance was not investigated in [13].

in the range of potentials in which there is no current through
the electrode—electrolyte interphase [14]. Curves of the capacitance in aqueous solutions also have the same form close
to the capacitance minimum [15, 16] (Fig. 100). At a frequency
of 80 kHz (Fig. 99, lowest curve) the measured capacitance in
methylformamide is close to the calculated capacitance of the
space charge region in the semiconductor (see § 12). With a
fall in frequency the capacitance increases: $C \sim \omega^{-1/2}$. When
the semiconductor surface is enriched in majority carriers
(i.e., at potentials more negative than the capacitance minimum), the excess capacitance can apparently be associated only
with fast surface states. As a rule the frequency dependence of
the capacitance observed experimentally cannot be described by
formula (43.23) for one level. At the germanium—methylformamide or germanium—water interphase there is apparently a
set of levels, whose density increases with distance from the
center of the forbidden band, as is shown by the potential dependence of the excess capacitance. (This is confirmed by
photopotential measurements. See § 18.)

In the region of the inversion layer (at potentials more positive than the potential of the capacitance minimum) the frequency dependence of the capacitance is also apparently due to
fast surface states. The fall in capacitance with an increase in
frequency may also be connected with slow diffusion of holes in
the quasineutral region, which is analyzed in § § 42 and 45. However, this effect is not clearly marked on Figs. 99 and 100 because the samples of germanium with a conductivity close to
the intrinsic value were used. In this material the equilibrium
bulk concentration of holes is very high and, therefore, even with
the maximum frequency used [14, 16] the diffusion current of
holes in the quasineutral region was quite high so that the capacitance of the electrode differed little from its static value
(at zero frequency). With samples whose conductivity is far
from the intrinsic value, the fall in the capacitance of the inversion layer with frequency is clearly marked even at a frequency
of 10^3-10^5 Hz (in the systems germanium—electrolyte [15, 17],
silicon—dielectric—metal [10], and silicon—electrolyte [18]).

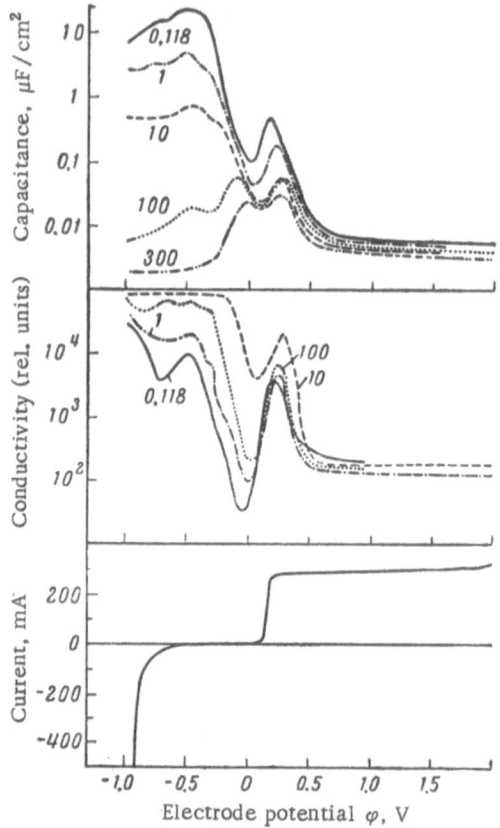

Fig. 100. Relation of the capacitance, parallel conductivity, and current (n-type germanium, 30 Ω · cm) to the
electrode potential φ in 0.1 N aqueous H_2SO_4 solution [16].
The parameter is the frequency (given on the curves in
kilohertz).

When an anode current is passing (less than the limiting
current of holes) the frequency dependence of the capacitance is
qualitatively similar to that calculated in § 45. As the current
approaches the limiting value, the resistance of the contact
passes through a minimum in accordance with the theory in [8]
(in Fig. 100, the resistance minimum corresponds to the conductivity maximum of the contact).

As Fig. 100 shows, with strong anode polarization of n-type germanium the capacitance falls somewhat with an increase in frequency. In some cases a linear relation is maintained here between the reciprocal square of the capacitance and the potential, as is characteristic of a depletion layer, but the slope of the $(1/C^2) - \varphi$ line changes with frequency. The same is also observed in the case of p-n junctions under conditions of the limiting current [19]. As in the cases examined above, the change in the capacitance with frequency at the limiting current may be connected with diffusion retardation of holes in the quasineutral region.

Under conditions of the limiting current the resistance also depends on the frequency. Figure 101 shows curves of the resistance (in the coordinates $R^2 - \varphi$) for cadmium sulfide under anode polarization [20]. As was shown in §§ 16 and 38, on a CdS electrode the density of the surface states is very low and a depletion layer arises on anode polarization so that the current density is determined by the generation of minority carriers in the depletion layer and on the electrode surface. Therefore, the elements C_I, C_t, $R_{n,t}$, and $R_{p,t}$ should be omitted from the equivalent circuit illustrated in Fig. 97b, i.e., the electrode is described by the capacitance of the depletion layer C_D and the resistance to the transfer of electrons through the depletion layer R_D connected in parallel with the capacitance of the Helmholtz layer and the bulk resistance of the semiconductor and solution connected in series with these. We can expect that R_D will be proportional to the thickness of the depletion layer and, consequently, the square root of the potential [equation (26.17)]. In actual fact, as Fig. 101 shows, with a frequency below 1 kHz the square of the resistance is related linearly to the potential with the $R^2 - \varphi$ lines intersecting the potential axis at the same point as the $(1/C^2) - \varphi$ lines, i.e., at the flat-band potential (§ 16). However, the reason for the change in the slope of the $R^2 - \varphi$ lines with a change in the frequency remains unknown.

The greatest frequency dependence of the capacitance is observed with strong cathode polarization of n-type germanium

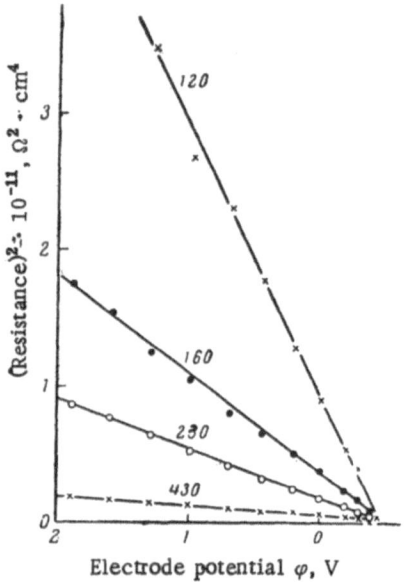

Fig. 101. Relation of the square of the differential
resistance of a cadmium sulfide electrode (0.8 Ω · cm)
to the electrode potential φ in K_2SO_4 solution [20]. The
parameter is the frequency (given in the figure in hertz).

and silicon (see Fig. 100) and anode polarization of p-type, which
corresponds to the forward direction of the current through the
interphase. Gobrecht and Meinhardt [16] associate the high
value of the capacitance and its frequency dependence with the
occurrence of an electrode reaction (pseudocapacitance) and also with the injection of minority carriers into the semiconductor (so-called diffusion capacitance [21]).

In conclusion, it should be noted that in most of the experimental investigations carried out up to now it has not been
possible to distinguish clearly between the different reasons for
the frequency dependence of the capacitance (fast surface states,
diffusion retardation of minority carriers, and an electrochemical reaction).

§48. Faraday Rectification in Semiconductor — Electrolyte System

When an alternating current passes through an electrochemical cell, in addition to an alternating voltage there also arises some constant shift of the electrode potential. This effect is called Faraday rectification and has been investigated very carefully both theoretically and experimentally for the system electrolyte—metal electrode [22-26]. Here we will examine Faraday rectification in a semiconductor—electrolyte system, basing the examination mainly on [27]. We will first show qualitatively how the passage of a sinusoidal current, which has no preferred direction, leads to the appearance of a constant potential difference in the system. It will be assumed that the passage of the current is associated with an oxidation—reduction reaction at the contact. By using equations (24.5) and (24.9) and neglecting the potential drop in the Helmholtz layer, we write an expression for the electric current passing through the contact:

$$i_s = i_p^0\left(1 - \frac{p_s}{p_s^0}\right) + i_n^0\left(\frac{n_s}{n_s^0} - 1\right) =$$

$$i_p^0(1 - e^{-\frac{e(\varphi_1 - \varphi_1^0)}{kT}}) + i_n^0(e^{\frac{e(\varphi_1 - \varphi_1^0)}{kT}} - 1).$$

(48.1)

Here the first term describes a reaction involving holes and the second term a reaction involving electrons of the conductivity band. We will assume for simplicity that $i_p^0 \gg i_n^0$. Then we obtain the following relation for the current:

$$i_s = i_p^0[1 - e^{-\frac{e(\varphi_1 - \varphi_1^0)}{kT}}].$$

(48.2)

As follows from equation (48.2), when a current is passed in a positive direction the electrostatic potential is shifted in a positive direction by $\tilde{\varphi} = \Delta\varphi_1' = \varphi_1 - \varphi_1^0$. During the subsequent half period, when the current is negative, the potential is shifted in a negative direction by $\Delta\varphi_1''$. By convention the oscillations of the current in the two directions are the same and, as follows from equation (48.2), are proportional to the change in the exponent $e^{-e\varphi_1/kT}$. The same increment in the exponent in positive and negative directions corresponds to different changes in the argument. It is readily seen that $|\Delta\varphi_1^n| < |\Delta\varphi_1^i|$, i.e., a shift in the potential arises in the system.

We then define the value $\tilde{A}$ by the relation

$$<\tilde{A}> = \frac{\omega}{2\pi} \int\limits_0^{2\pi/\omega} \tilde{A}(t)\,dt$$

$$(48.3)$$

and calculate the mean shift in the potential $\tilde{\varphi}_1$. Let the total current (41.1), which equals the sum of the current of free carriers and the displacement current, change in accordance with a harmonic law, i.e., in proportion to $e^{j\omega t}$. In the bulk of the semiconductor the displacement current is small and therefore the current of carriers also changes in accordance with the harmonic law.

By averaging equation (41.1) in accordance with the definition (48.3), we obtain the mean value of the total current:

$$<\tilde{I}> = <\tilde{i}> + \frac{1}{4\pi}<\frac{\partial\tilde{D}}{\partial t}> .$$

$$(48.4)$$

The mean value of the harmonic function $\tilde{I}$ equals zero. The mean value of the derivative of the periodic function with respect to time also equals zero, i.e., $<\partial\tilde{D}/\partial t> = 0$. Consequently $<\tilde{i}> = 0$. In particular, the following relation holds:

$$<\tilde{i}_s> = 0.$$

$$(48.5)$$

It will be assumed that the deviations from equilibrium are so small that $[e(\varphi_1 - \varphi_1^0)]/kT \ll 1$. Expanding the exponents in expression (48.1) into series with respect to $[e(\varphi_1 - \varphi_1^0)]/kT = e\tilde{\varphi}_1/kT$, and taking only terms of the second order of smallness we obtain for $\tilde{i}_s$ the expression

$$\tilde{i}_s = i_p^0\left(\frac{e\tilde{\varphi}_1}{kT} - \frac{e^2\tilde{\varphi}_1^2}{2(kT)^2}\right) + i_n^0\left(\frac{e\tilde{\varphi}_1}{kT} + \frac{e^2\tilde{\varphi}_1^2}{2(kT)^2}\right).$$

$$(48.6)$$

By averaging with respect to time relation (48.6) and using (48.5), we find

$$\frac{e}{kT}<\tilde{\varphi}_1> = \frac{i_p^0 - i_n^0}{i_p^0 + i_n^0} \frac{e^2}{2(kT)^2}<\tilde{\varphi}_1^2>.$$

(48.7)

To calculate $<\tilde{\varphi}_1^2>$ we use the value of $\tilde{\varphi}_1$ found in the harmonic approximation with an accuracy up to terms of the first order of smallness. In accordance with equation (41.11),

$$\tilde{\varphi}_1 = \frac{\tilde{I}}{\frac{1}{R} + j\omega C}.$$

(48.8)

Since the displacement current in the bulk of the semiconductor is small, then $\tilde{I} = i_B e^{j\omega t}$, where i_B is the amplitude of the current of carriers in the bulk of the semiconductor. Thus, relation (48.8) may be written in the form

$$\tilde{\varphi}_1 = \frac{i_B e^{j\omega t}}{\frac{1}{R} + j\omega C}.$$

(48.9)

By calculating the mean value of $\tilde{\varphi}_1^2$ in accordance with the definition (48.3), we find

$$<\tilde{\varphi}_1^2> = \frac{1}{2} \frac{i_B^2}{\frac{1}{R^2} + \omega^2 C^2}.$$

(48.10)

By substituting the value of $<\tilde{\varphi}_1^2>$ found in relation (48.7), we obtain

$$<\tilde{\varphi}_1> = \frac{i_p^0 - i_n^0}{i_p^0 + i_n^0} \frac{e}{4kT} \frac{i_B^2}{\frac{1}{R^2} + \omega^2 C^2}.$$

(48.11)

Then by taking into account the fact that the resistance R is given by

$$R = \frac{kT}{e(i_p^0 + i_n^0)} = \frac{kT}{ei^0},$$

(48.12)

where i^0 is the total exchange current, we rewrite relation (48.11) in the form

$$<\tilde{\varphi}_1> = \frac{i_p^0 - i_n^0}{i_0} \frac{kT}{4e} \frac{i_B}{(i^0)^2}^2 \frac{1}{1 + \left(\frac{kT\omega C}{ei^0}\right)^2} \cdot$$

$$(48.13)$$

Thus, the sign of the Faraday shift $<\tilde{\varphi}_1>$ is determined by the difference in the exchange currents $i_p^0 - i_n^0$.

No account was taken of the diffusion of minority carriers (holes) in the calculation given. In [27] the value of $<\tilde{\varphi}_1>$ was estimated taking into account this effect. The calculations were carried out for an n-type semiconductor on the assumption that diffusion retardation in the electrolyte may be neglected, i.e., we have the condition that $ec\sqrt{D\omega} \gg i_p^{lim}$ (here c is the concentration of ions participating in the oxidation−reduction reaction in the depth of the electrolyte and D is the diffusion coefficient of the ions). For germanium with a frequency $\omega = \tau_p^{-1}$ this relation normally holds beginning with a concentration c = 10^{-4} moles/liter. It may be shown that the ratio of the potential drop in the quasineutral region $<\tilde{\varphi}_L>$ to the potential drop in the space charge region $<\tilde{\varphi}_1>$ is given by

$$\frac{<\tilde{\varphi}_L>}{<\tilde{\varphi}_1>} = \frac{i^0}{i_p^{lim}} \frac{p^0}{n^0},$$

and, therefore, the potential drop in the quasineutral region may be neglected.

As a result of the calculations it was found that the mean value of the potential is again given by a formula of type (48.13). To the right-hand part of relation (48.13) were added terms of the order $(i_p^0)^2/(i^0 \cdot i^{lim})$. Since these terms are small in accordance with the assumptions made, the effects examined are appreciable only when the hole exchange current i_p^0 is very close to the electron exchange current i_n^0 so that the ratio $|(i_p^0 - i_n^0)/i^0|$ is of the order of $(i_p^0)^2/(i^0 \cdot i_p^{lim})$.

Examination of formula (48.13) shows that the sign of the mean value of the potential $<\tilde{\varphi}_1>$ depends only on the sign of the

difference in the exchange currents $i_p^0 - i_n^0$. Therefore, an experimental investigation of Faraday rectification makes it possible to determine the degree of participation in the reaction of
the two energy bands of the semiconductor at potentials close
to the equilibrium value.

It is interesting to compare the magnitude of Faraday rectification on a semiconductor electrode with the analogous value
for other systems. For a metal electrode we may write in
order of magnitude [24]

$$< \frac{e\widetilde{\varphi}}{kT} > \sim \left(\frac{\iota_B}{i_M^0} \right)^2,$$

where i_M^0 is the exchange current.

Comparison with relation (48.13) shows that on semiconductors we can expect a greater effect with the same current
amplitude as the exchange current i^0 is generally much less than
the exchange current on metals i_M^0.

Rough calculation gives the magnitude of Faraday rectification for p-n junctions:

$$< \frac{e\widetilde{\varphi_1}}{kT} > \sim \left(\frac{\iota_B}{i_p^{\lim}} \right)^2,$$

where $i_p^{\lim}$ is the limiting current of the p-n junction. Usually
this current considerably exceeds the exchange current i^0.

§ 49. Charging Curves

The relation between the charge and the potential of an
electrode when it is charged with a current of constant strength
is called a galvanostatic charging curve. We will briefly examine processes occurring during the charging of a semiconductor electrode by square current pulses of short duration. The
so-called "fast" charging curves* thus obtained have provided

---

*It is important that only the potential drop in the space charge region changes
during the plotting of a "fast" charging curve (in contrast to "slow" curves, see
§ 14). The Helmholtz potential drop remains constant as its relaxation time is
much greater than the duration of the charging pulse.

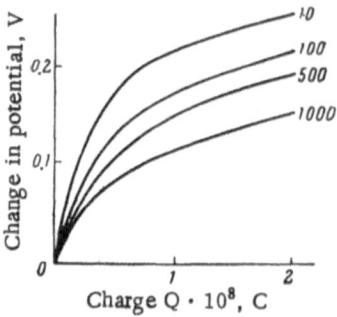

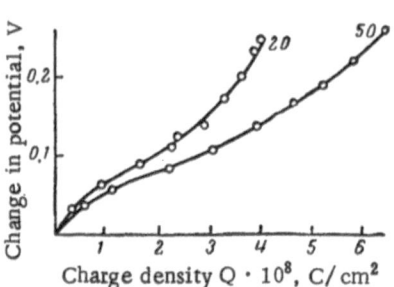

Fig. 102. Cathode charging curves of a germanium electrode (n-type, 5 Ω·cm) in a solution of KBr in methylformamide. Charging was carried out from the potential of the capacitance minimum. Duration of charging pulse (in microseconds) is given in the figure. The electrode area was 0.04 cm².

Fig. 103. Anode charging curves of a germanium electrode (n-type, 30 Ω·cm) in a solution of KBr in methylformamide [14]. The charging was carried out from the minimum point. The duration of a charging pulse (in microseconds) is shown on the curves.

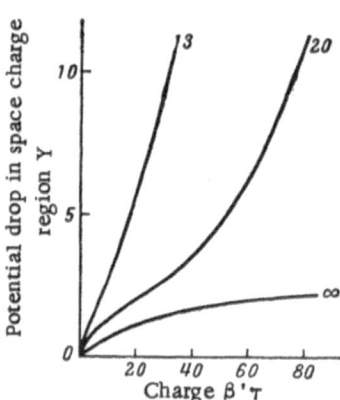

Fig. 104. Calculated dynamic charging curves of a semiconductor electrode [28]. The parameter is the value of β' (shown on the curves). The starting value of $\beta'\tau^0$ was taken as equal to 10; $p^0/n^0 = 100$.

a convenient method of investigating the properties of semiconductor electrodes in recent years.

The current i passed is consumed in changing the space charge (ΔQ_1), in charging the surface states (ΔQ_t), and in the electrochemical reaction. If we are considering a contact without leakage or if the charging time t ≪ RC (where R and C are the resistance and capacitance of the contact connected in parallel), then the Faraday current may be neglected and then it = $\Delta Q_1 + \Delta Q_t$.

With infinitely slow charging of the electrode, ΔQ_1 and ΔQ_t are related to the change in the potential $\Delta \varphi_1$ by equations (10.12) and (10.13), by means of which it is possible to calculate the static charging curve. If the charging time is finite, then it is necessary to take into account the movement of charges between the space charge region and the bulk of the semiconductor and also the deviation from equilibrium in the space charge region.

With an accumulative layer the charging of the electrode is connected with the movement of majority carriers, and if the charging time is much greater than the time to establish equilibrium in the space charge region t_d (see p. 289), then the relation of the space charge to the potential coincides with the static charging curve. An example is the upper curve in Fig. 102, which was obtained by Krotova for the germanium—methylformamide system by means of the pulse apparatus described in §23 with a pulse time of 10 μsec. With an increase in the pulse time the charging curves deviate from the upper curve (see Fig. 102) and, in particular, there is an increase in the charge which must be given to the electrode to change its potential to a given value. Here part of the current is apparently consumed in the charging of fast surface states, whose relaxation time is several tens of microseconds (so that with t = 10 μsec the charge in these states hardly changes). The difference between a curve plotted during time t and the upper curve, which was obtained in 10 μsec, is the relation of the charge to the fast surface states with a relaxation time of the order of t to the potential.

If the space charge is an inversion layer, then its change involves the movement of minority carriers from the neutral bulk to the space charge region, or the reverse. In a semiconductor with conductivity far from the intrinsic conductivity, the transfer of minority carriers proceeds by diffusion. The minority carriers necessary for charging the double layer come from the neutral volume and their concentration at the boundary of the Debye region p_1 falls. As a concrete example we will examine the charging of an n-type semiconductor by an anode current so that during charging there is an increase in a space charge consisting of minority carriers, i.e., holes. During the time of a pulse t the diffusion front moves away from the boundary of the space charge into the depth of the semiconductor by a distance $\sqrt{D_p t}$ (D_p is the diffusion coefficient of holes). When the electrode is charged with a high current (consequently, during a short time), the thickness of the diffusion region is small and the hole charge ΔQ moving from this region into the space charge region with the change in potential is found to be less than that calculated by equation (10.12). In this case, the charging curve deviates from the static curve, as is shown in Fig. 103 [14].

The dynamic charging curve for the case examined above was calculated in [28]. The electrode charging conditions are described by the dimensionless parameter:

$$\beta' = \frac{e\,(p^0)^2 D_p\,\sqrt{2Y*}}{I\pi L_1 N_D}\,,\tag{49.1}$$

where p^0 is the equilibrium concentration of holes in the bulk of the semiconductor, I is the charging current, L_1 is the Debye length, and $Y*$ is the dimensionless potential defined by the relation

$$p_1 e^{Y*} = N_D Y*.$$

Thus, β' takes into account simultaneously the magnitude of the charging current and the properties of the sample. (It is readily seen that the degree to which the charging process is nonequilibrium increases both with an increase in the charging current and with a decrease in the concentration of holes p^0 in the bulk of the semiconductor, i.e., with a decrease in β').

The results of the calculation are shown in Fig. 104 in the coordinates $Y-\beta'\tau$; here τ is dimensionless time defined by the relation $\tau = t/\delta$, where $\sqrt{\delta} = ep^0\sqrt{D_p}/(I\sqrt{\pi})$. The value of β' is proportional to the change in the space charge:

$$\Delta Q_1 = \frac{eN_D L_1}{\sqrt{2Y^*}} \beta'\tau. \qquad (49.2)$$

When $\beta' \to \infty$, the charging curve coincides with the static curve (Fig. 104, bottom curve); but with a decrease in β' the difference between the dynamic and static curves becomes appreciable (upper curves in Fig. 104).

From the calculation in [28] it follows that at the initial moment of time the charging current is totally electronic. The reason for this phenomenon is that for a diffusion current of holes to flow there must be a concentration gradient of holes at the boundary of the Debye and quasineutral regions, and this cannot arise instantaneously. However, after a certain time interval,

$$t_D = \left(\frac{n^0}{p^0}\right)^2 \frac{1}{2Y^*} \frac{L_1^2}{D_p} \qquad (49.3)$$

the charging current becomes mainly a hole current (when $\beta' \gg 1$) and there is an increase in the hole space charge. The thickness of the diffusion layer increases in time and the concentration gradient of holes falls; as a result the diffusion current of holes, which is proportional to their concentration gradient, becomes less than the charging current I. From this moment onward an electronic component again appears in the charging current. Electrons move into the depth of the semiconductor, leaving ionized atoms of the donor impurity. This means that the space charge increases not as a result of the influx of holes, but as a result of extension of the region in which the charge is formed by ionized donors.

Literature Cited

1. V.G. Levich, Yu.A. Vdovin, and V.A. Myamlin, Course in Theoretical Physics, Vol. 2, Fizmatgiz, Moscow (1962).
2. F. Berz, J. Electron. Control, 6 : 197 (1959).
3. F. Berz, J. Phys. Chem. Solids, 23 : 1795 (1962); 25 : 859 (1964).
4. Yu. Ya. Gurevich and V.A. Myamlin, Izv. Akad. Nauk SSSR, Ser. Khim., 1964:1776.
5. C.G.B. Garrett, Phys. Rev., 107:478 (1957).
6. Yu. Ya. Gurevich and V.A. Myamlin, Izv. Akad. Nauk SSSR, Ser. Khim., 1964:2237.
7. T.P. Birintseva and Yu. V. Pleskov, Izv. Akad. Nauk SSSR, Ser. Khim., 1965:251.
8. V.A. Myamlin and B.M Grafov, Izv. Akad. Nauk SSSR, Ser. Khim., 1963:1011.
9. V.A. Myamlin and Yu. Ya. Gurevich, Dokl. Akad. Nauk SSSR, 155 : 164 (1964).
10. K. Lehovec, A. Slobodskoy, and J.L. Sprague, Phys. Status Solidi, 3 : 447 (1963).
11. K. Lehovec and A. Slobodskoy, Solid–State Electronics, 7 : 59 (1964).
12. R.S. Nakhmanson, Fiz. Tverd. Tela, 6 : 1115 (1964).
13. W.H. Brattain and P.J. Boddy, J. Electrochem. Soc., 109 : 574 (1962).
14. M.D. Krotova and Yu. V. Pleskov, Phys. Status Solidi, 3 : 2119 (1963).
15. K. Bohnenkamp and H.-J. Engell, Z. Elektrochem., 61 : 1184 (1957).
16. H. Gobrecht and O. Meinhardt, Ber. Bunsenges., 67:151 (1963); H. Gobrecht, O. Meinhardt, and B. Reinicke, Ber. Bunsenges., 67:493 (1963).
17. O.G. Deryagina and E.N. Paleolog, Elektrokhimiya, 1 : 267 (1965).
18. R.M. Hurd and P.T. Wrotenbery, Ann. N. Y. Acad. Sci., 101 : 876 (1963).
19. A.A. Lebedev and V.M. Tuchkevich, Collection: Electron-Hole Junctions in Semiconductors, Izd. Akad. Nauk UzbSSR, Tashkent (1962), p. 220.

20. V.A. Tyagai, Izv.Akad.Nauk SSSR, Ser. Khim., 1964:34.
21. W. Shockley, Bell System Techn.J., 28:435 (1949).
22. K.S.G. Doss and H.P. Agarwal, Proc. Ind.Acad.Sci., 34(Sec.A):229 (1951).
23. K.B. Oldham, Trans. Faraday Soc., 53:80 (1957).
24. Yu.A. Vdovin, Dokl.Akad.Nauk SSSR, 120:554 (1958).
25. G.S. Barker, Trans. Symposium on Electrode Processes, Philadelphia (1959).
26. V.G. Levich and B.M. Grafov, Dokl.Akad.Nauk SSSR, 146:1372 (1962).
27. Yu.Ya. Gurevich and V.A. Myamlin, Dokl.Akad.Nauk SSSR, 155:1159 (1964).
28. V.A. Tyagai and Yu.Ya. Gurevich, Fiz.Tverd.Tela, 7:12 (1965).

Additional Literature

Gobrecht, H., M. Schaldach, F. Hein, R. Blaser, and H.-G. Wagemann, Dynamische Untersuchungen an der Plasengrenze Germanium–Elektrolyt. I: Eine Methode zur schnellen Messung der Komponenten eines komplexen Widerstandes und ihre Anwendung zur Ermittlung der differentiellen Kapazität. Ber. Bunsenges., 69:338 (1965).

Grove, A. S., E. H. Snow, B. E. Deal, and C. T. Sah, Simple physical model for the space-charge capacitance of metal oxide-semiconductor structures, J. Appl. Phys., 35:2458 (1964); Solid-State Electronics, 8:145 (1965).

Gurevich, Yu. Ya., and V. A. Myamlin, High-frequency capacitance of an electrolyte–semiconductor interphase Elektrokhimiya, 1:734 (1965).

Hofstein, S. R., and G. Warfield, Physical limitations on the frequency response of a semiconductor surface inversion layer, Solid-State Electronics, 8:321 (1965).

Lehovec, K., and A. Slobodskoy, The semiconductor surface impedance under conditions of flat bands. IEEE Trans. on Electron Devices, ED-12:121 (1965).

Nakhmanson, R. S., Estimation of recombination in the space charge region close to a semiconductor surface. Fiz. Tverd. Tela, 7:3439 (1965).

Chapter IV
Corrosion of Semiconductors

§ 50. Basic Principles

In the broad sense, corrosion is the oxidation of a material without the passage of an electric current. In this chapter we examine the corrosion of semiconductor materials in aqueous solutions, * and most attention will be paid to the electrochemical aspects of this process.

Since there is no current in the outer circuit, the oxidation of the solid must be accompanied by the reduction of other components of the system. For example, simultaneously with the dissolution of germanium there may occur the liberation of molecular hydrogen from the solution or the reduction of a dissolved oxidant. The overall equation of the process has the form

$$M + X \rightarrow M^+ + X^-, \tag{50.1}$$

where M is the semiconductor material, M^+ is its oxidized form (for example, an ion in solution), X is the oxidant, and X^- its reduced form.

---

* This process is also called chemical etching of semiconductors (in contrast to anodic etching). This name cannot be regarded as well-chosen, since the processes of "chemical" etching often have an electrochemical nature, as will be shown below.

337

The solution of the semiconductor and the reduction of the oxidant may actually occur in one act so that equation (50.1) reflects the microscopic mechanism of the process. In this case we talk of the chemical mechanism of the corrosion or the dissolution of the solid.

However, most often the corrosion has an electrochemical nature. At the solid–liquid interface there are two reactions proceeding simultaneously and at the same rate, namely, anodic and cathodic reactions:

$$M \to M^+ + e^-,$$
$$X + e^- \to X^-. \tag{50.2}$$

On samples with a uniform surface they are localized on the same sections of the surface, while in the opposite case they may be separated in space. These reactions are called conjugate reactions. In the electrochemistry of metals they are regarded as completely independent of each other. This means that by changing the external conditions it is possible to change the rate of one of them arbitrarily without affecting the other. Knowing the kinetic parameters of both reactions (the exchange currents and transport coefficients or simply the coefficients a and b of the Tafel relation) it is possible to determine unequivocally the potential and the corrosion rate from the condition

$$i^a = - i^c = |i_{corr}|, \tag{50.3}$$

where i^a and i^c are the densities of the anode and cathode currents (see, for example, [1]).

The peculiarities of corrosion processes on semiconductors as compared with metals are connected with the fact that in the first case the exchange between the solid and the solution proceeds through mobile charges of both signs, namely, free electrons and holes. Therefore, a condition of type (50.3) is inadequate. In the dissolution process there must be not only an overall balance of electric charges, but also a balance with respect to the different energy bands of the semiconductor. The recombination and generation of free carriers must be taken into account in this balance.

As an example, we will examine the scheme for the dissolution of germanium. The equation for the anode reaction may be written arbitrarily in the form (see § 32)

$$Ge + 2.4\, e^+ \rightarrow Ge\,(IV) + 1.6\, e^-,$$

where Ge (IV) denotes a compound of tetravalent germanium.

In the general case the cathode reaction also proceeds simultaneously through the conductivity band and through the valence band. In the limiting cases (the participation of only free electrons or only holes in the cathode process) the overall equation for the corrosion process has the form

$$Ge + 2.4\, e^+ + 2.4\, e^- + 4X \rightarrow Ge\,(IV) + 4X^-$$

or   (50.4)

$$Ge + 4X \rightarrow Ge\,(IV) + 1.6\, e^+ + 1.6\, e^- + 4X^-.$$

In the first case carriers of both signs participate in the reaction. It is readily seen that on p-type and on n-type samples the reaction rate is determined by the supply of minority carriers to the interphase. Therefore, it cannot exceed the limiting current of minority carriers.

In the second limiting case, the free carriers are reaction products and therefore the rate of the reaction is not limited by their supply to the sample surface. Essentially the anodic and cathodic reactions are not independent here. The cathode reaction involves valence electrons and thus "supplies" the anode reaction with holes. Therefore, the rate of the anode process on an n-type semiconductor depends on the "injection coefficient" of the reduction process.

In the general case for n-type germanium we may write

$$i^a = i_p \cdot \alpha',$$
$$i_{ps} = -\, i^c \cdot \gamma + i_p^{\lim}, \tag{50.5}$$

where γ is the fraction of valence electrons in the cathode current, α' is the multiplication factor of the reaction current of anode solution of germanium, and i_{ps} is the hole current at the surface.

Depending on the relation* between α' and γ, the following cases are possible:

a) $\gamma = 0$ (only free electrons participate in the cathode process). Then $i^a = i_{corr} = i_p^{lim} \cdot \alpha'$, i.e., the rate of corrosion is determined by the limiting current of holes.

b) $\gamma < 1/\alpha'$. The corrosion current $i_{corr} = i_p^{lim} / [(1/\alpha') - \gamma]$ is greater than in (a), but the process is controlled primarily by the supply of holes to the electrode surface.

c) $\gamma = 1/\alpha'$. In this case the rate of removal of holes by the anode process equals their rate of generation as a result of the cathode reaction. Therefore, the surface concentration of holes equals the equilibrium value and on the surface $i_{ps} = 0$.

d) $\gamma > 1/\alpha'$. The flow of holes formed by the reduction of the oxidant with the participation of valence electrons is greater than the consumption of holes in the anode solution of germanium. From the surface of the electrode into the bulk there flows a current of excess holes

$$i_{ps} = i_{corr}\left(\gamma - \frac{1}{\alpha'}\right) \tag{50.6}$$

and a current of free electrons equal to it. These nonequilibrium carriers recombine in the quasineutral region at a distance

---

*Strictly speaking, α' and γ are not independent values. According to §24 they are determined by the surface concentrations of free electrons and holes. However, in practice, for reactions which occur during the corrosion of the semiconductor materials which have been investigated in greatest detail, the question of the dependence of these parameters on the surface potential remains open. Therefore, we regard α' and γ here as values which are independent of each other. It should also be remembered that even for a definite process these parameters are not always constant. For example, in the anode solution of germanium, α' falls with an increase in the hole injection current [2, 3].

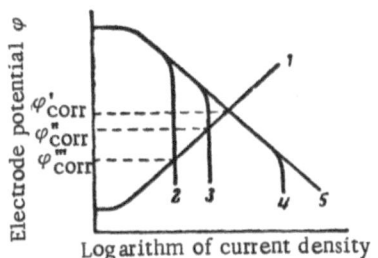

Fig. 105. Diagram illustrating the peculiarities of the dissolution of an n-type semiconductor in presence of an oxidant, which is reduced with participation of the valence electrons. 1) Overvoltage curve of cathode process; 2) overvoltage curve of anode process with $\gamma = 0$; 3) the same with $\gamma < 1/\alpha'$; 4) the same with $\gamma > 1/\alpha'$; 5) the same for a p-type semiconductor.

from the surface of the order of the diffusion length. Consequently, in the corrosion of a semiconductor, as under illumination, electron—hole pairs are formed. The difference lies in the fact that in the first case the nonequilibrium carriers arise only at the surface, and in the second case they arise in the bulk of the semiconductor at the depth of the absorption of light.

The rate and potential of the dissolution depend on the rate of the anode process, as is shown in Fig. 105.*

§ 51. Effect of Light on Corrosion [4-7]

On illumination, nonequilibrium electrons and holes are formed and their concentration at the surface increases, and this leads to the acceleration of reactions in which minority carriers participate. Moreover with an increase in the concentration of free carriers there is a decrease in the potential in the space charge region in the semiconductor and an increase in the potential drop in the Helmholtz region, which is accompanied by a change in the rates of all electrochemical reactions. Here we will examine the particular case where the exchange of charges between the semiconductor and the solution both in the dark and in the light proceeds solely through one band (as a definite case,

---

*In Fig. 105 we assumed that with a change in the electrode potential there is also an appreciable change in the potential drop in the Helmholtz region. Otherwise the rate of the cathode reaction, which proceeds through the valence band, does not change on polarization and does not exceed the hole exchange current (see § 24).

through the valence band), while the redistribution of potential
at the interphase on illumination may be neglected. Under conditions of quasiequilibrium, we may write the following expressions for the densities of the anode hole current in the dark and
in the light:

$$i_p^{dark} = -k \cdot p_1 \cdot e^{-e\varphi_1^{dark}/kT},$$

$$i_p^{ill} = -k(p_1 + \Delta p)\, e^{-e\varphi_1^{ill}/kT},$$

(51.1)

where p_1 and $(p_1 + \Delta p)$ are the concentrations of holes in the
space charge region in the dark and on illumination, respec-
tively, and φ_1^{dark} and φ_1^{ill} are the potential drops in the space
charge region under the same conditions.

The equations (51.1) show that had the potential drop not
changed on illumination, then the rates of the anode process
would have increased because of the increase in the preexponential term. However, the rate of the reverse process, i.e.,
the cathode current into the valence band, is independent of φ_1
and is determined only by the potential drop in the Helmholtz
region, which we took to be constant. From condition (50.3) it
follows that the corrosion rate does not change on illumination.
Therefore, the potential drop in the space charge region φ_1
changes so as to compensate for the increase in p_1:

$$\Delta\varphi_1 = \varphi_1^{ill} - \varphi_1^{dark} = -\frac{kT}{e}\ln\frac{p_1}{p_1 + \Delta p}$$

(51.2)

The potential drop in the space charge region changes toward
positive values and consequently (see footnote on p. 32) the
electrode potential becomes less positive. The magnitude of
the photopotential is determined by the ratio of p_1 and Δp. In
p-type samples the holes are the majority carriers, their concentration changes little on illumination, and the photoeffect is
small. In n-type samples the light concentration of holes may
be high in comparison with the dark concentration and, therefore, the potential changes markedly on illumination. The concepts developed also hold if the nonequilibrium carriers are
produced not by illumination, but by injection, as a result of a
cathode reaction involving valence electrons (p. 340, case d).

In this way it may be shown that in the case of the exchange of charges between the conductivity band of the semiconductor and the solution the photopotential (which is defined here as the change in electrode potential on illumination) is positive and is large on p-type samples and small on n-type samples.

The general case when exchange between the solid and the solution proceeds simultaneously through both bands was analyzed by Lazorenko-Manevich [5] and Kuznetsov and Dogonadze [7]. The Dember potential should be taken into account in all cases (§ 17).

Under conditions where the semiconductor and electrolyte exchange charges the photopotential is usually called the slow photopotential in contrast to the instantaneous photopotential which was examined in § 17. In the first case in the calculation we use the condition $i = 0$, i.e., $i^a = -i^c$ [equation (50.3)] and in the second case, $Q_1 = const$. The rate of establishment of the steady value of the slow photopotential is greater, the greater the exchange currents of free electrons and holes in the light.

§ 52. Relation of Corrosion Rate to Properties of Solid and Solution

Above we examined only the electrochemical stage of corrosion. Under actual conditions, it is often complicated by the adsorption of reagents at the interphase and the desorption of reaction products. An essential stage of dissolution is also the transport of material between the sample surface and the bulk of the solution. Let us examine the effects of various factors on the individual stages of the process.

a. Crystallographic Orientation of Surface. In the most important semiconductor materials the bond is covalent. In crystals with a covalent bond the different crystallographic directions and faces are not equivalent. Depending on the orientation of the crystal surface, there is a change in the number of atoms per unit surface and also in the character of the bond of the surface atoms with the bulk of the crystal.

Thus, for example, in semiconductors of group IV with a diamond-like cubic lattice (germanium and silicon), at the (111) face the surface atoms are attached to three atoms of the second layer and have one unpaired electron. On the (110) face each atom is attached to two neighbors and to one atom from the second layer and also have one unpaired electron. Finally, on the (100) surface the atoms are attached to the crystals by two covalent bonds and have two unpaired electrons each. The adsorption energy and overvoltage of the electrochemical stage depend on the energy state of the surface atoms and are therefore functions of the crystallographic orientation.

b. Impurities in the Crystal and on the Surface. The amount of doping impurity in pure semiconductor materials is very small and, therefore, it cannot affect the dissolution by changing the chemical nature of the semiconductor. However, effects of an electrical character may occur. For example, the rate of dissolution in most etching agents is independent of the type of conductivity of the sample. However, if a p-n junction is immersed in a solution of an oxidant which is reduced with the participation of valence electrons, then electrons and holes are injected into the sample during the dissolution process. These are separated in the electric field of the p-n junction and charge the p-type region relative to the n-type region. Therefore, the p-type region of the crystal is an anode and dissolves more rapidly than the n-type region. The same occurs when a sample is illuminated in a nonetching solution. Essentially we are dealing here with a photoelectric effect (see § 51).

Impurities present in the etching agent may be deposited on the crystal surface and create galvanic microcells.

c. Dislocations. Excess energy is associated with crystallographic defects and, in particular, dislocations. Therefore one would expect that the activation energy of adsorption and the chemical stage of corrosion would be lower at the points of emergence of dislocations at the surface, while the rate of solution would be higher than on the surface of a perfect crystal. Moreover, the segregation of impurities often occurs at dislocations and therefore the electrophysical properties of the material close to dislocations differ from the properties of the rest of the crystal. Finally, the recombination rate is higher at dislocations. Therefore the corrosion rate close to dislocations is increased if it is limited by the limiting current of minority carriers.

The effect of the dissolution conditions on its kinetics was examined in the reviews of Irving [8] and Holmes [9].

§ 53. Germanium

a. Steady Potential of Germanium. Germanium does not dissolve at an appreciable rate in solutions which do not contain oxidants, although this reaction is possible thermodynamically (§ 8). The most accurate measurements of the steady potential of germanium* in relation to the solution pH under particularly pure conditions were made by Lovrecek and Hockris [10]. It was found that the steady potential does not depend on the type of conductivity of germanium, the crystallographic orientation of its surface, or the concentration of GeO_2 in solution and may be represented by curve a in Fig. 5. Comparison of the experimental curve with the calculated φ^0−pH diagram made it possible for the authors of [10] to conclude that the steady potential of germanium is not the equilibrium potential but the corrosion (hybrid) potential. The cathode reaction is the liberation of hydrogen from water. Its high overvoltage explains the very low rate of corrosion of germanium in nonoxidizing solutions. The anode process is the formation of the brown modification of GeO at pH 0-4 and the yellow modification at pH 6-12 in accordance with (8.18) and (8.19). The formation of GeO_3^{2-} anions may possibly occur at pH > 12.5.†

---

*A series of earlier papers [11-15] was also devoted to the measurement of the steady potential of germanium. The effect of traces of copper in the solution and in the germanium on the reproducibility of the steady potential of germanium was investigated by Sparnaay [16].

†In comparing the experimental and theoretical φ^0 − pH relations, Lovrecek and Bockris assumed that the anode reaction is not accompanied by an overvoltage. This point of view was criticized in [17, 18].

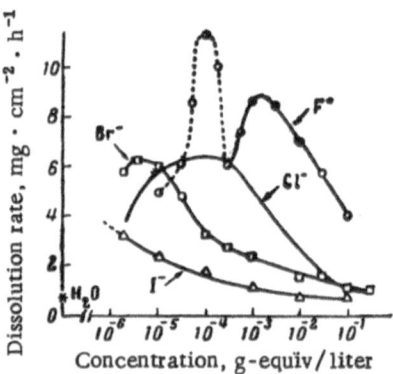

Fig. 106. Effect of anions on the rate of dissolution of germanium in potassium salts (pH 6), saturated with oxygen [20].

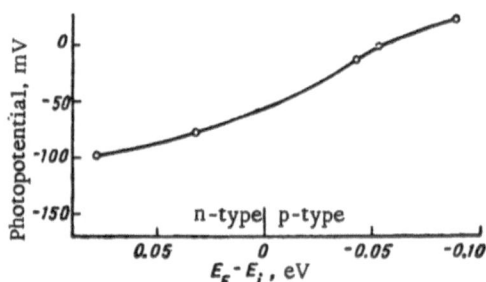

Fig. 107. Change in steady potential of a germanium electrode on illumination as a function of the position of the Fermi level in the bulk of the sample [21]. Solution 0.1 N KCl + 0.001 M GeO$_2$, saturated with oxygen. Steady illumination.

b. Corrosion in the Presence of Dissolved Oxygen. Under these conditions the rate of dissolution is determined by the cathode reaction, namely the reduction of molecular oxygen. According to Harvey and Gatos [19-21], a change in the electrophysical properties and illumination of the samples do not affect the reaction rate. It should be noted that the rate of the anode reaction does not exceed 15 μA/cm^2, i.e., it remains below the limiting current of minority carriers. The relative rates of solution on the (100), (110), and (111) crystallographic faces are 1.00, 0.89, and 0.62, respectively, i.e., they change approximately the same as the relative densities of free bonds on the surface of germanium (1.00, 0.71, and 0.58, respectively). The rate of corrosion is related to the partial pressure of oxygen above the solution P by the equation

$$v = AP/(1 + BP),$$

where A and B are constants. The activation energy of the process is about 20 kcal/mole. All this makes it possible to conclude that the rate of corrosion is determined by the adsorption of oxygen on germanium or the electrochemical stage of the process. *

It was found that the composition and concentration of the foreign electrolyte have an appreciable effect on the rate of dissolution of germanium. In solutions of halides the rate of dissolution of germanium passes through a maximum at a definite concentration of the anion and this concentration falls in the series $I^- < Br^- < Cl^- < F^-$ (Fig. 106). This effect is not connected with the solubility of oxygen or the electrical conductivity of the solution and is apparently explained by the fact that the adsorption of anions facilitates the reduction of oxygen. However, with too great a concentration of the anions competition begins between the anions and oxygen for packing onto the surface and the rate of dissolution again falls [20]. Thus, the concentration corresponding to the maximum is a function of the adsorption power of the anion.

The change in the photopotential with a change in the Fermi level (Fig. 107) in combination with the independence of the dissolution rate of illumination agrees with the concept of the exchange between germanium and the electrolyte predominantly through the valence band.

---

*With a very low oxygen content in solution the slow stage is the diffusion of oxygen to the sample surface.

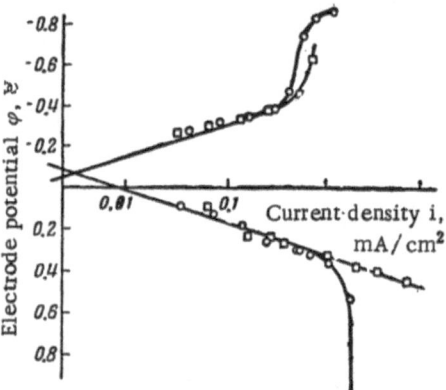

Fig. 108. Corrosion diagram for germanium in H_2SO_4 solution (pH 1), saturated with air [22]. ○) n-Type, 13 Ω · cm; □) p-type, 21 Ω · cm.

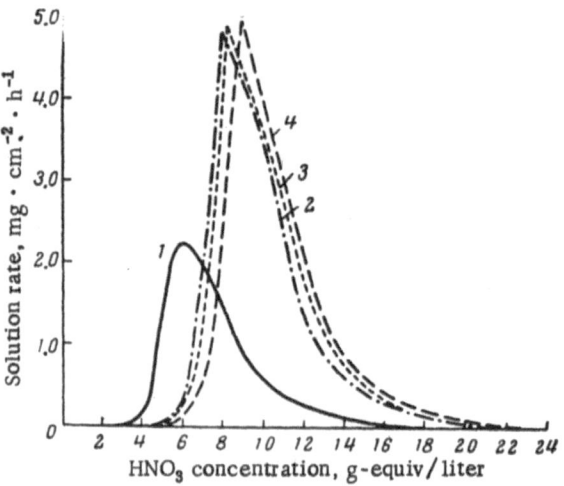

Fig. 109. Rate of dissolution of germanium in relation to the nitric acid concentration [23]. Stirrer rotation rate: 1) 0; 2) 194 rpm; 3) 390 rpm; 4) 800 rpm.

The rate of dissolution of germanium may be found graphically from the intersection of the extrapolated curves of the overvoltages of the anode and cathode reactions, as is shown in Fig. 108 [22]. The dissolution current determined in this way equals $3.2 \cdot 10^{-6}$ A/cm^2 and the value found by direct measurement of the amount of germanium passing into solution is $4.7 \cdot 10^{-6}$ A/cm^2. The good agreement between these values indicates that corrosion proceeds by an electrochemical mechanism.

c. Kinetics of Corrosion in Nitric Acid Solutions. Nitric acid is included in the composition of many etching agents for germanium and silicon. The kinetics of the solution of germanium in nitric acid solutions were found to be very complex [23-27]. With an increase in the acid concentration the solution rate passes through a maximum, which lies at $c_{max} = 6$ g-equiv/liter in an unstirred solution, and is displaced toward higher concentrations with stirring (Fig. 109). If the solution contains no traces of nitrogen oxides or nitrites, then rapid dissolution begins only after a certain induction period. At a concentration less than c_{max} the rate of dissolution is proportional to the products of the concentrations of nitric and nitrous acids in solution. The relation between the potential and the dissolution rate is represented by the Tafel relation and when the potential is shifted toward positive values to the left of the maximum (see Fig. 109), the rate increases, while to the right of the maximum it falls. The following scheme has been proposed for the cathode reduction of HNO$_3$:

$$H^+ + NO_3^- \rightleftarrows HNO_3 \tag{53.1}$$

$$HNO_3 + HNO_2 \rightarrow N_2O_4 + H_2O \tag{53.2}$$

$$N_2O_4 \rightleftarrows 2NO_2 \tag{53.3}$$

$$2NO_2 + 2e^- \rightleftarrows 2NO_2^- \tag{53.4}$$

$$2NO_2^- + 2H^+ \rightleftarrows 2HNO_2 \tag{53.5}$$

Overall equation $3H^+ + NO_3^- + 2e^- \rightarrow HNO_2 + H_2O.$ (53.6)

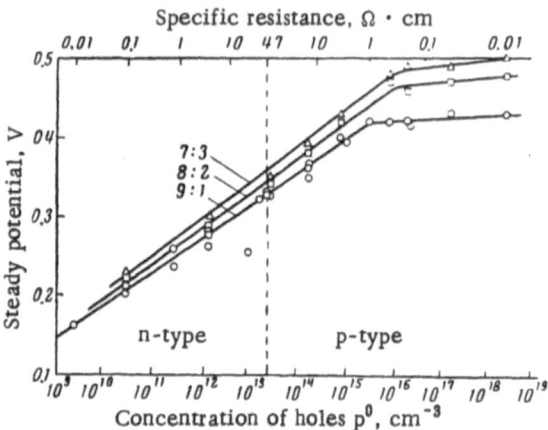

Fig. 110. Relation of the steady potential of germanium in a mixture of nitric and hydrofluoric acids to the concentration of holes p^0 in the bulk of the sample [28]. The ratio of the concentrations of HNO_3 and HF is shown on the curves.

Stage (53.2) is the slow stage and determines the dissolution rate. The autocatalytic mechanism also explains the characteristic dependence of its rate on stirring.

At a nitric acid concentration above 6 g-equiv/liter (in a stirred solution) the germanium is passivated as a result of the formation of a layer of GeO_2, whose thickness varies from 150 Å to several microns. The rate at which passivation appears is increased with an increase in the nitric acid concentration. The addition of HF prevents passivation because of solution of the oxide layer.

d. Dissolution in the Presence of Oxidants Which are Reduced with the Participation of Valence Electrons. As will be shown below, nitric acid which was examined in the previous section is an oxidant which is reduced with the participation of valence electrons. The effect of the semiconductor properties of the sample on the corrosion potential and rate in nitric acid solutions was investigated in [28-31].

The etching rate is independent of the type and the magnitude of the conductivity of the sample. The steady potential is shifted regularly toward more positive values with an increase in the bulk concentration of holes p^0 right up to $p^0 \approx 10^{16}$ cm^{-3}. A further increase in p^0 hardly affects the potential * (Fig. 110). An analogous relation is observed for the dissolution of silicon in HNO_3 + HF mixtures [28] and germanium in $Ce(SO_4)_2$, $FeCl_3$, and $K_3Fe(CN)_6$ solutions (Fig. 111) [32]. We can readily see the analogy between the dependence on the specific resistance of the semiconductor of the photopotential in a weak etching solution (see Fig. 107) and the potential difference between n- and p-type samples in the etching agents listed above (see Figs. 110 and 111). In both cases the shift in potential is small for p-type samples and increases with an increase in the conductivity of n-type samples.

Although the apparent reasons for these shifts in potential are different, the nature of the phenomenon observed is the same in the two cases. The change in potential is apparently connected with the injection of electron−hole pairs into the semiconductor: in the first case, as a result of the illumination of the sample, and in the second as a result of the reduction of an oxidant with the participation of valence electrons. When $\gamma > 1/\alpha'$ (case d, p. 340) a stream of nonequilibrium holes and free electrons flows from the surface of the sample into the bulk of it.† Then, by using equations (50.6) and (51.2), and assuming that $\Delta p \sim i_p$, we may calculate the relation of the corrosion potential to the corrosion rate and the specific resistance of the semiconductor, which is observed experimentally.

This explanation is supported by the following experimental data:

1) The limiting current of anodic dissolution of n-type semiconductors in injecting etching agents is higher than in nonin-

---

*The potentials given in Figs. 110 and 111 are relative to a saturated calomel electrode.

†We will call etching agents in which this effect occurs injecting etching agents.

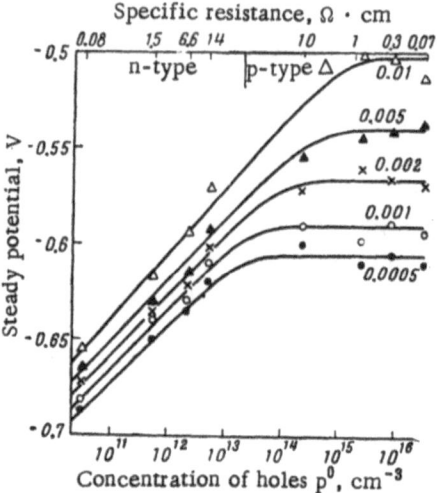

Fig. 111. Relation of the steady potential of germanium in
$K_3Fe(CN)_6$ solutions of various concentrations in 0.1 N NaOH
+ 1 N NaNO$_3$ to the concentration of holes p^0 in the bulk of
the sample [32]. The concentration of $K_3Fe(CN)_6$ in moles
per liter is given on the curves.

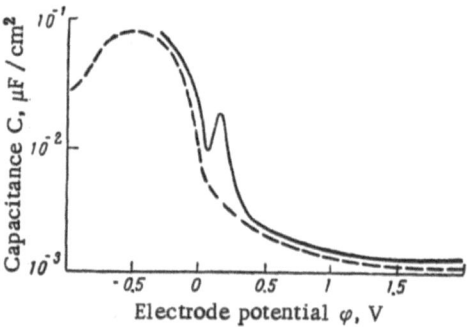

Fig. 112. Relation of the capacitance C of germanium (n-
type, 1 Ω · cm) to the electrode potential [35]. Broken
curve) in 0.1 N H_2SO_4 solution; solid curve) in the same solution with $K_3Fe(CN)_6$ added.

jecting etching agents by a value which is proportional to the dissolution rate [see equation (50.6)]. Turner [33] proposed the use of the measured limiting anode current in injecting etching agents for following the dissolution rate.

2) When a p-n junction is immersed in an injecting etching agent in the dark, there arises on it a potential difference as on illumination in a nonetching solution or in a gas atmosphere. The power developed for germanium and silicon diodes in SR-4 etching solution reaches 0.3 mW on a sample 1 cm long [28].

The appearance of the potential difference results in one of the regions of the junction dissolving faster than the other, even though samples of n- and p-type germanium separately corrode at the same rate in this solution [34].

3) On anode polarization of n-type germanium in solutions containing no oxidants there arises in it a depletion layer as is shown, for example, by the monotonic decrease in the differential capacitance with anode polarization (§ 35). When potassium ferricyanide is added to the solution the $C - \varphi$ curve assumes a form characteristic of equilibrium conditions, i.e., a minimum appears on it (Fig. 112) [35]. Under these conditions the concentration of holes beyond the space charge region apparently approaches the equilibrium value. An analogous phenomenon occurs on silicon in potassium ferricyanide solution {the limiting current of anode solution of n-type silicon is also increased when $K_3Fe(CN)_6$ is added to the solution [36]}.

In the case of germanium with intrinsic conductivity the introduction of ferricyanide into the solution increases the capacitance of the electrode on both the hole and electron sides. This confirms the above conclusion that during dissolution both holes and free electrons are injected into germanium [35].

4) Finally, the current of nonequilibrium carriers into the bulk of germanium during its dissolution in solutions of $K_3Fe(CN)_6$ and $KMnO_4$ was found by direct experiment with the electrode system described in §33 [37]. The back current on the indicator p-n junction increases when an oxidant is introduced into the solution and this increase in the indicator current increases with an increase in the dissolution rate (for ex-

ample, with an increase in the concentration or when the solution is stirred).

Thus, the role of many oxidants which are used in the most important mixtures for chemical etching of semiconductors consists of "supplying" the anode reaction with holes by their reduction through the valence band of the semiconductor.

e. Etching in Hydrogen Peroxide. Like nitric acid, hydrogen peroxide is one of the most common components of etching agents for germanium. It is used in both weakly alkaline and in acid (with HF added) solutions. The etching rate depends on the pH and the hydrogen peroxide concentration [38-42]. With a constant H_2O_2 concentration the rate is maximal in 0.1 N NaOH solution, and with a constant pH the relation to the hydrogen peroxide concentration passes through a maximum. The apparent activation energy is 10-17 kcal per mole, indicating that the slow stage is not a diffusion process.

The dissolution rate is independent of the type of conductivity and specific resistance of the germanium. It falls on anode or cathode polarization of the electrode, indicating the electrochemical nature of the process. At the same time, the experimentally determined dissolution rate was found to be several times that found by extrapolation of the anode and cathode curves of the overvoltage on p-type germanium. It was considerably higher than the limiting current density of anode solution of n-type germanium [42].

The fact that the limiting anode current on n-type germanium is independent of the H_2O_2 concentration indicates that the reduction of hydrogen peroxide is not accompanied by the injection of holes into the semiconductor. Therefore, the rate of the electrochemical dissolution of n-type germanium under these conditions cannot substantially exceed the density of the limiting current of holes. The corrosion of germanium in hydrogen peroxide apparently proceeds simultaneously by electrochemical and chemical routes and the rates of these processes are comparable.

The kinetics of etching of germanium in an $H_2O_2 - HF - H_2O$ mixture were investigated by Camp [43]. The electrophysical properties of a germanium surface in contact with hydrogen peroxide solutions were studied in [44, 45].

In the production of some types of semiconductor apparatuses etching is used on the germanium itself before the assembly of the apparatus and also on the prepared apparatus before sealing. In the latter case a multicomponent system is immersed in the etching agent and it contains various metals in addition to germanium. Solution under the influence of galvanic couples is superposed on the dissolution of germanium. As was shown in [46], germanium is the anode of such a couple in alkaline solutions of hydrogen peroxide and dissolves. The most effective cathode is copper, which is used for leads in several types of diode. Indium and tin are strongly polarized and do not have a substantial effect on the dissolution of germanium. The operation of a germanium — copper galvanic couple may be described by means of a polarization diagram, since we know the anode overvoltage curve for germanium, the cathode curve for copper, and the corrosion current density in the couple [46].

Among the other oxidants for germanium investigated, we should mention ozone and the halogens. The rate of dissolution in the first case is determined by the diffusion of ozone in the solution to the sample (the apparent activation energy is about 2 kcal/mole) [41]. In the case of aqueous solutions of bromine there are apparently also diffusion kinetics for the dissolution and in the case of iodine solutions (in the presence of KI), the kinetics are chemical [47].

Finally, an important case of corrosion is the electrochemical displacement by germanium of more noble metals from solutions of their salts. If ions of more electropositive metals are in solution, these metals are deposited on the surface of germanium and the latter passes into solution. This process is the basis of the chemical deposition of metals on, semiconductors (see Chapter V).

§ 54. Silicon

Silicon dissolves at an appreciable rate in solutions containing fluorides or alkali. In other cases it is passivated by a film of insoluble oxides.

a. Solutions Containing Hydrofluoric and Nitric Acids. The role of nitric acid in the etching process was analyzed above. It consists of the injection of holes into the semiconductor. The investigation of the kinetics of dissolution in HNO_3-HF mixtures [48, 49] showed that with a low concentration of nitric acid the reaction rate is proportional to this concentration and is determined by an autocatalytic process of type (53.6). Under these conditions the addition of nitrous acid accelerates the reaction. In concentrated nitric acid the etching is limited by the diffusion of hydrofluoric acid to the surface of the sample. The etching proceeds at the maximum rate with a ratio of HF and HNO_3 concentrations of 3 : 1. In this intermediate region the addition of water or acetic acid affects the dissolution rate, probably as a result of a change in the degree of dissociation of the nitric acid. A possible mechanism for dissolution in pure hydrofluoric acid was examined in [50].

The rate of etching in an HNO_3-HF mixture is independent of the semiconductor properties of the silicon. However, as Archer [51, 52] showed, the rate of growth and composition of surface films differ on samples with different electrophysical properties. These films have a thickness of 20-200 Å and consist mainly of finely dispersed silicon and its hydrides.

b. Dissolution in Alkaline Solutions. The steady potential of silicon in alkaline solutions is independent of the semiconductor properties of the sample and is related to the alkali concentration by the equation*

---

*The potential was measured at 50°C relative to a saturated calomel electrode. In equation (54.1) the potential is expressed in volts and the concentration in gram-equivalents per liter.

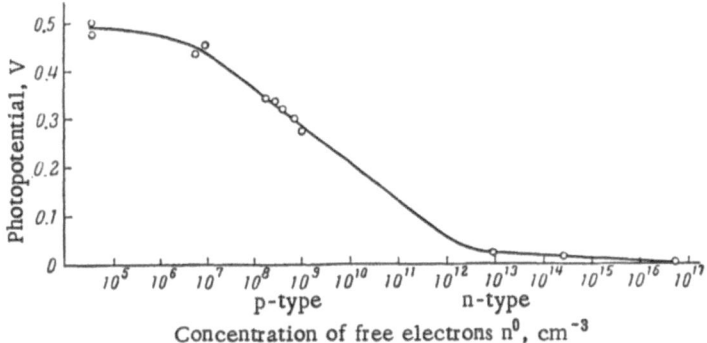

Fig. 113. Change in the steady potential of a silicon electrode in 10 N NaOH solution on illumination in relation to the concentration of free electrons in the bulk of the samples [53]. Steady illumination.

$$\varphi = -1.42 - 0.17 \log c_{NaOH}. \qquad (54.1)$$

It is much more positive that the thermodynamic equilibrium potential and is apparently the corrosion potential. The relation between the steady photopotential and the concentration of free electrons in the sample is given in Fig. 113 [53]. In accordance with what was stated in § 51, it may be explained by the fact that the surface of silicon is enriched in free electrons in comparison with the bulk and the exchange between the semiconductor and the solution proceeds through the conductivity band (cf. Fig. 107 for germanium).

Silicon dissolves vigorously in alkaline solutions with the liberation of hydrogen. As Izidimov showed [54-57], the etching rate is independent of the type of conductivity and the specific resistance of the silicon. With an increase in the alkali concentration it passes through a maximum (in 1 N NaOH), while the steady potential changes monotonically in accordance with the equation (54.1). Thus, there is no direct relation between the steady potential and the dissolution rate. Cathode polarization hardly affects dissolution. However, with anode polarization the dissolution rate falls and the process ceases at the potential of the maximum on the anode polarization curve

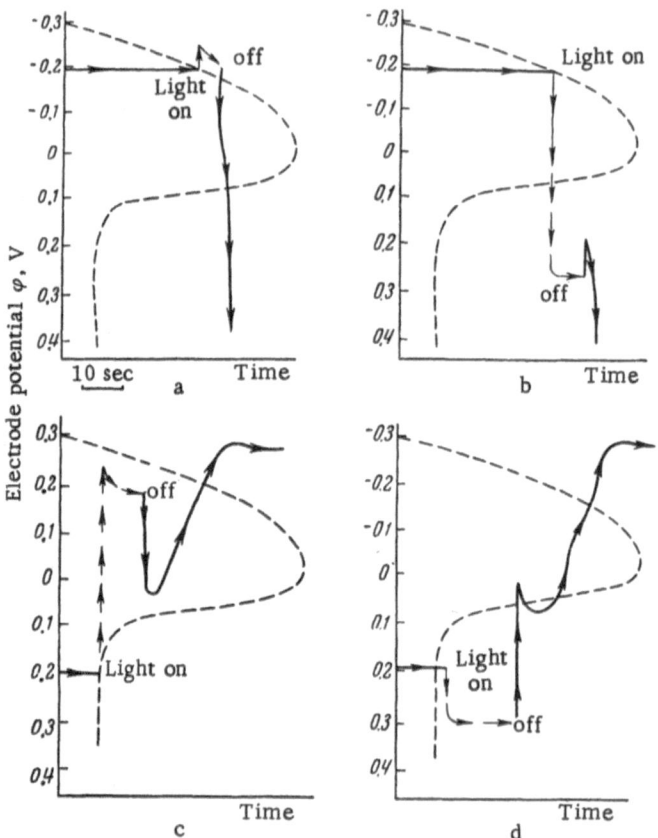

Fig. 114. Photopassivation (a and b) and photoactivation (c and d) of a silicon electrode in NaOH solution [57]. a,c) n-Type; b,d) p-type. The broken curve shows the anode polarization curve.

(Fig. 87). It is interesting that total suppression of dissolution is achieved with an anode current which is two orders less than the dissolution current at the steady potential. Extrapolation of the cathode overvoltage curve to the steady potential also gives a corrosion current which is two orders less than that measured experimentally. All this indicates that the dissolution of silicon in an alkali proceeds simultaneously by chemical and electrochemical mechanisms with the former predominating.

The cessation of dissolution on anode polarization is apparently connected with the formation of an oxide film on the surface of the silicon. The anode oxide film dissolves less readily than surface oxides formed as intermediate products in the dissolution process. Thus, the potential indirectly affects the etching process by a chemical mechanism.

One of the most interesting phenomena in the field of dissolution of semiconductors is the photopassivation and photoactivation of silicon [56, 57]. The photopassivation process is as follows. If silicon is illuminated at a potential more negative than the potential of the maximum on the anode polarization curve (with a constant anode current density), after the light has been switched off the silicon potential spontaneously shifts toward positive values. Thereupon dissolution ceases, i.e., the silicon is passivated (Fig. 114a, b).

On the other hand, if passive silicon is illuminated, when the light has been switched off its potential shifts toward negative values and finally assumes a value between the potential of the maximum on the $i - \varphi$ curve and the steady potential; the silicon is activated (Fig. 114c, d). If the initial potential of the passivated silicon (determined by the anode current density chosen) is very positive, then after the light has been switched off there is incomplete activation, i.e., the potential becomes less positive, but does not reach the potential of the maximum on the $i - \varphi$ curve and the system remains in the passive region.

It was found that photopassivation requires a higher light intensity, the lower the anode current density selected (i.e., the less positive the initial potential of the silicon). The action of the light appears to supplement the action of the anode current, which is insufficient for complete passivation. Close to the maximum on the $i - \varphi$ curve the silicon is passivated even after weak illumination. On the other hand, at the steady potential (i.e., without a current) the silicon is passivated by illumination only if an artificial oxide film is already present on its surface.

All the factors which facilitate the formation of an oxide film on silicon, i.e., lowering the temperature, reducing the alkali concentration, stopping the stirring of the solution, extend the range of potentials in which photopassivation is observed, and, on the contrary, prevent photoactivation.*

There is as yet no detailed theory of photopassivation and photoactivation. According to [57] the reason for photopassivation is the acceleration by light of the cathode reaction of electrochemical corrosion (i.e., the liberation of hydrogen or the reduction of oxygen), which proceeds through the conductivity band. Since the external current is kept constant, then the anode reaction is simultaneously accelerated and the potential is shifted to the region where the electrode is passive. Photoactivation results from the fact that light accelerates the anode oxidation of silicon, which involves holes. Under galvanostatic conditions this leads to a shift in the electrode potential in a cathode direction. Izidinov [57] relates the predominance of the effect of light on the cathode or anode process to the properties of the surface oxygen, which is an acceptor or a donor at potentials which are more negative or more positive than the potential of the maximum on the $i - \varphi$ curve, respectively.

This explanation is not exhaustive. It is difficult to understand the basic lack of an appreciable difference between n- and p-type silicon in the photopassivation process. The illumination of samples whose conductivity is far from the intrinsic value produces little change in the concentration of majority carriers and, consequently, the rate of electrode processes in which they participate. The concentration of minority carriers changes substantially on illumination. If the reason for photopassivation is the acceleration of one of the stages of the electrochemical reaction as a result of the injection of free carriers, then one would expect differences in the kinetics of photopassivation, connected with the type of conductivity of the samples.

---

*Photopassivation has also been observed in an $HNO_3 - HF$ mixture with a low HF concentration, i.e., under conditions where an oxide layer may exist on the surface [57].

In the explanation of photopassivation and photoactivation it is apparently necessary to take into account the fact that in the oxidation of the surface which is associated with passivation there is a substantial change in the potential drop in the Helmholtz layer (by 0.2-0.3 V) (see § 39).

§ 55. Binary Compounds

The corrosion and surface properties of intermetallic compounds of the type A^3B^5 have been studied most fully and this applies above all to indium antimonide. A series of papers by Gatos et al. [58-63] was devoted to the peculiarities of their dissolution. The crystallographic lattices of these compounds are of the zinc-blende type and consist of atoms of two sorts. * This has two basic consequences.

First, not all directions in the crystal are equivalent. Let us consider the (111) face. Each surface atom (for example, A) is attached by three bonds to the layer of the atoms B beneath. Each of these in its turn is attached to the surface layer by three bonds, but to the layer beneath by only one bond. This structure is shown schematically in Fig. 115. Therefore, if the surface layer of the atoms A is removed by etching, in the uncovered layer of B there are three free bonds at each atom. This configuration is highly unstable. Therefore, the removal of material occurs immediately in two layers and after etching the surface layer again consists of atoms of type A.

For this reason the opposite ($\overline{1}\overline{1}\overline{1}$) face of the crystals always consists of atoms of type B and this situation is maintained during etching. Thus, the crystal has polarity in the < 111 > direction.

Secondly, atoms A and B differ in their chemical properties and this leads to a difference in the macroscopic properties of the faces of A and B. The electronic structure of the

---

*For brevity we shall subsequently refer to the atoms of groups 3 and 5 by A and B, respectively.

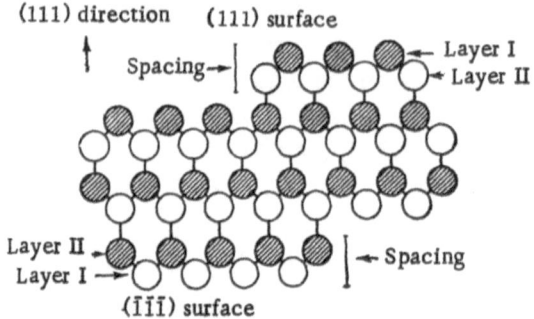

Fig. 115. Schematic diagram of the crystal lattice of binary compounds with a zinc-blende structure [62].

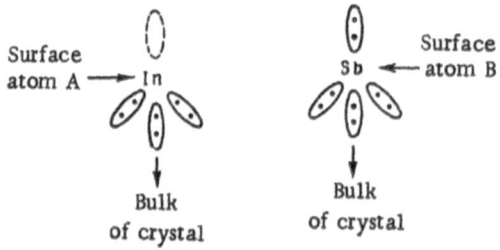

Fig. 116. Schematic diagram of the electronic structure of surface atoms of groups 3 (A) and 5 (B) in compounds of the type A^3B^5 [62].

surface atoms of Groups 3 and 5 is shown schematically in Fig. 116. In the atoms A all the valence electrons are used in covalent bonds with other atoms of the crystal. The atoms B have one free bond. Therefore, they are more reactive than the atoms A.

An investigation of the etching of indium antimonide and other compounds of the type A^3B^5 in mixtures of hydrofluoric and nitric acids and also in solutions of various oxidants (H_2O_2, Fe^{3+}, MnO_4^-, $Cr_2O_7^{2-}$, Ce^{4+}) showed that the (111) surface faced with atoms of Group 5 (B) has a higher reactivity than the surface of A. It is etched 10-20 times as fast and its steady potential is more negative. As a result of etching the surface of B is polished, while the surface of A is normally covered with etching pits, which coincide with dislocations.

In the presence of inhibitors (amyl-, butyl-, and propyl-amines) the rate of etching of the surface of B falls to values characteristic of the surface of A (on which inhibitors have no effect) and the difference in the character of the surfaces of the crystal formed as a result of etching is eliminated.

The features described are observed if the etching rate is determined by the chemical stage. In dilute solutions of oxidants the dissolution shows diffusion kinetics. Different A^3B^5 compounds dissolve at the same rate and this is determined by the rate of diffusion and convection in the solution [58, 64]. At a concentration of the oxidant of $\geq 10^{-2}$ g-equiv/liter, the diffusion kinetics change into chemical kinetics (Fig. 117).

In acid solutions in the absence of oxidants the compounds A^3B^5 are hydrolyzed with the formation of volatile hydrides. The rate of hydrolysis increases in the series GaSb < GaAs < InP. No appreciable hydrolysis is observed in alkaline solutions. The surface of intermetallic compounds is passivated in concentrated nitric acid [58].

The semiconductor properties were found to have an effect on the corrosion mechanism for the system gallium arsenide—alkaline solution of $KAu(CN)_2$ [65]. If the surface of an n-type sample is illuminated nonuniformly, the illuminated parts dissolve while gold is deposited on the unilluminated part of the surface. In the case of p-type gallium arsenide, on the contrary, etching occurs on the dark surface and gilding on the illuminated surface (until the film of gold becomes impenetrable to light). Here the anode and the cathode reactions in corrosion, i.e., anode solution and cathode deposition of gold, are separated in space. This is connected with the fact that a depletion layer arises in gallium arsenide at the interphase with the solution (see §38). In n-type samples the space charge consists of ionized donors and in p-type samples, acceptors. The nonequilibrium electrons and holes generated by light are separated by the field of the depletion layer and the direction of the field in each case is such that the minority carriers move to the illuminated surface, while the majority carriers move into the bulk of the crystal. Thus, a potential difference arises be-

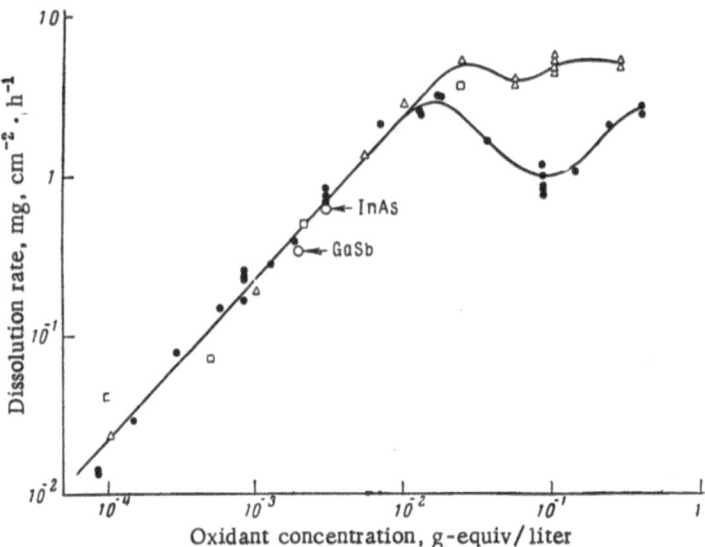

Fig. 117. Relation of the dissolution of indium antimonide and other compounds of the type A^3B^5 in solutions of oxidants to the oxidant concentration [58]. ●) Ce^{4+}; △) Fe^{3+}; □) $V(OH)^{4+}$.

tween the illuminated surface and the rest of the crystal. The illuminated and unilluminated parts of the sample are short circuited through the electrolyte and a galvanic couple is formed with steady illumination.

Among compounds of the type A^2B^6, the corrosion of cadmium sulfide in neutral solutions of $CdSO_4$ has been investigated [66]. The corrosion of cadmium sulfide on illumination may be described by the equations

$$Cd^{2+} + 2e^- \rightarrow Cd$$

$$\frac{CdS + 2e^+ \rightarrow Cd^{2+} + S}{CdS + 2e^- + 2e^+ \rightarrow Cd + S} \tag{55.1}$$

with the anode process involving holes and the cathode process, free electrons. Decomposition products (metallic cadmium and elementary sulfur) accumulate on the surface. The proposed scheme explains the photochemical decomposition of cadmium sulfide in the presence of water. Equations (55.1) show that this decomposition is accompanied by the recombination of nonequilibrium electrons and holes at the surface, since they participate in the electrochemical reactions.

§ 56.   Practical Application of Chemical Etching

Chemical etching is used widely in the production of semiconductor apparatuses. Its main task is to remove material whose crystal structure has been disrupted as a result of preliminary treatment (cutting and polishing). The thickness of this layer is normally several tens of microns [67, 68]. The concentration of structural defects and impurities in it is considerably higher than in a deeper layer of the semiconductor with a perfect crystal lattice. Moreover, etching is also used to determine the crystallographic orientation of the surface of a sample and the degree of perfection of its crystal structure, and also to prepare the surface before measurement of the electrophysical properties.

The composition of the most common etching agents for germanium, silicon, and binary semiconductor compounds, the etching conditions, and the character of the etched surface are given in Holmes' review [9] and also in the reviews [69, 70]. Some other etching agents, which are largely for the compounds A^3B^5 and A^2B^6, are described in [71-77]. The character of the surface oxides formed as a result of the etching of germanium was examined in [78].

Literature Cited

1.   A.N. Frumkin, V.S. Bagotskii, Z.A. Iofa, and B.N. Kabanov, Kinetics of Electrode Processes, Izd.Mosk. Gos.Univ. (1952).

2.   Yu.V.Pleskov, Dokl.Akad.Nauk SSSR, 132 : 1360 (1960).

3.   H. Gerischer and F. Beck, Z.Phys.Chem., Frankfurt, 24 : 378 (1960).

4. J. F. Dewald, Surface Chemistry of Metals and Semi-conductors (ed. H. C. Gatos), Wiley, New York (1960), p. 205.

5. R. M. Lazorenko-Manevich, Zh. Fiz. Khim., 36: 2066 (1962).

6. G. Feuillade, Compt. Rend., 252: 1288 (1961).

7. A. M. Kuznetsov and R. R. Dogonadze, Izv. Akad. Nauk SSSR, ser. khim., 1964: 1885.

8. B. A. Irving, Electrochemistry of Semiconductors (ed. P. J. Holmes), Academic Press, London-New York (1962), p. 256.

9. P. J. Holmes, Electrochemistry of Semiconductors (ed. P. J. Holmes), Academic Press, London-New York (1962), p. 329.

10. B. Lovreček and J. O'M. Bockris, J. Phys. Chem., 63: 1368 (1959).

11. J. I. Hall and A. E. Koenig, Trans. Electrochem. Soc., 65: 215 (1934).

12. M. L. Nichols and S. R. Cooper, Ind. Eng. Chem., Anal. Ed., 7: 350 (1935).

13. R. Schwarz, F. Heinrich, and E. Hollstein, Z. Anorg. Chem., 229: 146 (1936).

14. J. E. Land, J. Alabama Acad. Sci., 20: 37 (1948).

15. W. L. Jolly and W. M. Latimer, J. Am. Chem. Soc., 74: 5751 (1952).

16. M. J. Sparnaay, Surface Sci., 1: 102 (1964).

17. J. I. Carasso, M. M. Faktor, and H. Holloway, J. Phys. Chem., 65: 2260 (1961).

18. J. I. Carasso and M. M. Faktor, Electrochemistry of Semiconductors (ed. P. J. Holmes), Academic Press, London-New York (1962), p. 205.

19. W. W. Harvey and H. C. Gatos, J. Electrochem. Soc., 105: 654 (1958).

20. W. W. Harvey and H. C. Gatos, J. Electrochem. Soc., 107: 65 (1960).

21. W. W. Harvey and H. C. Gatos, J. Appl. Phys., 29: 1267 (1958).

22. E. N. Paleolog, N. D. Tomashov, and A. Z. Fedotova, Zh. Fiz. Khim., 34: 1027 (1960).

23. M.C. Cretella and H.C. Gatos, J. Electrochem. Soc., 105:487 (1958).

24. H.C. Gatos, W.W. Harvey, and M.C. Lavine, Rev. Met. (Paris), 60:1149 (1958).

25. R.L. Myuller, T.P. Markova, and S.M. Repinskii, Vestn. Leningr. Gos. Univ., 16:106 (1959).

26. B. Schwartz and H. Robbins, J. Electrochem. Soc., 111:196 (1964).

27. B.A. Irving, J. Electrochem. Soc., 109:120 (1962).

28. D.R. Turner, J. Electrochem. Soc., 107:810 (1960).

29. J. Dewald, Collection: Semiconductors (ed. N. Hannay) [Russian translation] IL, Moscow (1962), p.619.

30. H.C. Gatos, Surface Chemistry of Metals and Semiconductors, Wiley, New York (1960), p.381.

31. P.P. Konorov and O.V. Romanov, Fiz. Tverd. Tela, 5:3039 (1963).

32. H. Gerischer and F. Beck, Z. Phys. Chem., Frankfurt, 23:113 (1960).

33. D.R. Turner, J. Electrochem. Soc., 108:561 (1961).

34. O.G. Deryagina, E.N. Paleolog, and N.D. Tomashov, Zh. Prikl. Khim., 35:1276 (1962); H. Gerischer, Z. Naturforsch., 19a:553 (1964).

35. H. Gobrecht and O. Meinhardt, Ber. Bunsenges., 67:142 (1963); O.G. Deryagina and E.N. Paleolog, Elektrokhimiya, 1:267 (1965).

36. R.M Hurd and P.T. Wrotenbery, Ann. N.Y. Acad. Sci., 101:876 (1963); R.M. Hurd and N. Hackerman, Electrochim. Acta, 9:1633 (1964).

37. Yu.V. Pleskov, Zh. Fiz. Khim., 35:2576 (1961); W.W. Harvey and M.C. Finn, Surface Sci., 2:456 (1964).

38. R.E. Smolyanskii, V.M Gurevich, A.M. Raikhlin, and M.I. Lukasevich, Zh. Tekhn. Fiz., 28:2135 (1958).

39. G.S. Supin, Zh. Prikl. Khim., 32:478 (1959).

40. R.L. Myuller, A.V. Danilov, T.P. Markova, V.N. Mel'nikov, A.B. Nikol'skii, and S.M. Repinskii, Vestn. Leningr. Gos. Univ., Ser. Fiz. Khim., 1960:80.

41. K.J. Miller, J. Electrochem. Soc., 108:296 (1961).

42. G.A. Kataev and L.N. Rozanova, Trudy Tomsk. Gos. Univ., 154:122 (1962).

43. P.R. Camp, J. Electrochem. Soc., 102:586 (1955).
44. A.A. Yakovleva, T.I. Borisova, and V.I. Veselovskii, Zh. Fiz.Khim., 36:2541 (1962).
45. P.P. Konorov and O.V. Romanov, Fiz. Tverd. Tela, 4: 1655 (1962).
46. O.G. Deryagina, E.N. Paleolog, and N.D. Tomashov, Zh. Fiz.Khim., 34:1952 (1960).
47. R.L. Myuİler and N.A. Baglai, Vestn. Leningr.Gos. Univ., Ser. Fiz. Khim., 1960:88.
48. H. Robbins and B. Schwartz, J. Electrochem. Soc., 106:505 (1959); 107:108 (1960); 108:365 (1961).
49. D.L. Klein and D.J. D'Stefan, J. Electrochem. Soc., 109:37 (1962).
50. G. Feuillade, Compt. Rend., 252:3958 (1961).
51. R.J. Archer, J. Electrochem. Soc., 104:619 (1957).
52. R.J. Archer, J. Phys.Chem.Solids, 14:104 (1960).
53. M. Seipt and H. Fischer, Anal. Real.Soc.Espan.Fis. Quim., B56:443 (1960).
54. S.O. Izidinov, T.I. Borisova, and V.I. Veselovskii, Dokl. Akad. Nauk SSSR, 133:392 (1960).
55. S.O. Izidinov, T.I. Borisova, and V.I. Veselovskii, Zh. Fiz.Khim., 36:1246 (1962).
56. S.O. Izidinov, T.I. Borisova, and V.I. Veselovskii, Dokl. Akad. Nauk SSSR, 145:598 (1962).
57. S.O. Izidinov, Dissertation. L. Ya.Karpova Physico-chemical Institute (1963).
58. H.C. Gatos and M.C. Lavine, J. Electrochem.Soc., 107: 427 (1960).
59. H.C. Gatos and M.C. Lavine, J. Electrochem. Soc., 108: 645 (1961).
60. M.C. Lavine, H.C. Gatos, and M.C. Finn, J. Electro-chem. Soc., 108:974 (1961).
61. M.C. Lavine, A.J. Rosenberg, and H.C. Gatos, J.Appl. Phys., 29:1131 (1958).
62. H.C. Gatos and M.C. Lavine, J. Phys.Chem.Solids, 14: 169 (1960).
63. J.W. Faust and A. Sagar, J.Appl. Phys., 31:331 (1960).

64. R.L. Myuller, G.M. Orlova, and Ts'ut Chin-Hua, Zh. Org.Khim., 31:2461 (1961).

65. R.W. Haisty, J. Electrochem. Soc., 108:790 (1961).

66. V.A. Tyagai, Izv.Akad.Nauk SSSR, Ser. Khim., 1963: 1556.

67. T.M. Buck and F.S. McKim, J. Electrochem. Soc., 103:593 (1956).

68. D. Baker and H. Yemm, Brit.J.Appl.Phys., 8:302 (1957).

69. P.Wang, Sylvania Technologist, 11:50 (1958).

70. J.L. Richards and A.J. Crocker, J.Appl.Phys.,31:611 (1960).

71. C.S. Fuller and H.W. Allison, J. Electrochem.Soc., 109:880 (1962).

72. E.P. Warecois, M.C. Lavine, A.N. Mariano, and H.C. Gatos, J.Appl.Phys., 33:690 (1962).

73. T.H. Yeh and A.E. Blakeslee, J. Electrochem.Soc., 110:1018 (1963).

74. R.V. Bakradze and I.A. Rom-Krichevskaya, Kristallo-grafiya, 8:238 (1963).

75. R.L. Petrusevich, E.S. Sollertinskaya, and O.I. Pavlova, Fiz.Tverd.Tela, 4:1378 (1962).

76. R.L. Petrusevich and E.S. Sollertinskaya, Kristallo-grafiya, 8:243 (1963).

77. A.D. Trakhtenberg and S.M. Fainshtein, Fiz.Tverd. Tela, 1:373 (1959).

78. E.A. Efimov, I.G. Erusalimchik, and G.P. Sokolova, Zh.Fiz.Khim., 36:765 (1962).

Additional Literature

Chu, T. L., and J. R. Gavaler, Dissolution of silicon and junction delineation in silicon by the CrO_3-HF-H_2O system, Electro-chim. Acta, 10:1141 (1965).

Dogonadze, R. R., and A. M. Kuznetsov, Some steady processes in the system semiconductor—electrolyte solution, Elektrokhi-miya, 1:1008 (1965). Etching of Semiconductors, Izd. Mir

(1965) (with bibliographies of Soviet and foreign work on the etching of semiconductors from 1959 to 1965).

Faust, J. W., Etching of intermetallic III–V compounds, in: Compound Semiconductors, Vol. 1 (eds. R. K. Willardson and H. L. Goering). Reinhold, New York (1962), p. 445.

Gatos, H. C., and M. C. Lavine, Chemical behavior of semiconductors: etching characteristics, in: Progress in Semiconductors, Vol. 9 (ed. A. F. Gibson and R. E. Burgess). Heywood, London (1965), p. 1.

Lainer, L. V., V. I. Lainer, and Z. A. Baranova, Chemical polishing and etching of single crystals of silicon to show dislocations, Zh. Prikl. Khim., 38:2473 (1965).

Lowen, J., Gallium arsenide surface preparation, J. Electrochem. Soc., 112:1057 (1965).

Orlova, G. M., S. K. Erofeev, and N. V. Romanova, Kinetics of the chemical etching of a gallium arsenide single crystal in alkaline and hydrochloric acid solutions of hydrogen peroxide, in: Chemistry of Solids, Izd. Leningr. Gos. Univ. (1965), pp. 218, 227.

Packard, R. D., Notes on the chemical polishing of gallium arsenide surfaces, J. Electrochem. Soc., 112:871, 1057 (1965).

Repinskii, S. M., Mechanism of the solution of germanium in hydrogen peroxide, Zh. Fiz. Khim., 39:1674 (1965).

Rozanova, L. N., Solution of germanium in ammonia solutions of hydrogen peroxide, Zh. Prikl. Khim., 38:2334 (1965).

Wolkenberg, A., Electrochemical and photochemical properties of Group IV elements, Przemysl. Chem., 43:692 (1964).

Chapter V

Use of Electrochemical Methods for Investigating Properties and Treating the Surface of Semiconductor Materials

§ 57. Investigation of Bulk Properties

In Chapters I and II it was shown that the surface properties of semiconductor electrodes, i.e., the photopotential, the capacitance, etc., are definite functions of the bulk properties, for example, the concentration of free carriers, diffusion length, etc. Therefore, it is possible to use the measurement of the electrochemical parameters to check the bulk properties of semiconductor materials. Naturally all these characteristics may be determined by a nonelectrochemical method. However, in many cases electrochemical methods have great advantages. This refers primarily to the determination of the characteristics of samples which are intended for use in electrochemical investigations.

As an example we will examine the determination of the concentration of ionized donors by measurement of the differential capacitance. In investigating the relation of the flat-band potential of zinc oxide to the concentration of free carriers over the range of concentrations of $0.6 \cdot 10^{18}$–$0.9 \cdot 10^{18}$ cm^{-3}, Dewald

[1] used a crystal with a nonuniform distribution of the doping impurity. By progressively etching 1-μ layers from the surface of the crystal, he obtained each time a "new" electrode with a different concentration of donors in the surface layer. We should mention that the capacitance is determined by the concentration of donors in the space charge region whose thickness is $\approx 10^{-4}$ cm. This concentration, a knowledge of which was required for comparison with the flat-band potential, was determined each time from the slope of $1/C^2 - \varphi$ lines by using equation (12.11), i.e., from the same capacitance measurements as for calculating the flat-band potential. Naturally, another method of finding this concentration in a crystal with a varying specific resistance would have been extremely difficult. The differential capacitance method is particularly convenient for investigating deep donors which are incompletely ionized in the bulk of the crystal at room temperature. In this case their concentration cannot be found by simple measurement of the conductivity (see § 16).

Another method is the determination of low values of the diffusion length from the potential dependence of the photocurrent, which was described in § 40. This method is a variation of a nonelectrochemical method proposed for semiconductor-metal rectifying contacts and p-n junctions. The advantage of the electrochemical method is the fact that it does not require the preparation of a special contact on the sample or a p-n junction. Its part is played successfully by the semiconductor-electrolyte interphase. Moreover, the surface of the semiconductor investigated is not spoiled as a result of the electrochemical measurements.

We will examine methods of finding nonuniformity of germanium samples as regards specific resistance. The use of the electrodeposition of copper has been proposed for this purpose [2-5]. The deposition is carried out under conditions where the overvoltage of the reaction is low and the current density is determined by the series resistance of the sample. Since the potential on the cell is fixed, the current density is higher on sections with a lower specific resistance. The thickness of the copper deposit is greater where these sections emerge at the surface of the electrode. Thus, the degree of uniformity of the thickness of the copper deposit is an indicator of the uniformity of the material of the electrode with respect to specific resistance. To avoid concentration polarization with respect to copper ions in the electrolyte (as a result of which there is a decrease in the "contrast factor" of the copper deposit and the unresolving power of the method), the electrodeposition is carried out under pulsed conditions. This method can be applied to semiconductor instruments such as photodiodes with thin bases [6, 7], as well as to massive samples. It has also been used in the case of silicon [8] and sulfide minerals with semiconductor properties [9]. The electrodeposition of gold has also been used for finding defects on the surface of germanium [10].

Instead of electrodeposition it is also possible to use an anode solution of the material of the electrode and to estimate the local current density from the amount of material removed. The etching of silicon is carried out in a mixture of hydrofluoric and acetic acids $(1:1)$ [11, 12] or in a solution of 0.04 N KNO_3 + 0.01 N NH_4F in methylformamide [13]. Germanium is etched in a solution of KOH or Na_2SO_4 [14, 15].

In the case of n-type germanium, the etching is carried out with an alternating current. As was shown in § 32, with a fall in the specific resistance there is a fall in the breakdown voltage of the surface barrier, which arises at the semiconductor—electrolyte interface. Therefore, with a sufficiently high voltage on the cell there is breakdown at the points with the maximum concentration of donors. At these places, the current density is considerably higher than on the rest of the surface of the electrode and depressions are formed (pitting). Another reason for local breakdown is the presence of various structural defects such as dislocations [16]. In polycrystals, breakdown normally occurs at the boundaries of individual crystallites.

Gerischer [17] proposed the measurement of the corrosion potential of a semiconductor in "injecting" etching agents (see § 53, d) for determining the type of conductivity.

To conclude this section, we should mention a method of measuring the specific resistance of powdered silicon [18]. The powdered silicon is placed in a solution of hydrofluoric acid (which dissolves the SiO_2 from the surface of the silicon grains). The concentration of the acid is adjusted until the specific electrical conductivities of the silicon and solution are identical. *
In this case, the resistance of the cell is independent of whether there is pure hydrofluoric acid solution or the same solution with silicon powder added between the electrodes. Knowing the concentration we can calculate the electrical conductivity of the solution and hence that of the silicon.

§ 58.   Detection of p-n Junctions in Crystals

The rectifying properties of p-n junctions may be used conveniently to detect the exact position of boundaries between regions with n-type and p-type conductivity. Back bias (the n-type region is connected to the positive pole of the voltage source and the p-type region to the negative) is applied to the junction and it is immersed in a solution for the electrodeposition of copper ($CuSO_4 \cdot 5H_2O$, 20-200 g/liter; HF, 1%) or gold. There is hardly any current through the p-n junction and a potential drop (close to the potential drop of the source) arises at the junction with the p-type region negatively charged relative to the n-type region. Therefore, electrodeposition of copper occurs on the surface of the region with p-type conductivity, while the section with n-type conductivity acts as an anode. The boundary of the copper deposit marks the line at which the p-n junction emerges at the crystal surface [19].

If the junction is obtained by diffusion of the doping impurity from the crystal surface into the bulk, then the thickness of the diffusion region (i.e., the depth of alloying of the p-n junction) normally does not exceed a few microns. It is difficult

---

*The electrical conductivity is measured with an alternating current of sufficiently high frequency for the impedance of the silicon—solution interphase to be much less than the ohmic resistance of the silicon and solution.

to observe a copper deposit of this width on the side surface of the sample. Therefore, an oblique section (angle of 1-5°) is prepared, covering the n- and p-type regions (Fig. 118). A drop of solution is put onto the section and electrodeposition carried out as described above. By measuring the distance from the apex of the section to the boundary of the copper deposit, with a knowledge of the angle it is easy to determine the depth of alloying of the p-n junction, which may be only 0.5μ. This method has been used in the case of germanium [20, 21] and gallium arsenide [22].

A saturated solution of the dye methylene blue may be used for the same purpose [23]. On the surface of the p-type region the dye is reduced to the colorless and insoluble leuco base, while the n-type region remains blue. If the dye is then washed from the surface of the n-type region and the sample left in the air, the white leuco base is again ozidized by atmospheric oxygen and the surface of the p-type region becomes blue; the n-type region remains colorless.

The electrodeposition of copper and silver may also be used in the case of silicon [24-29]. Instead of polarization of the p-n junction, it is illuminated. The electrons and holes arising close to the junction are separated by the electric field of the junction with the electrons passing into the n-type region, charging it negatively relative to the p-type region, into which the holes are drawn. The principle of the development of a potential difference here is the same as in solar batteries. However, in our case, the "solar battery" is short circuited through the electrolyte solution and not through an external load. A current arises in the system with the n-type region as the negative electrode and metal is deposited on it. Solution or oxidation of the silicon occurs on the p-type region. The resolving power of the method is about 2μ (Fig. 119).

However, here there is the following complication. The electrode potential of silicon is very negative and therefore the silicon is stable in aqueous solutions only in a passive state. The hydrofluoric acid or hot alkali used in solutions for electro-

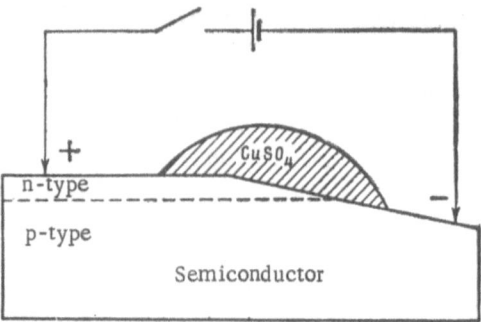

Fig. 118. Diagram of experiment for determining the depth
of alloying of a p-n junction by electrodeposition on an
oblique section. The broken line shows the plane of the p-n
junction.

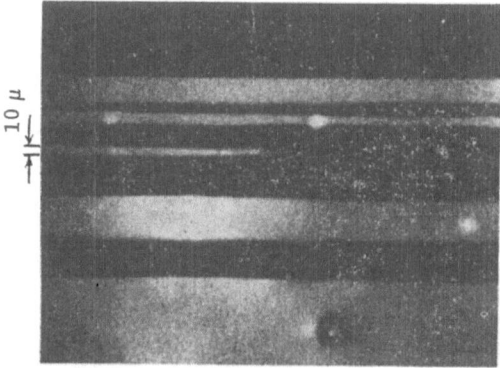

Fig. 119. Surface of silicon sample with several p-n junctions after the electrodeposition of copper [28]. Light sections indicate copper deposit (surface of n-type regions).

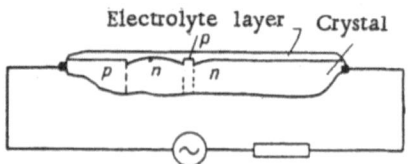

Fig. 120. Diagram of experiment to show up p-n junctions
by electrolytic etching [38].

deposition* activates the silicon; it displaces copper, silver, or gold from solutions of their salts and passes into solution. There occurs a peculiar corrosion of silicon, whose rate as a rule is limited by the cathode reaction, namely, the diffusion of metal ions to the electrode. Therefore, the deposition occurs uniformly over the whole of the sample surface, regardless of its electrophysical properties. However, in some cases, anode control of the corrosion is observed and then its rate is greater on the surface of sections of the semiconductor with a low specific resistance [26] or on a polished surface [28]. Thus, the deposition of metal occurs simultaneously by two mechanisms: by an "electrochemical" mechanism (due to the difference in the potentials of the n- and p-type regions on illumination) and by an electrochemical replacement mechanism. The rate of the reaction by the first route is determined by the potential difference between the regions with different types of conductivity and increases with an increase in the illumination intensity, while the rate of the second is determined by the difference in the electrode potentials of silicon and the metal selected for deposition. Since the electrochemical replacement does not show up the p-n junction, strong illumination is necessary to increase the fraction of the "electrochemical" deposition of the metal and thus increase the contrast factor of the deposit.

Finally, we should bear in mind the following situation. The cathodic reaction in the bath for the electrodeposition of the metal may involve valence electrons and hence inject holes into the semiconductor. In such a solution, a potential difference arises at a p-n junction, as on illumination (see, for example, [30]), so that the junction also shows up in the dark.

It is possible to detect a p-n junction by anode etching. Although the resolving power of this method is considerably lower, it does not spoil the surface of the sample and, moreover, the

---

* A bath used for copper plating has the following composition [28]: $CuSO_4 \cdot 5 H_2O$ (200 g/liter), HF (24%). Bath for gold-plating [25]: $KAu(CN)_2$ (10 g/liter), KOH (200 g/liter), temperature 30-70°C.

detection of junctions may be combined with etching them. It was shown above that when negative bias is applied to a junction immersed in an electrolyte solution, the n-type region is an anode and dissolves. The cathode is the p-type region or an auxiliary electrode in the solution. A solution of KOH or NaOH (10%) is used as the electrolyte. A step forms at the boundary of regions with different types of conductivity [31]. If the etching is carried out at high temperature ($\approx 80°C$) in a solution saturated with germanium oxides, then on the surface of the p-type region there appears a dark deposit, which may be germanium oxide or amorphous germanium. The surface of the n-type region is lustrous.

Another method [32] is based on the difference in the rate of anode solution of n-type and p-type germanium. The sample with the p-n junction is the anode. The p-type region dissolves with a very low overvoltage, while the rate of solution of the n-type region is limited by diffusion (or generation) of holes and as a rule it does not exceed 10^{-3} A/cm^2. Therefore, the p-type region is etched preferentially and a step is again formed at the p-n junction. The method has also been used for gallium arsenide [22] and silicon [33-35]. In the latter case, a solution of hydrofluoric acid with nitric acid added was used as the electrolyte. With a low current density, the surface of the p-type region is covered with a dark layer of amorphous silicon or the silicon oxide SiO, which adheres poorly to the surface.

The resolving power of the method is limited by the fact that a p-n junction connected in the forward direction injects holes into the n-type region. Therefore, the n-type region dissolves quite rapidly at a distance of the order of the diffusion length of holes from the junction and no sharp boundary with the p-type region is obtained [36].

If there are several junctions in the sample and it is difficult to apply a contact separately to each of them, it is possible to apply a voltage from a source only to the ends of the sample on whose surface is a thin layer of electrolyte. Then all the junctions connected in a back direction (i.e., half of all

the junctions in the sample) undergo etching with the n-type regions as anodes which dissolve, while the p-type regions are
cathodes. Reversing the direction of the current "operates" the
second half of the junctions. In practice, an alternating potential is applied to the ends of the sample [37, 38]. The system
for this method is illustrated in Fig. 120.

§ 59. Electrochemical Etching and Electropolishing

a. General Information. Electrochemical etch-
ing is used widely together with chemical etching for treating
the surface of semiconductor devices.

The advantage of the electrochemical method over chemical etching lies in the possibility of very accurate regulation of
the etching rate and control of the amount of material removed.
Moreover, the problem of local etching is solved relatively
simply by a local change in the current density. Since the details of the technology of etching of semiconductor devices are
industrial secrets and are not published in the open press, in
the present section we will give only the basic principles of various methods of etching and the most common electrolyte compositions proposed for this purpose.

The most common electrolyte for etching germanium is a
solution of potassium hydroxide or sodium hydroxide. The current density during etching is usually 10^{-2} A/cm^2 and in electropolishing, above 1 A/cm^2. This gives a surface with a very
low rate of surface recombination (≈ 100 cm/sec) [39, 40].* During etching, n-type samples are illuminated so that the current
remains below the limiting anode current; otherwise, the etching conditions and the structure of the etched surface are extremely sensitive to various local heterogeneities (see § 57).
Electropolishing occurs under conditions where the surface barrier breaks down and, as a rule, it is insensitive to the semicon-

---

*A low rate of surface recombination may be obtained by anode treatment of
germanium in solutions of antimony salts [41, 42].

ductor properties of the sample. Some problems in the technology of etching are examined in [43–45].

A whole series of other electrolytes for the etching and
electropolishing of germanium is described in the scientific and
patent literature [46–50]. Nonaqueous solutions (using formamide, ethylene glycol, and glycerol as solvents) are used in
addition to aqueous solutions. It is also possible to use molten
chlorides or fluorides of sodium and potassium at 700°C [51].

The etching and electropolishing of silicon is carried out
in solutions of hydrofluoric acid in water and sometimes in organic solvents [52–55]. The kinetics of electropolishing have
been investigated by Turner [56]. With a low current density
the surface of silicon is covered with a friable dark deposit of
amorphous silicon, which is formed as a result of the secondary reaction [56, 57]:

$$Si + 2HF + (2 - m)e^+ \rightarrow SiF_2 + 2H^+ + me^-,$$

$$2SiF_2 \rightarrow Si + SiF_4.$$

This deposit reacts relatively slowly with water, changing to silicon dioxide, which then dissolves in the hydrofluoric acid. Electropolishing begins when a critical current density is reached. During this stage the silicon surface is covered with a very
thin layer of oxides (probably less than 100 Å). The critical
current density of electropolishing is proportional to the concentration c of hydrofluoric acid and is related to the temperature ϑ by the equations

$$i_{cr} = c\,(1.06\vartheta - 16.6) \text{ when } \vartheta > 30°\,C;$$

$$i_{cr} = c\,(0.40\vartheta + 3.2) \quad \text{when } \vartheta < 30°\,C.$$

In aqueous glycerol solutions of hydrofluoric acid $i_{cr} \sim \eta_c^{-1/4}$,
where η_c is the viscosity of the solution. The rules presented
led Turner to the conclusion that i_{cr} is the current density at
which the electrode reaction changes from the chemical kinetic
region to the diffusion (or hybrid) kinetic region.

There have been very few investigations of the etching of other semiconductor materials. Silicon carbide is etched in hydrofluoric acid solutions [58] and lead telluride in a water—alcohol—glycerol solution of potassium hydroxide [59]. Gallium arsenide may be electropolished with a 10% solution of KOH or NaOH [22] or with Jacques electrolyte (a mixture of perchloric acid and acetic anhydride) [38]. The latter solution is also suitable for indium antimonide [60].

b. Electrochemical Milling (Shaping). We will examine some special etching methods. Methods for showing up p-n junctions, which were described in § 58, may be used for selective etching of individual regions (emitter, base, collector) of transistors. To avoid heating of the devices, the etching is carried out under pulsed conditions [61-64]. It should be remembered that modern semiconductor devices are often a few microns in size. Thus, for example, in the preparation of a tunnel diode between a p-n junction and an ohmic contact, a neck less than 10 μ in diameter and length is etched out. To avoid "overetching" the device it is necessary to adjust its characteristics directly during etching and to automatically stop the process. In an article with the characteristic heading "Tunnel Diode Controls Its Own Etching," Vulcan [65] describes an automatic apparatus for producing tunnel diodes. The etching is carried out with pulses with a frequency of 10 Hz, and in the intervals between these, pulses of a measuring current are passed through the diode, which is in a solution, to determine its volt-ampere characteristics. As soon as the electrical parameters of the diode reach given limits, etching is automatically stopped. The peculiarities of the etching of p-n junctions in degenerate germanium (as applied to tunnel diodes) are examined in [66].

Other shaping methods are based on a nonuniform distribution of anode current density across the surface of a homogeneous crystal. Their advantages lie in the fact that here the type of conductivity and other semiconductor properties of the sample are not of appreciable importance. These methods are equally applicable to relatively high-resistance and to degenerate samples and also to metals. It is only necessary that one of the two phases, i.e., the semiconductor or the solution, should have a high specific resistance relative to the other. In the first case (high-resistance sample), the localization of etching results from the fact that the current in the semiconductor is concentrated largely close to the ohmic contact. Therefore, the etching rate is maximal on those sections of surface which are closest to the contact [67].

Methods based on a nonuniform distribution of current in a poorly conducting solution are most common. A dilute solution of alkali (0.1%) or distilled water is used as the solution. A drop of solution is placed on the surface of the sample and a cathode (steel or tungsten points) touches the surface of the liquid. When the current is switched on, etching occurs in the immediate vicinity of the cathode. When a current of a few microamperes is reached, the local current density ≈ 100 A per cm^2. By this method it is possible to etch right through a germanium plate 0.4 mm thick. It is also possible to carry out the etching in a large volume of solution; for this, the side surface of the cathode is insulated, while the end of the point is placed close to the anode [68, 69]. If the electrode potential of the semiconductor is much more negative than the potential of the second (metal) electrode, external polarization is not necessary and etching occurs as a result of this "natural" potential difference (so-called contact etching).

Electrochemical corrosion of the sample occurs here and in the semiconductor−metal couple, the anode reaction is localized at the semiconductor, while the cathode reaction is localized at the second (metal) electrode which is in electrical contact with the semiconductor. This cell makes contact through the solution and if the solution (or semiconductor) has a high resistance, etching is concentrated in a narrow region close to the semiconductor−metal contact. Gold, silver, and nickel are used as cathodes for etching germanium and silicon, and gold is used for gallium arsenide, while the dilute etching agent SR-4 is used as the solution [70].

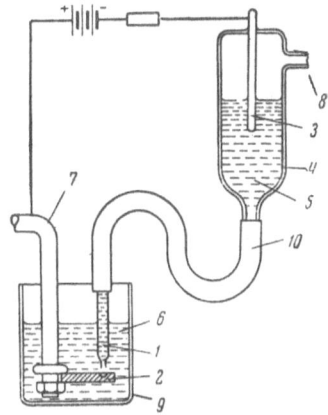

Fig. 121. Diagram of apparatus for electrochemical drilling of semi-conductor materials [71]. 1) Tube ending in capillary; 2) sample; 3) cathode; 4) reservoir of electrolyte; 5,6) electrolyte; 7) sample holder; 8) connection to system with compressed gas; 9) cell; 10) flexible tube.

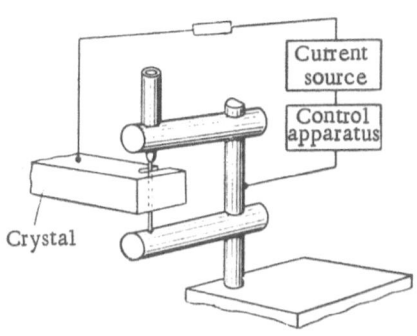

Fig. 122. Diagram of apparatus for electrochemical cutting of germanium [74].

To obtain a high local rate of etching, Uhlir [71] proposed the apparatus illustrated in Fig. 121. The end of a fine capillary (a few microns in diameter), which ends in the cathode section of the cell, is pressed to the surface of the anode. Liquid is fed through the capillary continuously with gas pressure so that it does not boil because of overheating. In this way it is possible to drill holes with a diameter of a few microns. Here it is not necessary to use an electrolyte with a high specific resistance.

If a semiconductor anode in the form of a thin rod is lowered into the solution, then the maximum current density is established at the boundary of the three phases (solution−solid−air) and a constriction is produced on the sample as a result of etching. It is possible to carry out the etching in such a way as to "overetch" the sample into two halves and to obtain a point with a radius of curvature of $\approx 1\ \mu$, which may be used, for example, to investigate field emission [72].

Finally, for photoelectrochemical shaping of n-type semi-conductors it is possible to use the fact that the limiting current density of anode etching is increased by illumination of the electrode surface. By nonuniform illumination of the anode, it is possible to achieve a local increase in the etching rate [67].

c. Electrochemical Cutting. In the production of instruments from large ingots of germanium, the latter are cut into pieces. Thereupon a large part of the ultrapure material is converted into filings. A considerable part of the remaining germanium is wasted as a result of the etching which always follows mechanical working (to remove the layer with a disrupted structure).

To increase the utilization coefficient of the semiconductor material, many methods have been proposed, such as the preparation of the pure material directly in the form of tablets of the required form and size, the use of dendritic ribbons, the deposition of thin layers of germanium from the gas phase on a solid backing, etc., and among these electrochemical cutting of ingots occupies an important position. The cathode is a thin tungsten wire about 80μ in diameter, arranged vertically (Fig. 122), about which flows the electrolyte (0.002% solution of KOH). The germanium anode is set in the immediate vicinity of the cathode. The specific resistance of the electrolyte is considerably greater than the resistance of the germanium, and this leads to localization of the etching in a narrow zone close to the cathode. As etching proceeds, the cathode is moved steadily into the anode (at a rate of the order of 0.15 mm/min with a current of 30 mA), leaving a narrow cut behind it. The width of the cut equals approximately twice the diameter of the cathode. The electrolyte consumption is 10 ml/min [73, 74]. In another method [75] the ingot is completely immersed in distilled water which acts as the electrolyte. The cathodes are two series of parallel wires which move so as to cut off from the ingot tens of rectangular samples simultaneously by etching. The germanium surface does not require special etching after cutting.

d. Precision Electropolishing. The so-called planar technique for producing transistors has become

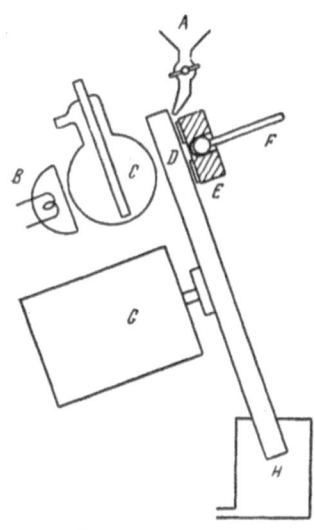

Fig. 123. Diagram of apparatus for precision electropolishing by Sullivan's method [77]. A) Electrolyte reservoir; B) light source; C) water filter; D) polishing disk; E) plastic disk with samples; F) spindle of disk with joint; G) electric motor; H) reservoir for collecting spent electrolyte.

common in recent years. Its principle is as follows. A thin plate of a semiconductor cut from an ingot is the blank for tens or hundreds of devices. They are all produced simultaneously as a result of a series of successive operations (masking of individual sections of the surface, diffusion of one or two impurities from the surface to a depth of several microns, the deposition of ohmic contacts, local etching, etc.). Finally, the plate is cut into tens or hundreds of parts, each of which is a complete device.

This method of producing devices requires absolutely flat plates several square centimeters in area, whose roughness is slightly greater than atomic dimensions. Ordinary electropolishing gives a mirror shine, but a wavy surface with "piled up" edges. A new method of electropolishing was proposed by Sullivan [76-79]. Plates of semiconductor are cemented to the surface of plastic disks with the ohmic contacts to the samples on the rear face of the disks. These disks are placed against the surface of the cathode — a polishing steel disk covered with paper or cloth. As the disks are arranged eccentrically with respect to the polishing disk, when the latter rotates they also rotate and the samples (anodes) slip relative to the surface of the polishing disk. The polishing plane is arranged at a sharp angle to the vertical, as shown in Fig. 123, or. horizontally. Electrolyte with a high resistance and a high viscosity drips from above onto the polishing disk. The high resistance is necessary so that the current density, which is determined here by the ohmic resistance of

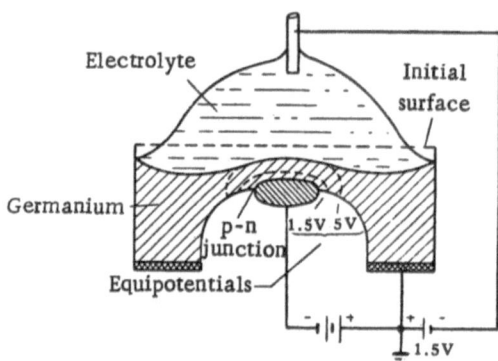

Fig. 124. Diagram of apparatus for preparing thin germanium
plates by anode etching by Froschle's method [83].

the solution, is greatest at projections on the semiconductor
surface. Therefore, the projections dissolve more rapidly and
the surface is automatically smoothed. The distance between
the anode and the cathode is set by the thickness of the viscous
layer, which, in its turn, is regulated by the viscosity of the
electrolyte. Since the hydrodynamic conditions in the liquid
are rigidly set and uniform to a high degree over the whole of
the anode surface, etching gives an ideally smooth surface. *

Electrolysis conditions: current density 1.6 A/cm^2, time
about 30 min. The electrolyte is a 0.1% aqueous solution of
KOH with 25% glycerol added (for germanium) or 0.5-3% HF solution with 30% glycerol added (for silicon). The solution consumption is 10-30 ml/min. The n-type samples are illuminated
through openings in the cathode. The result of the etching is a
surface up to 2 cm^2 in size with macro-unevenness no greater
than 1 μ/cm and microroughness no greater than 25 Å (according to [78], 0.04 μ/cm and 10-13 Å, respectively). The final
thickness of the sample is 25 μ and greater. The electrophysical properties of the surface and devices based on it are better
than with the normal etching method.

---

*The apparatus illustrated in Fig. 123 may also be used for chemical polishing, for
example, of gallium arsenide in a mixture of bromine and methanol [80].

e. Preparation of Thin Samples by Etching.

For the preparation of thin plates of n-type germanium of high area, the sample is illuminated during anodic etching with light which is absorbed in the bulk of the semiconductor (a light filter consisting of a germanium plate is placed between the light source and the sample for this purpose). The limiting anode current density is proportional to the number of light quanta absorbed and hence the thickness of the sample. Therefore the thickness of the plate is automatically leveled out during photoetching. Germanium layers 1-3 μ thick with thickness deviations no greater than 50% were obtained by this method [81, 82].

A completely different principle is the basis of a method of preparing p-n-p triodes with a thin base, * which was proposed by Froschle [83, 84] (see also [38]). Here, the etching is stopped automatically through the electric field of the p-n junction, which is on the opposite side of the plate (the dry side) from the n-type germanium (Fig. 124). Back bias is applied to the junction. The width of the space charge of the p-n junction is determined by expression (26.17). A space charge region also exists at the germanium—electrolyte boundary. The two charged regions are separated by the neutral bulk of the semiconductor and approach as etching proceeds. In the dark the limiting anode current density is a function of the generation of holes in the neutral bulk and falls as the thickness of this region decreases. To maintain a constant high rate of etching, and to make it uniform, the sample is illuminated on the solution side with light which is absorbed at the surface of the semi-

---

*The thickness of the base w determines the frequency characteristics of the transistor. Thus, the limiting frequency of amplification on the short-circuit current is

$$\omega \approx \frac{D}{w^2},$$

where D is the diffusion coefficient of minority carriers. In high-frequency transistors, the thickness of the base is no more than a few microns.

conductor. All the holes injected by light are used up at the
germanium — solution interphase in the anode solution reaction
and, therefore, the back current of the p-n junction is independent of the illumination.

When the thickness of the sample decreases sufficiently
for the two space charge regions to merge, etching automatically ceases. This occurs because the potential drop at the p-n
junction is made greater than that at the germanium — solution
interphase and hence in the space charge region of the p-n
junction there is always an equipotential surface (shown by a
broken line in Fig. 124) whose potential equals the potential of
the germanium — solution interphase. When the etching surface
reaches this equipotential, the etching current falls to zero.
All the holes injected by light are drawn into the p-type region
by the field of the p-n junction and the back current of the junction increases. The effect of illumination from the electrolyte
side on the back current of the p-n junction may be used for an
additional indication of the completion of etching.

Electrolysis conditions used: potential on p-n junction,
20 V; potential on electrochemical cell, 5 V; electrolyte, 5%
KOH solution; diameter of light spot, approximately equal to
the diameter of the hole etched out, ≈ 0.1 mm; etching time, 3
min. A hole with a mirror-smooth and level surface is obtained. Depending on the potential on the junction, the thickness
of the base is 2.5-25 μ ($\pm 1\mu$).

The apparatus of Rediker and Sawyer [85] for preparing
p-n-p triodes with a thin base differed from that illustrated in
Fig. 124 in that forward and not back bias was applied to the p-n
junction. The holes thus injected diffused to the germanium —
electrolyte interphase and fulfilled the same functions as illumination in Froschle's method. As there are no conditions here
for automatically stopping etching, Rediker and Sawyer used the
following method for controlling it. To the p-n junction were
alternately applied pulses of forward potential (for injection of
holes) and pulses of back potential (control), during which the
back current of the junction was measured. When the thickness
of the sample fell to the size of the space charge region (1-5 μ),

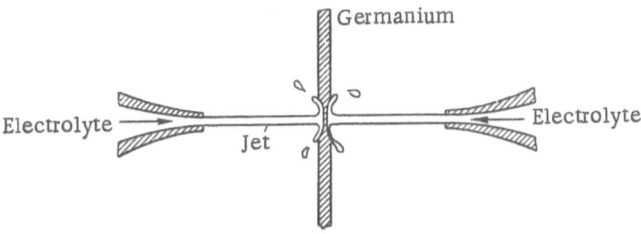

Fig. 125. Diagram of apparatus for jet etching [86].

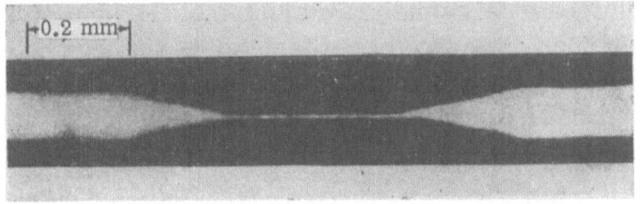

Fig. 126. Silicon sample after jet etching (cross section) [87].

the back current of the p-n junction fell sharply, * and a special relay switched off the etching current.

f.   Jet Etching.   The etching of germanium in a jet of electrolyte was first used by Tiley and Williams [86] for making surface barrier triodes. In these devices, the emitter and collector are not p-n junctions, but metal–semiconductor contacts, whose rectifying properties are due to an appropriate choice of the sign of the difference in the electronic work functions of germanium and the metal. The emitter and collector are deposited on opposite sides of a germanium plate (base) whose thickness is a few microns. Although surface barrier triodes have not become common, the method of jet etching with subsequent electrodeposition is used widely in the produc-

*This occurs because at the time the space charge region of the p-n junction reaches the surface of the germanium–solution interphase, the injection of minority carriers into the semiconductor because of the electrochemical reaction becomes possible.

tion of microalloy and diffusion transistors for obtaining the required geometry and to deposit metals which are then used in
alloying or diffusion processes. A review of the patent literature is given by Schnable and Lilker [87].

Figure 125 illustrates the principle of jet etching. A fine
jet of electrolyte is directed onto a sample of semiconductor
(anode) from a glass nozzle. The cathode is fused into the wall
of the nozzle. The specific resistance of the electrolyte is 2-3
orders higher than the specific resistance of the semiconductor
and, therefore, etching occurs only on the region of the surface
on which the jet falls and the diameter of the hole etched away
normally does not exceed the diameter of the jet (0.1-1 mm).
The distance from the sample to the nozzle is 1-2 mm. Two
jets in opposite directions and on opposite sides of the plate are
used to make triodes.

Jet etching itself does not require special electrolytes and
may be carried out with the solutions which are used in other
etching methods: in 0.1-0.01% solution of potassium hydroxide
or of nitrates for germanium [88-91] and in hydrofluoric acid
solution for silicon [87]. However, as has been stated already,
electrodeposition usually follows etching and these two operations are carried out more conveniently without changing the solution, but simply by changing the sign of polarization. Therefore, salts of the metals which are then to be deposited on the
sample are included in the etching electrolytes. For example,
the electrolyte for germanium contains indium, zinc, or cadmium, that for silicon contains gold, and that for gallium arsenide contains cadmium [92]. The consumption of electrolyte is
a few tens of milliliters per minute and the mean linear velocity
in the jet is several meters per second. The current strength
normally equals 1-10 mA per nozzle, which corresponds to a
current density of 10-100 A/cm^2 and above, i.e., etching occurs under electropolishing conditions and is complete in 1-2
min. Its result is a mirror-smooth surface. The n-type samples
are usually illuminated. The electrolyte column in the nozzle
and jet may be used as the "optical system" for conveying light
from the source to the sample surface. The mechanism of jet

etching of germanium and silicon is examined in the papers of
Schmidt [93, 94].

An important problem is the control of the thickness of
the base and automation of the termination of etching. The relation of the absorption of light in the sample to its thickness is
usually used for this purpose. On one side of the plate is arranged a source of modulated light and on the other, a photocell. * The intensity and spectral characteristics of the light
transmitted by the sample are known functions of its thickness
and may be used for controlling the latter [38]. Etching is ter-
minated automatically when the thickness of the remaining neck
is reduced to a given value $(1-5 \mu)$.

Two holes with flat smooth bottoms are obtained as a result of etching on opposite faces of the plate (Fig. 126).

Let us examine some special applications of jet etching.
In the preparation of field triodes, a jet of electrolyte is directed
onto the side surface of a germanium cylinder, perpendicular
to its axis. During etching the sample is rotated about the axis
so that an annular groove is etched out of it as a result. The
diameter of the remaining neck is about 0.2 mm. Then indium
is deposited on the etched surface and this forms a circular collar around this neck. In the operation of the triode the indium
acts as the field electrode, regulating the strength of the current along the axis of the cylinder [96].

Booker and Stickler [97] developed a method of jet electropolishing of large areas on germanium and silicon. The sample
and nozzle are set in reciprocating motion relative to each other
with the result that the whole surface of the sample is "scanned"
by the jet. The rate of relative motion is 25 mm/ sec horizontally and 0.08 mm/ sec vertically. The jet passes repeatedly
across the sample, leaving an ideally level surface; its roughness is no more than 25 Å and unevenness 0.5μ in 25 mm. Po-

---

*If an n-type semiconductor is etched, then instead of a photocell it is possible to
use as the "radiation receiver" the anode reaction itself, whose rate is proportional to the illumination [95].

lishing conditions for p-type germanium: 0.1 N H_2SO_4 solution
with a small amount of HF added, current strength 200 mA
(which corresponds to a current density of 6 A/cm^2 with a jet
diameter of 2 mm), and time 2 h. The electropolishing of n-
type germanium is carried out with half the current with strong
illumination. Silicon is polished in 0.2 N HF solution.

In recent years, the cathodic formation of germanium hydride
in accordance with equation (36.1) has been proposed for the
treatment of a germanium surface. Since this reaction proceeds at an appreciable rate only beginning with a current density of the order of 10 A/cm^2, * it is convenient to carry it out
under jet etching conditions, where the attainment of high current densities is no problem [99]. It was found that with a fall
in the current density the current efficiency of the formation of
GeH_4 falls sharply. As a result of this, there is an increase in
the "resolving power" of the etching, and holes with very sharp
edges and a flat bottom are obtained.

§ 60.   Electrodeposition of Semiconductors. Electrodeposition of Metals on Semiconductors

a.   Electrodeposition of Semiconductor Materials. †

In principle, the electrodeposition method may
be used for the following purposes: the production and refining
[102, 103] of semiconductor materials (in particular, compounds
of complex composition) and the deposition of semiconductor
layers in solid backings. Although such processes have not
found wide application up to now, we may hope that the latter
will find application in the technology of epitaxial semiconductor
films, ‡ which are now deposited mainly from the gas phase.

---

* The formation of GeH_4 on a cathode is also observed at lower current densities,
  but only when germanium ions are present in the electrolyte [98]; then, the
  material of the cathode does not participate in the reaction.

† The polarography of germanium and also the codeposition of germanium with
  zinc and other metals in electrometallurgy are not considered here. A review of
  the literature on these problems is given in the monographs [100, 101].

‡ I.e., single crystal films repeating the crystal structure of the backing. Such systems are used for producing semiconductor instruments.

Germanium of 99.4% purity may be obtained by electrolysis of molten mixtures of germanium dioxide with sodium fluoride or borosilicate at 1000-1100°C [104]. Electrolysis of a mixture of lithium silicate with germanium dioxide at 1250°C gives at the cathode solid solutions of silicon in germanium, whose composition depends on the ratio of the components in the melt [105]. Alloys of germanium and silicon with various metals may be obtained analogously [106].

Silicon is deposited on the cathode from a melt of alkali metal chlorides and fluorosilicates at 400-1000°C [107]. An indium–antimony alloy may also be obtained by electrolysis [108]. Finally, Bockris, Diaz, and Green [109] electrolyzed a $Na_2B_4O_7-GeO_2$ melt at 800-900°C and obtained on a graphite cathode a deposit of germanium in the form of dendrites, which may find application in the production of semiconductor devices.

In the electrolysis of aqueous solutions of germanium compounds, the main reaction at the cathode is the liberation of hydrogen. It is possible to deposit a very thin layer of germanium from sulfide solutions, but the quality of the deposit is poor [110, 111]. It is also possible that the overvoltage of hydrogen liberation on germanium is less than the overvoltage of germanium liberation, otherwise other reactions interfere with the deposition (for example, the formation of GeH_4).

On the other hand, dense lustrous layers of germanium may be deposited on various metals or graphite from nonaqueous solutions [112-115]. The best results are obtained in the electrolysis of a 7% solution of $GeCl_4$ in propylene glycol at 59°C with a current density of $0.4 \ A/cm^2$. Traces of water interfere with the formation of a good deposit.

In all the work listed, the electrodeposited material apparently did not have definite semiconductor properties and could not be used directly for preparing semiconductor devices without additional purification and treatment.

b. Preparation of Contacts to Semiconductor Materials by Electrodeposition. Electrodeposition is used for preparing both ohmic and rectifying contacts. In

the latter case the electrodeposition operation is normally followed by thermal treatment, during which the metal alloys or
diffuses into the semiconductor. In the surface layer of the latter there is formed a p-n junction which is responsible for the
rectifying action.

The systematic investigation of electrodeposited contacts
on germanium and silicon was first undertaken by Borneman et
al. [116,117] and Turner [118]. The deposition was carried out
by the jet method (see below) from aqueous and ethylene glycol
solutions. The compositions of the baths used are given in the
original papers and in the review [38]. It was found that most
of the metals investigated (In, Zn, Cd, Pb, Cu, Au, Ni, etc.)
give rectifying contacts on n-type germanium and ohmic contacts on p-type germanium. On the other hand, antimony forms
an ohmic contact on n-type and a rectifying contact on p-type
germanium [118,119]. No correlation was found between the
electric properties of the contacts and the difference in the electronic work functions of the semiconductor and metal which is
predicted by the theory of the metal−semiconductor contact
(see, for example, [120]). The electric field and the height of
the barrier at the contact are apparently set, not by the difference in the work functions, but by the surface electron states at
the interphase, which are maintained (or arise) during electrodeposition.

On the other hand, on silicon, at least qualitative corre-
spondence was found between the height of the barrier and the
difference in the work functions in the system metal−semicon-
ductor. Metals with a low work function (for example Zn, In,
Pb, and Sn) give an ohmic contact on electrodeposition on n-
type silicon and a rectifying contact on p-type silicon. Metals
with a high work function (including Au, Rh, and Pt) have the
opposite effect.

The adhesion of the deposited metal to the backing depends
to a large extent on the degree of oxidation of the semiconductor
surface. Firmly adhering deposits may be obtained only when
surface oxides are absent. On germanium, oxides may be reduced readily by cathode polarization during the electrodeposi-

tion, but on silicon this does not occur. Therefore, hydrofluoric acid is usually added to baths for the electrodeposition of metals on silicon so as to dissolve the SiO_2 [118, 121, 122].

As a rule, the kinetics of electrodeposition do not show a relation to the semiconductor properties of the samples. The peculiarities of these processes on semiconductors are caused only by their higher specific resistance than metals. Only in the deposition of nickel on silicon after chemical polishing of the surface was there observed a large difference in the polarizabilities of n- and p-type cathodes (Fig. 127). While on n-type silicon electrodeposition proceeds with a very slight overvoltage, on p-type samples the potential reaches 50-200 V even with a current density of the order of 10^{-4} A/cm^2. The potential of the cathode then falls suddenly to 1 V and then electrodeposition proceeds as on n-type material. Polishing the surface eliminates this difference. It is possible that free electrons are required for the beginning of electrodeposition (i.e., minority carriers in p-type silicon) [118, 123].

Let us examine some special methods for localization of electrodeposition on definite sections of the surface of a semiconductor. This is of great importance for the manufacture of miniature semiconductor devices. One of these methods is masking of a silicon surface with an oxide film which is obtained by thermal oxidation in air. By treatment with hydrohalic acids, "windows" of the required form are etched in the film at the places where contacts are to be applied and then electrodeposition is carried out [124].

Electrodes of complex form may be obtained by "electrodeposition on marks" [125]. A figure is drawn on the surface of the semiconductor with a point of a soft metal. Points of Pb−As, Pb−Sb, and Au−Sb alloys are used for n-type samples and points of Al, Au, and In, and Sn−In and Au−Ga alloys for p-type samples. Then the lead from the negative pole of a current source is clamped to the figure and the sample placed in a gilding bath. Gold is deposited on the metal marks. Electrodeposition is followed by alloying. The resolving power of the method is 1 μ and it may be applied to germanium, silicon, gallium arsenide, indium antimonide, and also, apparently, insulators.

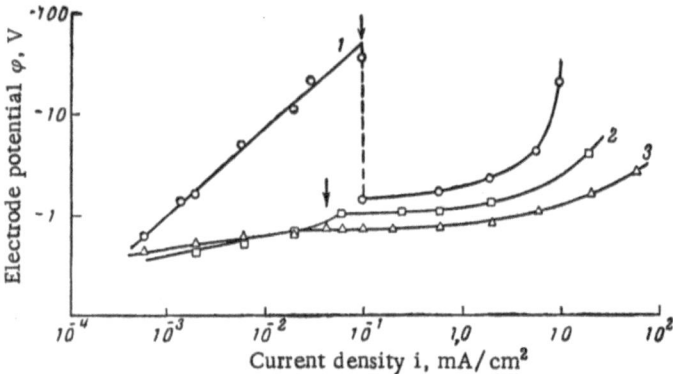

Fig. 127. Polarization curves of the electrodeposition of nickel on silicon [118].
1) n-Type; 2) p-type, polished surface; 3) p-type, surface after chemical polishing. Arrows mark the beginning of electrodeposition.

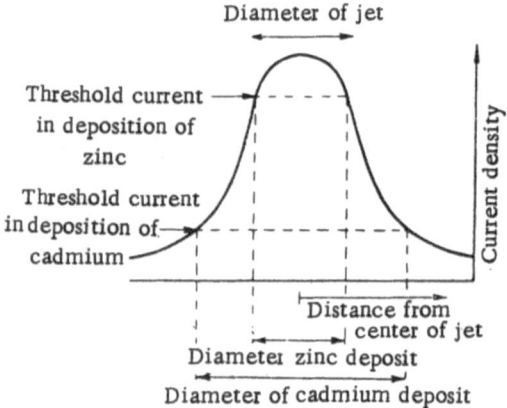

Fig. 128. Diagram showing the relation between the diameter of the metal deposit and the distribution of current density along the surface in jet electrodeposition
[87].

   In another method [126], the surface of the semiconductor
is scratched with a hard point. The formation of nuclei in elec-
trocrystallization is facilitated on the scratches and preferen-
tial formation of the deposit occurs. Microscratches (1-25 μ)
may also be used to localize etching.

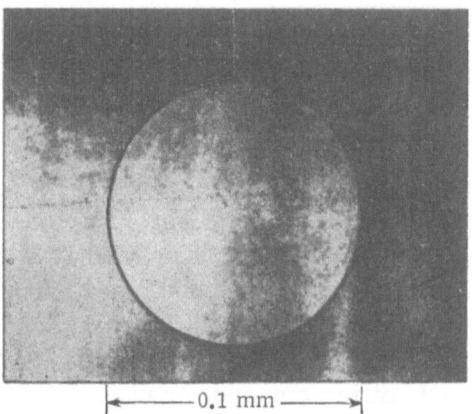

Fig. 129. Deposit of nickel on germanium obtained by jet
electrodeposition [128].

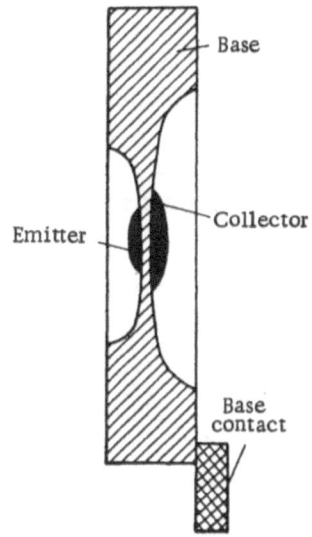

Fig. 130. Diagram of surface barrier triode (cross section).

The most common method
of obtaining deposits of small area
is jet electrodeposition, the principle of which was examined above.
It is carried out with electrolytes of
normal composition, but with a
low electrical conductivity (of the
order of 0.01 $\Omega^{-1} \cdot cm^{-1}$) at a high
current density (normally higher
than 1 A/cm^2). The distribution
of the current along the surface of
the sample at the point on which
the jet falls is shown in Fig. 128.
The current efficiency of electrodeposition depends on the current
density and reaches values close
to 100% only at a certain critical
current density which is different for different metals. There-
fore, the diameter of the deposit with a given diameter of jet
depends on the nature of the metal. The addition of surface-
active substances such as ethylenediaminetetraacetic acid to the
electrolyte is another method of regulating the current distribu-

tion in the jet and the size of metal deposits [127]. Round deposits, from 0.06 to 2.5 mm in diameter with sharp edges have been obtained by this method (Fig. 129). Rotating the jet about an axis parallel to it gives an annular contact.

The jet electrodeposition method has found wide application in the production of semiconductor devices. It is very "technological" and, therefore, it is readily standardized and automated. A bibliography of patent and technical literature on its applications is given in [128]. This method was used first in the production of surface barrier triodes [86] and is now used for microalloyed, diffusion, and other types of transistors. Figure 126 shows the section of a silicon sample after jet etching, and this is the blank for a microalloyed or surface barrier triode. After the production of the emitter and collector by jet electrodeposition of metals on the bottom of the etched holes, the device has the form illustrated in Fig. 130. Thermal treatment is normally carried out at the end of electrodeposition, but Williams [129] proposed the combination of deposition and alloying in one operation. A nonaqueous solution (for example, indium salts in ethylene glycol) is used for this and the deposition is carried out at a temperature above the melting point of indium (180 and 150°C, respectively).

In addition to electrodeposition, the following methods are used to prepare metal deposits on semiconductors:

1) Deposition by electrochemical displacement of a more noble metal from its salts by a less noble metal. This phenomenon was examined in Chapter IV. "Chemical" nickel plating with subsequent thermal treatment is used widely to apply ohmic contacts to silicon [130] and also to other semiconductor materials [131].

However, electrochemical displacement is often an undesirable side reaction in the electrochemical treatment of semiconductors if the solution contains traces of noble metals. Copper, silver, and other electropositive metals are readily deposited on germanium and silicon and this often impairs the characteristics of semiconductor devices. This must be considered particularly in jet treatment when the sample is in contact with a relatively large volume of well-mixed electrolyte [132].

2) Contact deposition as a result of the action of a macrogalvanic couple consisting of the semiconductor and a metal with a more negative electrode potential. This phenomenon also interferes frequently in the etching of prepared devices which contain electrodes and leads of very diverse metals (In, Sn, Cu, Au, Al, Ni, etc.).

3) "Nonelectric" (electroless) deposition by chemical reduction of metal ions in solution with a specially added reducing agent.

§ 61. Application of the Semiconductor — Electrolyte Interphase in Electronic Devices

The examples of electrochemical etching and electrodeposition listed above are only subsidiary, though very important operations in the production of semiconductor devices. The properties of the solid are used exclusively in the completed device. The importance of electrochemistry in semiconductor radio technology and electronics will increase considerably if the semiconductor—electrolyte system can be used directly in electronic devices.*

Attempts to find such applications have been made repeatedly, but up to now they have not led to practical results. The attention of investigators has been attracted primarily to the rectifying properties of the semiconductor—electrolyte interphase.† In 1954, Garrett and Brattain [134] proposed the use of the system n-type germanium—0.1 N KOH solution—silver oxide electrode as a "transistor with an internal power supply."

---

*Here we are not considering systems of the semiconductor—liquid dielectric—metal type such as the fieldistor [133].

†As was shown in § 26, a high rectification factor can be expected for this system, connected with the very low values of the exchange current at a semiconductor—electrolyte interphase.

The germanium electrode is a thin plate, on one side of which a p-n junction is formed by alloying with indium. The other side of the electrode is in contact with the solution (see Fig. 64b). The germanium (through the indium) and the silver oxide electrode are connected through a load resistor. The potential of silver oxide is more positive than that of germanium. As a result of this natural potential difference, the germanium is polarized anodically and dissolves, with the rate of the process determined by the diffusion of holes from the bulk of the semiconductor to its surface. When forward bias is applied to the p-n junction, a flow of holes arises in the germanium electrode from the junction to the boundary with the electrolyte and the limiting rate of anode solution, i.e., the current in the electrochemical cell, increases. Thus, the germanium electrolyte interphase acts as the hole collector in the transistor (see § 32). "Back bias" on the collector is the result of the difference in the electrode potentials in the germanium—silver oxide system and, therefore, the transistor requires no external power source. It has power amplification (since the impedance of the collector is higher than the impedance of the emitter) and also low current amplification ($\alpha \approx 1.5$). Its frequency characteristics are the same as for a junction transistor of the same geometry. However, the operating period of the device is short as the germanium electrode dissolves during its operation.

A brief communication [135] was published recently on the use of a silicon—sulfuric acid interphase as a detector of weak light signals. The operation of this device is based on the sensitivity of a rectifying silicon—electrolyte contact to illumination. In other words, this device operates as a solar battery, but its sensitivity is 10 times as great. This is explained by the lower exchange current in the silicon—electrolyte system than in the silicon—metal system or a p-n junction, and also the low ohmic resistance of the electrolyte in comparison with the resistance of the thin surface layer of a semiconductor in solar batteries. The device is as big as a pin's head and can operate over a wide range of temperatures. The interphase between KCl solution and a crystal of gallium arsenide or cadmium sulfide may be used for the same purpose [136].

The drawback of the systems described is their low stability due to electrochemical and corrosion processes at the semiconductor—electrolyte interphase. For the same reason it is probably unprofitable to use the peculiarities of the volt-ampere characteristics of semiconductor electrodes for producing new devices. The use of their capacitance properties seems more promising.

One way of using the semiconductor capacitance is to modulate the width of the space charge region in the semiconductor with an electric field applied to the semiconductor—electrolyte contact. The design of the field triode is based on this principle. An example is the field triode proposed by Dewald [137]. It consists of a single crystal of zinc oxide whose thickness is comparable to the size of the space charge region. To the ends of the crystal are applied indium ohmic contacts for passing a direct current. The central part of the crystal is surrounded by a solution in which zinc oxide is not corroded. In the solution is an auxiliary platinum electrode to which is fed the alternating signal, so that the zinc oxide is polarized relative to the solution. Thereupon, there is a change in the thickness of the space charge regions on opposite faces of the sample and hence in the thickness of the uncharged layer of semiconductor lying between them, through which the direct current passes. The thickness of this layer and the conductivity along the sample are thus functions of the external signal and this is used for amplification. An analogous device was made from cadmium sulfide [138].

Another method is based on a specific property of the semiconductor capacitance, namely, its nonlinearity. As was shown in Chapter I, the potential dependence of the capacitance in the case of a semiconductor with a narrow forbidden band is represented by a curve with a minimum, while for materials with a wide forbidden band the capacitance changes monotonically with potential. The nonlinear capacitance is used widely now in radio technology [139]. In the semiconductor—dielectric —metal systems normally used, the maximum value of the capacitance is limited by the capacitance of the layer of dielectric, which normally does not exceed $0.5 \, \mu F/cm^2$. In the case of a semiconductor−electrolyte contact, the role of the field electrode is played by the electrolyte, while the Helmholtz layer, whose capacitance is very great ($\approx 10 \, \mu F/cm^2$), acts as the "dielectric layer." Therefore, in the electrolytic system, it is possible to obtain very high nonlinearity of the capacitance. Thus, for example, with a change in the external potential, the capacitance of a cadmium sulfide electrode changes by 3 orders and reaches $5 \, \mu F/cm^2$ [140]. As this electrode is practically ideally polarized in K_2SO_4 solution (the current through the interphase does not exceed $10^{-6} \, A/cm^2$), this device can be expected to have a high stability and a long lifetime.

Literature Cited

1. J.F. Dewald, Bell System. Tech.J., 39:615 (1960).
2. P. Camp, J.Appl.Phys., 25:459 (1954).
3. R.C. Smith, J. Electrochem.Soc., 108:238 (1961).
4. F.E. Roberts, Solid-State Electron., 1:93 (1960).
5. A.A. Davydov and V.N. Maslov, Zh.Fiz.Khim.,37:778 (1963).
6. T. Masami, J.Phys.Soc.Japan, 15:2254 (1960).
7. T. Tokuyama, Solid-State Electron., 5:135 (1962).
8. G. Feuillade and S. Marette, J. Chim.Phys., 58:418 (1961).
9. I.N. Plaksin and R.Sh. Shafeev, Dokl.Akad.Nauk SSSR, 128:777 (1959).
10. W. Mehl and M.C. Coutts, J.Appl.Phys., 34:2120 (1963).
11. M.G. Mil'vidskii and A.V. Berkova, Zavodsk. Lab., 27:557 (1961).
12. B.M. Turovskii, Fiz.Tverd.Tela, 5:1750 (1963).
13. A. Goetzberger and C. Stephens, J. Electrochem.Soc., 109:604 (1962).
14. J.J. Oberly, Acta Met., 5:122 (1957).
15. V.N. Maslov, A.V. Ovodova, and L.V. Nabatova, Kristallografiya, 7:271 (1962); Zavodsk. Lab., 30:1362 (1964).

16. E.A. Efimov and I.G. Erusalimchik, Zh.Fiz.Khim., 38:589 (1964).

17. H. Gerischer, Z. Naturforsch., 19a:553 (1964).

18. D. Tjapkin and S. Joksimović-Tjapkin, Tehnika, 13; Electrotehnika, 7:E165 (1958).

19. G.L. Schnable, U.S. Pat. 2,893,929 (1959); C.A., 53: 18697b (1959); S. Ratcliffe and J.E. Hughes, Brit.J. Appl. Phys., 12:193 (1961).

20. R. Glang, J. Electrochem. Soc., 107:356 (1960).

21. G. Schwabe, Z. Angew. Phys., 14:297 (1962).

22. Yu.V. Pleskov, Dokl.Akad.Nauk SSSR, 143:1399 (1962).

23. J.T. Law and P.S. Meigs, U.S. Pat., 2,837,471 (1958); Ref.Zh.Met., 1959:22169.

24. P.J. Whoriskey, J.Appl.Phys., 29:867 (1958).

25. S.I. Silverman and D.R. Benn, J. Electrochem.Soc., 105:170 (1958).

26. P.A. Iles and P.J. Coppen, J.Appl.Phys., 29:1514 (1958).

27. P.A. Iles and P.J. Coppen, Brit.J.Appl.Phys., 11:177 (1960).

28. D.R. Turner, J. Electrochem. Soc., 106:701 (1959).

29. H. Robbins, J. Electrochem. Soc., 109:63 (1962).

30. D.R. Turner, J. Electrochem. Soc., 107:810 (1960).

31. E. Billig and J.J. Dowd, Nature, 172:115 (1953).

32. R.W. Jackson, J. Appl. Phys., 27:309 (1956).

33. C.S. Fuller, U.S. Pat., 2,740,700 (1956); Ref.Zh.Met., 1957:58131.

34. C.S. Fuller, J.Appl.Phys., 27:553 (1956).

35. B.A. Joyce, Solid-State Electronics, 5:102 (1962).

36. O.G. Deryagina, E.N. Paleolog, and N.D. Tomashov, Dokl.Akad.Nauk SSSR, 133:388 (1960).

37. J.J. Pankove, RCA Rev., 16:398 (1955).

38. J.J. Pankove, Electrochemistry of Semiconductors (ed. P.J. Holmes), Academic Press, London-New York (1962), p.290.

39. J.P. McKelvey and R.L. Longini, J. Appl.Phys., 25:634 (1954).

40. G. Wallis and S. Wang, J. Electrochem. Soc., 106:231 (1959).

41.   J.R. Haynes and W. Shockley, Phys. Rev., 81: 835 (1951).

42.   T.M. Buck and W.H. Brattain, J. Electrochem. Soc.,
       102: 636 (1955).

43.   P. Jacquet, Atomes, 8: 183, 208 (1953).

44.   C.H. Pool, U.S. Pat. 2,763,608 (1956); Ref. Zh. Khim.,
       1958: 43995.

45.   M.V. Sullivan, U.S. Pat. 2,983,655 (1961); Ref. Zh. Khim.,
       1962: 12K207.

46.   B.I. Él'kin, Problems in the Metallurgy and Physics of
       Semiconductors. Izd. Akad. Nauk SSSR, Moscow (1957),
       p. 142.

47.   J.I. Carasso and E.A. Speight, Brit. Pat. 861,679 (1961);
       Ref. Zh. Met., 1961: 10G368.

48.   B.H. Claussen, Brit. Pat. 807,297 (1959); Ref. Zh. Met.,
       1959: 24281.

49.   W. Müller and Quadlitz, Ger. Pat. 823,763 (1951); C.A.,
       47: 7461h (1954).

50.   I. Epelboin and M. Froment, J. Phys. Radium (Phys.
       Appl., Suppl), 18(3): 60A (1957).

51.   P. Brouillet and I. Epelboin, Compt. Rend., 237: 895
       (1953).

52.   D.R. Turner, J. Electrochem. Soc., 105: 402 (1958);
       U.S. Pat. 2,871,174 (1959); Ref. Zh. Khim., 1960: 14203.

53.   N.M. Alpatova, A.I. Gorbanev, Yu.M. Kessler, and
       L.G. Lozhkina, Dokl. Akad. Nauk SSSR, 142: 1073 (1962).

54.   L.V. Maslova, O.A. Matveeva, and V.F. Afanas'ev,
       Fiz. Tverd. Tela, 3: 2699 (1961).

55.   J.W. Faust, U.S. Pat. 2,861,931 (1958); Ref. Zh. Khim.,
       1960: 14201.

56.   D.R. Turner, Surface Chemistry of Metals and Semicon-
       ductors (ed. H.C. Gatos), Wiley, New York (1960), p. 285.

57.   M.M. Koltun, Zh. Fiz. Khim., 38: 723 (1964).

58.   J.W. Faust, Silicon Carbide. Pergamon Press (1960),
       p. 403.

59.   M.K. Norr, J. Electrochem. Soc., 109: 433 (1962).

60.   J.F. Dewald, J. Electrochem. Soc., 104: 244 (1957).

61.   I.A. Lesk and R.E. Gonzalez, J. Electrochem. Soc.,
       105: 469 (1958).

62. F.W. Dehmelt, Ger. Pat. 1,001,077 (1957); C.A., 53: 1671b (1959).

63. L.D. Armstrong and P. Kuznetzoff, U.S. Pat. 2,850,444 (1958); Ref. Zh. Met., 1959:26654.

64. A. Amaya, U.S. Pat. 2,890,159 (1959); C.A., 53:16718b (1959).

65. V.A. Vulcan, Control Engineering, 9:132 (1962); A.M. Goodman, Rev. Sci. Instr., 35:642 (1964).

66. L. Varettoni, E.T. Casterline, and K. Glicksman, Electrochem. Technol., 2:21 (1964).

67. A. Uhlir, Bell System. Tech.J., 35:333 (1956).

68. W. Rindner and R.C. Ellis, J. Electrochem. Soc., 109:537 (1962).

69. Z. Majewski and I. Klamka, Arch. Elektrotech., 4:379 (1955).

70. W. Rindner and J.M. Lavine, Solid-State Electron., 2:190 (1961).

71. A. Uhlir, Rev. Sci. Instr., 26:965 (1955).

72. N.M. Alpatova, Radiotekhn. i Elektron., 5:1351 (1960).

73. S. Sheff, J. Electrochem. Soc., 108:60C (1961).

74. J.F. Barry and N.C. Seeley, U.S. Pat. 2,827,427 (1958); see [38].

75. V.I. Savchenko, Author's Certificate No. 125721 (1959).

76. M.V. Sullivan, Bell Lab. Record., 39:107 (1961).

77. M.V. Sullivan, D.L. Klein, R.M. Finne, L.A. Pompliano, and G.A. Kolb, J. Electrochem. Soc., 110:412 (1963).

78. T.M. Donovan and B.O. Seraphin, J. Electrochem. Soc., 109:877 (1962); D. Baker and J.R. Tillman, Solid-State Electron., 6:589 (1963).

79. Engineering, 191:909 (1961); A.F. Bogenschutz, Telefunken-Ztg., 36:151 (1963).

80. M.V. Sullivan and G.A. Kolb, J. Electrochem. Soc., 110:585 (1963); A. Reisman and R. Rohr, J. Electrochem. Soc., 111:425 (1964); W.G. Oldham, Electrochem. Technol., 3:57 (1965).

81. W.A. Albers and J.E. Thomas, Bull. Am. Phys.Soc., 3:219 (1958).

82. W.A. Albers, V.E. Noble, R.P. Poplawsky, and J.E. Thomas, U.S. Dept. Comm., Office Tech. Serv., PB Rept. 161867 (1960); C.A., 56:13664a (1962).

83. E. Fröschle, Telefunken-Röhre, No. 35:63 (1958).

84. E. Fröschle, Ger. Pat. 1,044,289 (1958); C.A., 1960: 603.

85. R.H. Rediker and D.E. Sawyer, Proc. IRE, 45:944 (1957).

86. J.W. Tiley and R.A. Williams, Proc. IRE, 41:1706 (1953).

87. G.L. Schnable and W.M. Lilker, Electrochem. Technol., 1:202 (1963).

88. V. Miles, M.V. Sullivan, and J.H. Eigler, J. Electrochem. Soc., 103:132 (1956).

89. D.D. Evers, U.S. Pat. 2,767,137 (1956); Ref. Zh. Khim., 1958:43994.

90. W.E. Bradley, U.S. Pat. 2,846,346 (1958); Ref. Zh. Met., 1959:26649.

91. J.S. Lamming and G.M. Wells, Brit. Pat. 806346 (1958); Ref. Zh. Khim., 1959:71992.

92. R. Barrie, F.A. Cunnell, J.T. Edmond, and I.M. Ross, Physica, 20:1087 (1954).

93. P.F. Schmidt and D.A. Keiper, J. Electrochem. Soc., 106:592 (1959).

94. P.F. Schmidt and M. Blomgren, J. Electrochem. Soc., 106:694 (1959).

95. Z. Gragoun and E. Sipek, Bergakademie, 10:78 (1958).

96. K. Hoselitz and T.B. Watkins, Brit. Pat. 807042 (1959); Ref. Zh. Met, 1959:26648.

97. G.R. Booker and R. Stickler, J. Electrochem. Soc., 109:1167 (1962).

98. M. Green and P.H. Robinson, J. Electrochem. Soc., 106:253 (1959).

99. A. Topfer, U.S. Pat. 2,913,383 (1959); Ref. Zh. Khim., 1960:93034.

100. T.A. Kryukova, S.I. Sinyakova, and T.V. Aref'eva, Polarographic Analysis, Goskhimizdat, Moscow (1959).

101. E.A. Efimov, I.G. Erusalimchik, Electrochemistry of Germanium and Silicon, Goskhimizdat, Moscow (1963).

102. H. Kaneko and K. Masumoto, J.Japan.Inst.Metals, 22 : 553 (1958).

103. R. Monnier and J.C. Giacometti, Helv.Chim.Acta, 47:345,2203 (1964).

104. M.J. Barbier-Andrieux, Ann.Chim., 10:754 (1955).

105. M.J. Barbier-Andrieux, Compt. Rend., 242:2352 (1956).

106. G. Rosenberger, Ger. Pat. 939,100 (1956); C.Z., 1958:6642.

107. D.R. Stern and Q.H. McKenna, U.S. Pat. 2,892,763 (1959); C.A., 53:16770h (1959).

108. G. Busch, R.Kern, and E. Steigmeier, Proc. 9th CITCE Meeting, London (1959), p. 425; A.D. Styrkas, Zh.Prikl.Khim., 37:2431 (1964).

109. J.O'M. Bockris, J. Diaz, and M. Green, Electrochim. Acta, 4:362 (1961).

110. C.G. Fink and V.M. Dokras, J. Electrochem.Soc., 95:80 (1949).

111. S.I. Sklyarenko, I.I. Lavrov, and S.V. Yakobson, Izv. Vysshikh. Uchebn. Zavedenii Tsvetn.Met.,1962:129

112. G. Szekely, J. Electrochem. Soc., 98:318 (1951).

113. Brit. Pat. 711,065 (1954); C.A., 48:13493i (1954); Machinery (London), 83:824 (1953); Ref.Zh.Khim., 1954:45240.

114. V.V. Ostroumov, Zh.Prikl.Khim., 37:1483 (1964); V.V. Ostroumov and G.V. Anan'eva, Zh.Prikl.Khim., 37:1612 (1964).

115. A.N. Sysoev and N.N. Gavyrina, Zh.Prikl.Khim., 34:2001 (1961).

116. E.H. Borneman, R.F. Schwarz, and J.J. Stickler, J.Appl.Phys., 26 : 1021 (1955).

117. E.C. Wurst and E.H. Bornemann, J. Appl.Phys., 28:235 (1957).

118. D.R. Turner, J. Electrochem. Soc., 106:786 (1959).

119. R.A. Ehrhardt, U.S. Pat. 2,753,299 (1956); C.A., 50:12704i (1956).

120. A.R. Plummer, Electrochemistry of Semiconductors (ed. P.J. Holmes). Academic Press, London-New York (1962), p. 61.

121.   M.C. Waltz, U.S. Pat. 2,814,589 (1957); Ref.Zh.
        Khim., 1959 : 32036.
122.   W. Mehl, H.F. Gossenberger, and E. Helpest, J.
        Electrochem. Soc., 110 :239 (1963); V.V. Ostroumov,
        Elektrokhimiya, 1 :304 (1965).
123.   D.P. Zosimovich and V.D. Nemtsov, Ukr.Khim.Zh.,
        30 : 59 (1964).
124.   J. Roschen, U.S. Pat. 2,906,647 (1959). C.A., 54 :
        3012b (1960).
125.   W. Rindner and J.M. Lavine, Solid-State Electron.,
        5 : 85 (1962).
126.   W. Rindner and J.M. Lavine, J. Electrochem. Soc.,
        108 : 869 (1961).
127.   E.M. Zimmerman, U.S. Pat. 2,873,232 (1959); C.A.,
        53 : 8893a (1959).
128.   W.J. Hillegas and G.L. Schnable, Electrochem.
        Technol., 1 : 228 (1963).
129.   R.A. Williams, U.S. Pat. 2,945,789 (1960); Ref.Zh.
        Khim., 1961 : 16K190.
130.   M.V. Sullivan and J.H. Eigler, J. Electrochem. Soc.,
        104 : 226 (1957).
131.   D.R. Turner and H.A. Sauer, J. Electrochem. Soc.,
        107 : 250 (1960).
132.   G.M. Krembs and M.M. Schlacter, J. Electrochem.
        Soc., 111 : 417 (1964); J.W. Faust, J. Electrochem.
        Soc., 112 : 114 (1965).
133.   O.M. Stuetzer, Proc.Inst.Radio Engrs., 40 : 1377
        (1952).
134.   C.G.B. Garrett and W.H. Brattain, Phys.Rev., 95 :
        1091 (1954).
135.   Electronics, 1961(18) : 11.
136.   K.E. Plain and R.H. Bube, J. Electrochem. Soc.,
        111 :751 (1964).
137.   J.F. Dewald, J. Electrochem. Soc., 105 : 105C (1958).
138.   R. Williams, J. Phys.Chem. Solids, 22 : 129 (1961).
139.   L.S. Berman, Nonlinear Semiconductor Capacitance.
        Fizmatgiz, Moscow (1963).
140.   V.A. Tyagai, Author's Certificate No. 166074 (1964).

Additional Literature

Antonov, A. S., B. P. Osipenko, and L. G. Yuskeselieva,
  Mechanism of showing up junctions in silicon p-i-n de-
  tectors by electrochemical deposition of copper, Zh. Fiz.
  Khim., 39 : 2252 (1965).

Berkova, A. V., M. G. Mil'vidskii, and V. B. Osvenskii,
  Detecting nonuniformities in the distribution of impurities in gallium arsenide crystals, Zavodsk. Lab., 31:
  1095 (1965).

Daw, A. N., and R. N. Mitra, Electroless plating of
  metals on III-V group semiconductors, Solid-State
  Electron., 8 : 697 (1965).

Duffek, E. F., E. A. Benjamini, and C. Mylroie,
  Anodic oxidation of silicon in ethyleneglycol solutions,
  Electrochem. Technol., 3 : 75 (1965).

Evans, E. J., Technique for trim etching diffused resistors in silicon, Rev. Sci. Instrum., 36 : 1248 (1965).

Hobson, M. C., and H. Leidheiser, Increased rate of
  InSb formation on the surface of antimony under the electrochemical treatment. Trans. Met. Soc. AIME, 233:
  482 (1965).

Ivanov, V. G., Preparation of germanium emitters and
  production of a field emission image of pure germanium,
  Radiotekhn. i Elektrón., 10 : 576 (1965).

Kochegarov, V. M., V. D., Samuilenkova, and G. Ya. Semyachko,
  Electrochemical deposition of electric contacts on the surface of n- and p-type germanium, Zh. Prikl. Khim., 38:1300
  (1965).

Kochegarov, V. M., and L. N. Kolesov, Electrochemical
  deposition of electric contacts on the surface of p-silicon,
  Zh. Prikl. Khim., 38 : 1396 (1965).

Konoplya, L. N., A. F. Kravchenko, and V. P. Sirotkina,
  Nonrectifying contacts on gallium arsenide, Pribory i
  Tekhn. Éksperim., 1965(3) :243.

Leritsi, A., T. Nemet, P. Sebeni, and E. Tikhani,
  Device for protecting micrononuniformities in semiconductors, Pribory i Tekhn. Éksperim., 1965(3) : 201.

Mackintosh, I. M., P. F. Schmidt, and M. W. Lapkin, Integrators prepared using electrochemical methods, Proc. Inst. Elect. Electron. Engrs., 52:1447 (1964).

McNeill, W., L. L. Gruss, and D. G. Husted, The anodic synthesis of CdS films, J. Electrochem. Soc., 112:713 (1965).

Monnier, R., and P. Tissot, Étude du compartement et de l'electrolyse de solution d'oxide de germanium dans des fluorures fondus. Helv. Chim. Acta, 47:2203 (1964). New method of electropolishing germanium slices, Electrochem. Technol., 4:73 (1966).

Schmidt, P. F., T. W. O'Keeffe, J. Oroshnik, and A. E. Owen, Doped anodic oxide films for device fabrication in silicon (II). Diffusion sources of composition and diffusion results. J. Electrochem. Soc., 112:800 (1965).

Shchigolev, P. V., and Z. B. Safonova, Electropolishing of silicon, Elektrokhimiya, 1:1077 (1965).

Seo, J., Japanese Pat. N2258 (1958); C. A., 53:11067 (1959).

Wolkenberg, A., Electrochimical and photochemical properties of the group IV semiconductors. Phototelectrochemical cell with n-type germanium. Solid-State Electron., 8:581 (1965).

Zesimovich, D. P., and V. D. Nemtsov, Polarization of silicon electrode in electrodeposition of indium and antimony. Ukr. Khim. Zh., 32:20 (1966).

Reviews on Electrochemistry of Semiconductors

H. Göhr, Proc. 7th CITCE Meeting, London (1957), p. 379.

Yu. V. Pleskov, Khim. Nauka i Promy., 3:443 (1958).

J. F. Dewald, Semiconductors (ed. N. Hannay). Cambridge Univ. Press, New York (1959), p. 713.

M. Green, Modern Aspects of Electrochemistry (ed. J. Bockris). London (1959).

J. Mieluch, Wiadom. Chem., 13:679 (1959).

The Surface Chemistry of Metals and Semiconductors (ed. H. C. Gatos). Wiley, New York (1960).

R. Sh. Nigmetova, Tr. Inst. Khim. Akad. Nauk KazSSR, 6:178 (1960).

M. Green, J. Phys. Chem. Solids, 14:77 (1960).

H. Gerischer, Advances in Electrochemistry and Electrochemical Engineering, Vol. 1 (ed. P. Delahay). Interscience, London-New York (1961), p. 139.

Electrochemistry of Semiconductors (ed. P. J. Holmes). Academic Press, New York (1962).

V. A. Myamlin and Yu. V. Pleskov, Usp. Khim., 32:470 (1963).

Yu. V. Pleskov, Vestn. Akad. Nauk SSSR, No. 10:18 (1963).

E. A. Efimov and E. G. Erusalimchik, Electrochemistry of Germanium and Silicon. Goskhimizdat, Moscow (1963).

R. Buvet and J. Perichon, Bull. Soc. Franc. Electriciens, 4:199 (1964); Electrochim. Acta, 9:587 (1964).

H. -U. Harten, Die Grönzflache Halbleiter-Elektrolyt. Festkörperprobleme, No. III (ed. F. Sauter). Akademie-Verlag, Berlin (1964), p. 81.

Most Immediate Problems in the Electrochemistry of Semiconductors*

V. A. Tyagai and Yu. V. Pleskov

(Institute of Electrochemistry, Academy of Sciences of the USSR)

In the ten years of the existence of semiconductor electrochemistry, the basic principles of the electrochemical behavior of semiconductor materials and their fundamental differences from metals have been elucidated, at least qualitatively [1-5]. At the same time, a series of important problems, partly specific to semiconductors and partly common to the electrochemistry of metals and semiconductors, still await their solution. The aim of the present article is a systematic account of what are, in the opinion of the authors, the most important unsolved problems in the electrochemistry of semiconductors.

Distribution of Potential at a Semiconductor – Electrolyte Interface

The study of the distribution of the galvanic potential drop between the space charge region in a semiconductor and the

---

*This article originally appeared in Elektrokhimiya, Vol. 1, No. 10, October 1965, pp. 1167-1173, and is a comment on the contemporary state and most important problems in the electrochemistry of semiconductors.

ionic part of the electrical double layer (in concentrated solutions – the Helmholtz layer) is one of the most important stages in the investigation of the properties of the suface of a semiconductor in contact with an electrolyte. The magnitude of the surface deflection of the bands, determined as a result of these experiments, is the basic parameter of the statistical theory of the properties of the surface of semiconductors. Large amounts of experimental data on this problem have been accumulated by now and the main results may be divided into two groups.

1) The first group includes investigations for determining the magnitude of the potential drop in the ionic double layer and for establishing the relation between this value and the crystallographic orientation of the semiconductor surface and the chemical composition of the surface compounds formed. As in the case of metal electrodes, with the adsorption of ions on the surface compounds there is a considerable change in the potential drop in the Helmholtz layer and at the interphase. For example, in the case of germanium, silicon, and gallium arsenide, the potential drop in the Helmholtz layer is determined mainly by the formation of surface oxides by the reaction of the semiconductor with the solution and the change in their composition on polarization of the electrode [6, 7]. The same apparently occurs on oxide and sulfide semiconductors [8, 9]. The main experimental problem here is the determination of the sign and magnitude of the charge of the actual surface of the semiconductor electrode. The charge of the electrode surface is of great importance in discussing the kinetics of electrochemical processes at a semiconductor – electrolyte interface. Since traditional methods of semiconductor electrochemistry, i.e., measurement of surface conductivity, differential capacitance, etc., make it possible to measure the space charge, but not the surface charge, here we have to use other methods such as adsorption measurements, measurement of the hardness of electrodes, or measurement of the wetting angle of gas bubbles. Attempts have been made to determine the zero point potential of germanium in aqueous solution by the wetting angle method [10], but they were relatively unsuccessful. More definite conclusions on the magnitude of the charge on a germanium surface were drawn from

a comparison of the sizes of hydrogen bubbles in the cathode liberation of hydrogen from acid and alkaline solutions [11]. As yet there has been no direct measurement of the surface charge or the zero points of semiconductor single crystals.

Another important problem is the establishment of the type of bond between the surface atoms of a semiconductor and an adsorbed substance. This bond may be covalent and dipolar; as a result of adsorption there is formed a surface dipole in which the charges of one sign lie on the surface atoms of the semiconductor and charges of the opposite sign, in the layer of specifically adsorbed atoms (or molecules). The adsorption of substances whose molecules have their own dipole moment leads to a similar result. In the other limiting case (for example, in the chemisorption of charge particles), the charge of one sign is concentrated in the adsorption layer, while the charge of the other sign is formed by electrostatically adsorbed ions from the solution and also free charges in the solid.* It should be pointed out that the investigation of surface oxides on germanium indicates that here there are both types of potential drop, i.e., "dipole" and "ionic," which are comparable in magnitude and may be changed independently by a change in the composition of the solution or polarization of the electrode [6].

2) The second group of papers is connected with checking the theoretical concepts on the structure of the space charge region in a semiconductor. Direct measurement of the magnitude of the space charge in relation to the surface potential, made by means of "fast" charging curves, and also the investigation of surface conductivity, the differential capacitance, the surface recombination rate, and the photoeffect showed that for a series of semiconductor materials in contact with an electrolyte the existing theory agrees satisfactorily with experiment over a definite range of surface potentials.

However, in recent years data have appeared which indi-

---

*In the electrochemistry of metals these two limiting cases are apparently realized on platinum and mercury electrodes in the presence of iodine ions.

cate the inadequacy of this theory. Thus, Dewald [8] observed that with a very high surface charge the capacitance of a zinc-oxide electrode is appreciably below the calculated value. The same result was obtained in measurements of the charging curves of cadmium sulfide electrodes [12].

This discrepancy between theory and experiment was obtained in fields where the change in potential energy at a distance of the order of the wavelength of thermal carriers is comparable with the mean thermal energy of the particles; therefore, here, quantum effects become substantial. One of them is the tunneling of mobile particles into a classically forbidden region of space; as a result the concentration of particles is increased in comparison with the value calculated without allowance for the tunneling effect. This phenomenon is analogous to that examined by Keldysh [13] where the forbidden band of a semiconductor is contracted in a strong electric field for optical absorption processes; it must lead to an increase in the differential capacitance. On the other hand, with sufficiently high deflections of the bands close to the surface there is the possibility that the usual assumption of the quasicontinuous character of the spectrum of energy levels in the bands may not hold. This effect should appear as an increase in the width of the forbidden band at the surface and as a reduction in the concentration of mobile charges; as a result, the space charge region is wider and the differential capacitance lower.

In the examination of fixed charges on the surface of a semiconductor and close to it (ionized impurity atoms, surface states, and adsorbed ions), the discrete character of their arrangement in the double layer region may be important.* The case where a considerable part of the charges is concentrated in the surface levels is very important; apparently over a certain range of values of the charge trapped by surface states it is necessary to take into account the screening of each center separately by mobile current carriers [15].

---

*The problem of taking into account the discrete character of the charge in the double layer is also unsolved in the electrochemistry of metals [14].

One of the peculiarities of the semiconductor – electrolyte system is the fact that under some conditions the charge of the semiconductor plane of the double layer may change considerably as a result of the penetration into the electrode of charged particles from the electrolyte. As a result, the properties of the space charge region may differ from those calculated on the basis of the bulk characteristics of the electrode. Thus, it is known that protons and cations of small radius diffuse readily into oxide layers [16]. There is also the possibility of alloying of the surface layer of the semiconductor with the formation of intermediate compounds. However, it should be noted that effects of this type have not yet been studied on single-crystal electrodes.

Surface Electronic States and Adsorption at the Semiconductor – Electrolyte Interface

The concept of "fast" and "slow" surface states is often used in the electrochemistry of semiconductors to explain the slow relaxation of the electrical double layer on semiconductor electrodes. It is possible to pick out two basic problems connected with surface states.

1) The experimental determination of the physicochemical nature of surface states remains an unsolved problem. In particular, no unequivocal relation has been found between the production of a surface charge and the adsorption of foreign substances on the surface of a semiconductor (for example, in the case of copper on a germanium surface). It should be noted that the properties of the surface states in the case of a "dry" semiconductor surface and a surface in contact with an electrolyte sometimes differ markedly, and this may be connected with a fundamental difference in the structure of the double layer in these two cases. As a rule, the "natural" charge at a germanium – electrolyte interface (formed as a result of standard etching and keeping in an electrolyte solution) cannot be described satisfactorily by a system of discrete surface levels. At a germanium – electrolyte interface the levels corresponding to "fast" surface states are apparently split into a band [17]. In the few cases where it has been possible to

observe discrete surface levels [18], the microscopic structure
of the interphase still remains unclear.

As regards "slow" states, it is normally assumed that on
a "dry" semiconductor surface the levels lie inside the oxide
phase or at its inner boundary. However, on the surface of germanium in aqueous solutions, for example, there is no oxide
phase and only adsorbed oxygen is present. Here, slow relaxation may be associated not with "slow" surface electronic
states, but with polar germanium − oxygen bonds or with polar
solvent molecules adsorbed on the electrode, the effective
dipole moment of which may be a monotonic function of the polarization of the electrode. In discussing the role of oxygen in
producing a surface charge we should also take into account the
fact that adsorbed oxygen may exist on the surface of a semiconductor in several forms, which may be interconverted [6].

2) An important problem is the statistical description of
processes changing the surface charge with a change in the external conditions (for example, the surface potential). The number of electrons in surface levels is usually written in the form
$N_t \cdot f$, where N_t is the concentration of levels and f is the Fermi
distribution function:

$$f = \left[\, 1 + \exp\left(\frac{E_t - E_i + e\varphi_s}{kT}\right)\right]^{-1} .$$

Here, E_t is the energy of the level, E_i is the energy of the middle of the forbidden band, and φ_s is the surface potential. This
description is inadequate in the case of surface levels arising
as a result of adsorption, when it is also necessary to take into
account the change in the concentration of adsorption particles
on polarization of the electrode. In the first approximation it is
possible to describe the degree of adsorption by means of the
corresponding adsorption isotherm, as is done in considering
metal electrodes, but the problem of choosing the type of isotherm remains unsolved. Finally, since the concentration of
surface states at the semiconductor − electrolyte interphase
may be extremely high, it is necessary to take into account the
interaction of charges trapped on the surface. This situation

may substantially affect the statistics of filling of surface levels
and the relation of adsorption to the electrode potential.

Kinetics of Electrode Processes on Semi-conductor Electrodes

In the study of the kinetics of charge transfer through the
interphase, the most important problem is the determination of
the nature of the slow stage of the process. As for metal electrodes, in the case of a reaction on a semiconductor surface the
slow stage may be the discharge stage, adsorption, diffusion of
the reaction components, the formation and breakdown of a
solid phase, or a chemical stage. Naturally, progress in
studying the kinetics of electrochemical processes is closely
connected with the development of theoretical concepts on the
reaction mechanism. It should be emphasized that at the present time there are detailed microscopic theories only for the
simplest cases; these are the theory of concentration polarization
[4] and the theory of electron transfers at an electrode– electrolyte interphase [19]. The theory of concentration polarization is well confirmed by experiment for semiconductor electrodes; diffusion limiting currents of free carriers in a semiconductor have been observed and measured [1, 4]. The theory
of electron exchange between a semiconductor and an electrolyte has been confirmed only qualitatively; the electron and
hole components of the total reaction current have been distin-
guished and it has been shown that the magnitude of the electron component in the total current increases with an increase
in the surface concentration of free electrons [20]. The problem in experimental investigations in this field is to check the
theory quantitatively, i.e., to measure the electron and hole
components of the total current (in particular, the exchange
current) in relation to the surface potential, in relation to the
surface potential, the width of the forbidden band, and other
properties of the electrode. Unfortunately, most of the data
already available in the literature cannot be used directly for
this purpose, as in these experiments no investigation was made
of the distribution of the overvoltage between the ionic double
layer and the space charge region in the semiconductor. Impor-

tant information on the kinetics of oxidation-reduction reactions
on semiconductor electrodes may be provided by the investigation of Faraday rectification [21].

Electrochemical reactions accompanied by the formation
and rupture of chemical bonds are more complex for a detailed
microscopic analysis. In this region we require further accu-
mulation of experimental data, of which there are very few as
yet. Only the liberation of hydrogen on germanium and the
anode solution of semiconductors have been studied relatively
well. There remain unelucidated the nature of "current multiplication" during the anode solution of germanium and, in particular, the relation of the current multiplication factor to the
crystallographic orientation of the surface and other properties
of the electrode. The electrodeposition of metals, the kinetics
of adsorption of substances from solution, and the insertion of
hydrogen and alkali metals into semiconductor cathodes have
hardly been studied.

In contrast to metals, the surface concentration of mobile
current carriers in semiconductors is low and is a function of
the electrode potential. With thermal equilibrium the relation
between the surface and bulk concentrations of mobile particles
is represented by a Boltzmann distribution (for a nondegenerate
electron gas in sufficiently weak fields); however, the use of
this distribution for deriving electrochemical kinetic equations
requires further justification. In the simplest case, the effect of
a current flowing on the distribution of particles may be taken
into account by solving the transfer equation (with appropriate
limiting conditions [22]):

$$I = eD(neE/kT + \partial n/\partial x),$$

where I is the current, E the electric field, n the concentration
of particles, and D the diffusion coefficient. In this form the
approach is equivalent, on the one hand, to the problem of a
nonequilibrium diffuse double layer in an electrolyte [23] and on
the other hand, to the approach used in the diffusion theory of
rectification at a metal-semiconductor contact [24].

A well-known difficulty is that connected with the inadequate definition of the term "surface concentration" for particles with a finite wavelength. In a kinetic equation it is normal to introduce the concentration of electrons (or holes) close to the geometric interface (in a layer a few angstroms thick), while the wavelength of thermal current carriers is hundreds of angstroms.

Photoelectrochemical Processes

Photoelectric sensitivity is a characteristic feature of semiconductor electrodes. In contrast to various forms of photochemical process in solution which involve free radicals or excited states of ions, here the primary act of the photoeffect is the optical generation of nonequilibrium current carriers. This leads to a redistribution of the potential drop close to the interphase and a change in the surface concentration of electrons and holes, which participate in the electrode reaction. The study of the contribution of these two factors to the change in the rate of an electrochemical conversion is the main problem in this field.

In the general case, the picture of the photoeffect may be very complex; it is determined, on the one hand, by the nature of the slow stage of the reaction, and on the other, by the properties of the space charge region in the semiconductor (mainly by the deflection of the bands, the mobility of the current carriers, and the intensity of recombination). By an appropriate choice of the experimental conditions it is possible to simplify the actual picture substantially. Thus, for example, changes in the ionic charge are eliminated when pulsed illumination is used and it is possible to measure the change in the deflection of the bands on illumination by using the existing theory developed for a semiconductor–gas interface. A series of interesting data in this direction have already been obtained for many single-crystal semiconductors. The steady photoeffect has been studied to a much lesser degree [25].

Of great interest is the further investigation of photopassivation and photoactivation [26], photodesorption [27], and also the photocatalytic activity of electrodes when the simultaneous

occurrence of different electrochemical reactions involving
current carriers of both types becomes possible on irradiation.
The study of photoelectrochemical phenomena may promote the
understanding of the mechanism of photosynthesis, as it was
found recently that a grain of chlorophyll has semiconductor
properties [28]. Finally, investigations in this field are of
particular importance in connection with the problem of direct
conversion of radiation energy into electrical energy.

The material presented above refers mainly to single-
crystal semiconductor materials, which up to now have been the
subject of the actual electrochemistry of semiconductors. The
important problem remains of extending the theoretical concepts and experimental investigation methods to polycrystalline
semiconductors, in particular, oxide films on so-called barrier
metals and to oxide and sulfide electrodes, and also to use the
concepts of the electrochemistry of semiconductors for investigating the phenomena of the passivation of metals.

Naturally we can conclude that the electrochemistry of
semiconductors has reached that stage of its development where
before it stand, to a considerable extent, the same problems as
before the electrochemistry of metals. On the other hand, the
electrochemistry of semiconductors has a series of specific
problems and also has new methods for solving them.

The authors are grateful to B. N. Kabanov, Yu. Ya. Gurevich, and R. M. Iazorenko-Manevich for valuable comments.

Literature Cited

1.   V. A. Myamlin and Yu. V. Pleskov, this volume.
2.   E. A. Efimov and I. G. Erusalimchek, Electrochemistry
      of Germanium and Silicon, Goskhimizdat, Moscow (1963).
3.   H. Gerischer, Advances in Electrochemistry and Electro-
      chemical Engineering, Vol. 1 (ed. P. Delahay), Intersci-
      ences Publisher, New York (1961).
4.   J. Dewald, Semiconductors, (ed. N. Hannay) Cambridge
      Univ. Press, New York (1959).

5. The Surface Chemistry of Metals and Semiconductors (ed. H. C. Gatos), New York–London (1960).

6. Yu. V. Pleskov, Elektrokhimiya, 1:4 (1965).

7. T. P. Birintseva and Yu. V. Pleskov, Izv. Akad. Nauk SSSR, Ser. Khim., 251 (1965).

8. J. F. Dewald, Bell. System Tech. J., 34:615 (1960).

9. V. A. Tyagai, Izv. Akad. Nauk SSSR, Ser. Khim., 34 (1964).

10. M. Sparnaay, Surface Sci., 1:213 (1964).

11. R. M. Lazorenko-Manevich, Dissertation, Moscow (1963).

12. V. A. Tyagai, Elektrokhimya, 1:377 (1965).

13. L. V. Keldysh, Zh. Éksp. i Teor. Fiz., 34:1138 (1958).

14. B. V. Érshler, Zh. Fiz. Khim. 20:679 (1946).

15. J. Bardeen, Surface Sci., 2:381 (1964).

16. P. D. Lukovtsev, Trans. 4th Conference on Electrochemistry, Izd. Akad. Nauk Moscow (1959).

17. M. D. Krotova and Yu. V. Pleskov, Phys. Status Solidi, 3:2119 (1963).

18. P. J. Boddy and W. H. Brattain, J. Electrochem. Soc., 109:812 (1962).

19. R. R. Dogonadze and Yu. A. Chizmadzhev, Dokl. Akad. Nauk SSSR, 150:333 (1963).

20. Yu. V. Pleskov, Zh. Fiz. Khim. 35:2576 (1961).

21. Yu. Ya. Gurevich and V. A. Myamlin, Dokl. Akad. Nauk SSSR, 155:1159 (1964).

22. V. A. Tyagai, Elektrokhimiya, 1:387 (1965).

23. V. G. Levich, Dokl. Akad. Nauk SSSR, 67:309 (1949); 124:869 (1959).

24. A. I. Ansel'm, Introduction to the Theory of Semiconductors, Moscow-Leningrad (1962).

25. A. M. Kuznetsov and R. R. Dogonadze, Izv. Akad. Nauk SSSR, Ser. khim., 1885 (1964).

26. S. O. Izidinov, T. I. Borisova, and V. I. Veselovskii, Dokl. Akad. Nauk SSSR, 145:598 (1962).

27. R. M. Lazorenko-Manevich, Electrochim. Acta, 10:141 (1965).

28. E. K. Putseiko, Dokl. Akad. Nauk SSSR, 150:343 (1963).

Index